AGRICULTURE ECONOMICS

AGRICULTURE ECONOMICS

Editor

Vasanti Khurana

scitus academics

Agriculture Economics
Edited by **Vasanti Khurana**

This Book is Printed on Acid-Free Paper. ♾

ISBN: 9781681179056
Library of Congress Control Number: 2017948933

SCITUS Academics LLC,
616, Corporate Way, Suite 2, 4766,
Valley Cottage, NY 10989

www.scitusacademics.com

Notice

Printed in United States of America.

Preface

Agricultural economics is a branch of applied economics that takes the tools of both micro and macroeconomics and uses them to solve problems in a specific area. With food inflation soaring and agricultural disputes at the heart of the collapse of the latest round of world trade talks, the subject has seldom been so topical. It focused on maximizing the crop yield while maintaining a good soil ecosystem. Throughout the 20th century the discipline expanded and the current scope of the discipline is much broader. Agricultural economics today includes a variety of applied areas, having considerable overlap with conventional economics. Agricultural economics does not study only the behavior of a farmer at the farm level. That is, in a way, the micro analysis. Agricultural problems have a macro aspect as well. Instability of agriculture and agricultural unemployment are the problems which have to be dealt with, mainly at the macro level. At the micro level, we need to understand the relevant production functions and the relationship between labor and capital. At the macro level, the subject studies the way governments decide how to support farmers. In developed economies, policy has tended to be devised so as to support and protect farmers – subsidies for prices and agricultural import tariffs under the European Union's Common Agricultural Policy for example – often with significant effects on trade flows and trading relations. In contrast, developing nations seek ways to ensure adequacy of supply of food for their people and to gain access to world markets to earn export revenue. Agricultural economics plays a role in the economics of development, for a continuous level of farm surplus is one of the wellsprings of technological and commercial growth.

This book 'Agriculture Economics' presents coverage on agricultural finance, policy, marketing, farm and agribusiness management, rural sociology, and agricultural law. Research into farm and agribusiness management led to mathematical planning systems and statistical computation of farm-enterprise data, and interest has been drawn to decision-making behaviour studies of farm managers. It deals with the study of the production and consumption of food in both developed and developing countries along with analysis of the policies that shape the world's largest industry.

Table of Contents

Chapter 1 Competitive strategies among Ontario farms marketing direct to consumers 1

Chapter 2 Determinants of household's consumption preference for processed cocoyam in Enugu State, Nigeria 35

Chapter 3 Categorizing households into different food security states in Nigeria: the socio-economic and demographic determinants 49

Chapter 4 Demand relationships in orange exports to Russia: A differential demand system approach focusing on Egypt 77

Chapter 5 Farming household food storage, consumption and sales decision making under price risk in northern Uganda 101

Chapter 6 Farmgate versus market centre sales: A multi-crop approach 121

Chapter 7 Climate sensitivities and farmland values in Nepal: A spatial panel Ricardian approach 143

Chapter 8 Spatial market integration and price transmission for papaya markets in Ethiopia 173

Chapter 9 Impacts of corn price and imported beef price on domestic beef price in South Korea 189

Chapter 10 Determinants of participation in fertilizer subsidy programme among rice farmers in Ogun State, Nigeria 209

Chapter 11 Values-Based Supply Chains as a Strategy for Supporting Small and Mid-Scale Producers in the United States 221

Chapter 12 The Common Agricultural Policy 2014–2020: scenarios for the European agricultural and rural systems 247

Chapter 13 The social pillar of sustainability: A quantitative approach at the farm level 255

Chapter 14 Do the initial land endowments lift people out of chronic and transient poverty? Evidence from central and Northern Mozambique 281

Chapter 15 Impacts of Rural Labor Resource Change on the Technical Efficiency of Crop Production in China 305

Chapter 16 Effects of local institutions on the adoption of agroforestry innovations: Evidence of farmer managed natural regeneration and its implications for rural livelihoods in the Sahel 323

Chapter 17 Off-farm participation decision and its impact on crop yield in Northern Ethiopia 369

Chapter 18 Determinants of small-scale mango farmers' market channel choices in Kenya: An application of the two-step Cragg's estimation procedure 387

Index 405

CHAPTER 1

Competitive strategies among Ontario farms marketing direct to consumers

Eric T. Micheels[1*] and Andreas Boecker[2]

[1] *Department of Agricultural and Resource Economics, University of Saskatchewan, 51 Campus Drive, Saskatoon, SK S7N 5A8, Canada.*

[2] *Department of Food, Agricultural and Resource Economics, University of Guelph, Guelph, ON N1G 2W1, Canada*

ABSTRACT

This paper examines the importance of two resources, namely a market orientation and an entrepreneurial orientation, in new product development and satisfaction with firm performance among direct and alternative marketers. Using a partial least squares structural equation model and data from a 2013 survey, this paper models the effect of a market orientation and entrepreneurial orientation on satisfaction with current performance and expected future performance for agricultural firms in Ontario, Canada. Our results show that a market orientation and entrepreneurial orientation are important factors in determining rates of product and marketing innovations among farm-based agribusinesses. These findings further demonstrate that a market orientation is a powerful resource as it enables firms to become aware of opportunities to provide superior value for consumers, especially when operating in non-traditional markets.

Keywords: Direct marketing, Entrepreneurial orientation, Market orientation, Marketing innovation, Product innovation

1. BACKGROUND

The number of farmers' markets has been growing across Canada in the past decade (Dukeshire et al. 2014). A recent study of consumers at farmers' markets in Toronto, Ontario, found that consumers use farmers' markets not only to fulfil specific grocery needs but also to support local agricultural firms (Dodds et al. 2013). Hence, sales in Ontario farmers' markets by primary producers were estimated to be between $385 million and $577 million annually in 2010, up from an estimated $73 million in 1999 (Cummings et al. 1999; Regional Analytics and Planscape 2011). Innovative and entrepreneurial agricultural firms may aspire to improve performance by marketing their production in a manner that takes advantage of these trends. For example, farms in Ontario market their production direct to consumers through not only farmers' markets but also on-farm markets and roadside stands, community-supported agriculture and boxing schemes, and on- and off-farm events. It is estimated that these additional outlets account for 25% of direct sales while farmers' markets account for 75% (Schumilas 2010). Given the increased use of direct and alternative marketing channels in Ontario, a greater understanding of the factors that drive current and expected success within these alternative marketing channels seems warranted.

The move to direct and alternative markets may signal that firms are not satisfied with current levels of performance achieved through more traditional channels and are therefore searching for other means to improve performance (Levinthal and March 1993; Mount 2012). In the search for new profit opportunities, firms may modify product offerings or production practices to differentiate their production from that of other firms. For smaller firms, innovations in how the product is marketed may be more scale-appropriate and more likely to lead to improved performance (Moser et al. 2011). However, as the number of farmers' markets has increased (Dukeshire et al. 2014) along with the number of farms selling a portion of their production through farmers markets (Uematsu and Mishra 2011), these farm businesses need to further differentiate themselves from competition for continued success. In this sense, farm-based businesses who have moved into direct and alternative marketing channels may now face greater levels of competition compared to early adopters within these channels. This would suggest that the resources and capabilities that have been important in determining success in other small businesses may now be important in determining success of farm-based direct marketing ventures. However, as these farm businesses often start out as traditional farms, the level of human and relational capital that has been invested in developing resources to identify new opportunities in product or market spaces may be rather limited. As these markets become increasingly competitive, firms that are more quickly able to identify and respond to market needs may have improved performance outcomes.

Research across different industries has shown that firms that are aware of opportunities and are able to respond to these opportunities are able to improve performance (Renko et al. 2009; Rauch et al. 2009). However, since firms and their resource endowment, including managerial skill sets, are heterogeneous, firms may develop and employ diverse strategies in the search for profit. For research in this area, this raises the need for addressing the identification of critical resources—managerial skills on the one hand and adaptive strategies on the other—as well as determining the interrelations among them. In the managerial research literature, a number of constructs have been developed and applied empirically to capture the managerial resources that facilitate efficient responses to change.

The purpose of this paper is to examine how two of these constructs, market orientation and entrepreneurial orientation, affect the focus of innovation on the farm and the satisfaction with current and expected performance. Recent work in this area examined the role of these constructs and other factors on new product success and the number of marketing channels utilized by direct marketers (Mirzaei et al., 2016). Slater and Narver (1995, pp 67) define market orientation as "the culture that (1) places the highest priority on the profitable creation and maintenance of superior customer value while considering the interests of other key stakeholders; and (2) provides norms for behaviour regarding the organizational development of and responsiveness to market information." While cultures such as this are important, the delivery of superior value to the customer requires action on the part of the firm. That action may stem from an entrepreneurial orientation, which Lumpkin and Dess (1996, pp. 136) define as "the processes, practices, and decision-making activities that lead to new entry." The entry described may be entry into new marketing channels or new product markets and may be a response to increasing consumer demand for locally produced products (Adams and Adams 2011; Hu et al. 2012; Gracia 2012; Kim et al. 2014).

We developed a survey that measures market orientation, entrepreneurial orientation, and product and marketing innovations of farmers in Ontario over the past 3 years. Product innovations we examine are new products and new packaging of existing products, while marketing innovations include the use of new marketing channels and new promotional strategies as well as changes in pricing. Using the data from the survey, we build a partial least squares structural equation model (PLS-SEM) to determine how these latent variables are interrelated and affect expected future performance and satisfaction with current performance. This work contributes to the literature on performance analysis of agricultural firms marketing their production through direct and alternative market channels by simultaneously examining critical managerial skills and adaptive strategies in their impact on performance.

In the sections that follow, we review the prior literature surrounding direct marketing, market orientation, and entrepreneurial orientation. We use this literature to develop hypotheses and a structural model that will be used to test the hypotheses. The "Methods" section outlines the data and methods used in the study. The "Results and discussion" section describes the results of the structural equation model used to test the hypotheses. Finally, the paper concludes with a discussion of the results, the implications of these results to managers and other stakeholders, and an outlook on further research that would address some of the limitations of this study.

2. METHODS

2.1 Previous literature and model development

Considerable research on consumer preferences for local food and the drivers of increased demand has been conducted (Adams and Adams 2011; Cranfield et al. 2012; Feldmann and Hamm 2015; Thilmany et al. 2008; Weatherell et al. 2003), but research on the manner by which firms respond to and benefit from meeting these needs is somewhat limited. In a recent study, Low and Vogel (2011) find that local food marketers in the USA utilize both direct-to-consumer outlets (farmers markets, roadside stands, etc.) as well as intermediated outlets (direct-to-grocer, direct-to-distributor, etc.), and the value of production marketed though these alternative channels has been increasing since 1992. Low and Vogel (2011) further find that while all local food marketers utilize direct-to-consumer channels, farms with sales of $250,000 or more make greater use of intermediated channels. This is supported by the finding of Sun et al. (2014) that larger wineries in emerging wine regions in the USA market a larger proportion of their production through intermediated channels.

A key characteristic of these alternative marketing channels is that the primacy of the information that is shared between market participants leads to highly relational markets. Research within New Zealand farmers' markets, for example, found that sellers place considerable value on the ability to try out new product formulations or packaging redesigns and quickly gauge customer feedback (Guthrie et al. 2006). Although agricultural firms involved in direct marketing may not have vast innovation budgets that high-tech firms possess, Rosenbusch et al. (2011) found that even for smaller, resource-scarce firms, the benefits of differentiation, such as increased loyalty, price premiums, and entry barriers, outweighed the costs and the risks involved in the innovative activity.

While the studies referenced above provide valuable information on specific aspects and trends in local food marketing, they do not investigate how managerial resources affect innovation choices and performance. In the remainder of this section, we therefore present a literature review for each of the two managerial resource constructs most relevant to this research, market orientation (MO) and entrepreneurial orientation (EO), from which hypotheses are derived and a structural equation model developed. A challenge for this literature review, however, was that a number of studies have used scales that were similar but not necessarily based on the original scales by Slater and Narver (1995) and by Lumpkin and Dess (1996) to measure market and/or entrepreneurial orientations. In some studies, the differentiation between the two constructs has thus not been straightforward but required interpretation.

2.2 Market orientation

While small- and medium-sized enterprises (SMEs) in food production may rely on innovativeness and new product development to meet consumer needs (Trienekens et al. 2008; Gellynck et al. 2012), the success of this strategy depends on the firm's ability to become aware of and respond to these needs before the competition. For example, Webb et al. (2010) find that market-oriented firms are better equipped to recognize opportunities and to capitalize on them through innovation. This ability stems from the market-oriented firm's capability in generating and acquiring market-focused intelligence based on interactions with lead customers in order to provide the organization with the information needed to develop new products that are valued by the market (Slater and Narver 2000). Moreover, it has been shown by Gunday et al. (2011) that innovation performance (the ability to develop and introduce new products) improves market performance, as measured through sales or market share.

Market-oriented firms have been shown to be proactive in developing new products and services to serve their markets (Cambra-Fierro et al. 2011; Nasution et al. 2011; Beck et al. 2011). Firms that are able to integrate market awareness and appropriate responses into innovation programmes are able to achieve stronger linkages with customer value (Nasution et al.

2011). In a study of family firms in the Netherlands and Belgium, Beck et al. (2011) find that the type of market orientation (responsive, proactive, emerging) fully mediated the level of innovation that family firms undertook, even when controlling for other factors. Moreover, Tan and Liu (2014) find that firms with a responsive market orientation are expected to be successful in innovation strategies that build on current customers and capabilities. Lamore et al. (2013) suggest that firms with a proactive market orientation succeed by identifying unmet and unarticulated market needs while firms with a more responsive market orientation succeed when they are able to meet current customers already expressed needs.

It may seem that agricultural firms who market through direct and alternative markets are not applicable in this literature. While agricultural firms may bear little resemblance to high-tech firms or others who seem to be leaders in innovation, these firms are innovators according to the definition of innovation by Nelson and Winter (1982) who define innovation as a change in routine. As much of agriculture consists of routine (i.e. seeding, harvest, marketing) even incremental changes in the acquisition of inputs, the nature of production or the manner in which firms market the product could be seen as an innovation in this sense. However, in order for these changes to be successful, these firms must understand the nature of their current and prospective customers as well as the strengths and weaknesses of current and potential competitors. Based on the findings from previous conceptual and empirical studies presented above, we derive the following hypotheses:

H1: An increase in the market orientation of the firm is associated with an increase in the level of product innovations within the firm.

H2: An increase in the market orientation of the firm is associated with an increase in the level of marketing innovations within the firm.

H3: An increase in the market orientation of the firm is associated with higher levels of satisfaction with current performance.

H4: An increase in the market orientation of the firm is associated with an increase in the expectations of future performance.

2.2.1 Entrepreneurial orientation

A body of literature has also examined how entrepreneurial orientation in management can improve firm performance. Pérez-Luño et al. (2011) find that entrepreneurial firms, particularly those who are proactive and are willing to take on additional risk, are more likely to generate new ideas and processes as well as adopt successful ideas that other firms have implemented. Evidence of successful imitation strategies has been found directly in some studies (Lee et al. 2000; Lee and Zhou 2012), as well as in an attempt to leverage the competitive aggressiveness of entrepreneurial firms (Lumpkin and Dess 1996; Wang 2008; Rhee et al. 2010).

Using measures from Covin and Slevin (1989) and Wiklund (1999), Grande et al. (2011) find that farm firms involved in traditional agricultural production (as well as value-added processing who are more entrepreneurial) have better performance than firms who are not as entrepreneurial. Moreover, Knudson et al. (2004) suggest that entrepreneurial innovators, such as organic food farms, are determined to innovate and bring new products to market in order to boost firm performance. In addition, de Lauwere (2005) finds that more entrepreneurial farmers, referred to as social farmers and new growers in the study, had more optimistic expectations for future performance as well as higher actual farm incomes.

Grande et al. (2011) develop a model of farm performance with firm size, the use of networks, the financial position of the firm, and the entrepreneurial orientation of the firm as explanatory variables. Their results show that firms who score higher on the entrepreneurial orientation scale also achieve higher profits and more aggressive sales growth measures. Nasution et al. (2011) find evidence in a study of hotel managers that firms who score higher on proactiveness and risk-taking scales are more innovative in terms of product and process innovations developed. Similarly, Tajeddini (2010) finds that hotels with an entrepreneurial orientation are more innovative and are better able to meet goals relating to market and financial performance. These results are similar to studies in other industries, and the positive effect on profit and sales may be caused by the ability of these firms to recognize opportunities in the market and be proactive in their response to these opportunities (Lumpkin and Dess 2001; Moreno and Casillas 2008; Rauch et al. 2009).

Given the increased level of competition at farmers' markets (Beckie et al. 2012), Boehlje and Bröring (2011) suggest that first-mover advantages in new markets may create entry barriers for subsequent firms as leading firms are able to develop valuable relationships with key customers. However, given the capital intensive nature of agricultural production, firms may be reluctant to invest in new products or processes. This reluctance may be further embedded due to the perceived lack of dynamism in agriculture. However, findings by Wiklund and Shepherd (2005) suggest that an EO could increase performance regardless of capital availability or market dynamism. From the above presented findings and considerations, we derive the following hypotheses:

H5: An increase in the entrepreneurial orientation of the firm is associated with an increase in the level of product innovations within the firm.

H6: An increase in the entrepreneurial orientation of the firm is associated with an increase in the level of marketing innovations within the firm.

H7: An increase in the entrepreneurial orientation of the firm is associated with an increase in the level of satisfaction with current performance.

H8: An increase in the entrepreneurial orientation of the firm is associated with an increase in the level of expected performance.

2.2.2 Innovation

Previous studies have shown that innovativeness mediates the link between a market orientation and firm performance (Han et al. 1998; Tajeddini et al. 2006; Hong et al. 2013). While awareness of market conditions and the willingness to take on additional risk are important success factors, firms must initiate and implement appropriate changes in routines to improve firm performance. Research has shown that the use of market information is a critical success factor in the complex internal decisions of agribusiness companies to select innovation projects for funding (Roucan-Kane et al. 2011). Köhler et al. (2012) find that greater use of market-driven search increases the rate of success when firms imitate product or process innovations of rival firms. As firms that market their production through direct and alternative channels have close contact with all customers, we would expect that they would have greater access to relevant market information, e.g. through customer surveys and product trials, and would thus be in a better position to react to this information and improve performance. Hence, we propose the following hypotheses:

H9a: An increase in the level of product innovations within the past three years is associated with an increase in satisfaction with current performance.

H9b: An increase in the level of product innovations within the past three years is associated with an increase in expectations regarding future performance.

Researchers have developed various typologies of innovation (Garcia and Calantone 2002; Rowley et al. 2011). Innovations can be thought of along product and technological lines or according to how radical the change is. For example, the innovation undertaken by a firm could be new to the world or it could be an adoption of an existing innovation with the change being new to the firm (i.e. Garcia and Calantone 2002). The innovation may also encompass a change in how the products are marketed (Boehlje et al. 1995; Boehlje 2002).

Marketing innovations entail the identification and development of new markets or ways of marketing. Regardless of whether the innovation is related to new or existing products, finding the right consumers for these products is also an important capability. Di Stefano et al. (2012), using a meta-analysis of innovation studies, find evidence that matching market needs to new products is an important capability that may ultimately ensure the success of the new product. Recent studies have shown that marketing innovations (measured as changes in product distribution, pricing, and promotion of products) significantly contribute to satisfaction with performance (Camisón and Villar-López 2011; Gunday et al. 2011). Ngo and O'Cass (2012) find evidence that suggests that a firm's marketing capability and their innovation- and customer-related performance are significantly related to the firm's market orientation. Expecting that these relationships would also hold for farms selling through direct and alternative channels in Ontario, we propose the following hypotheses:

H10a: An increase in the level of marketing innovations within the past three years is associated with an increase in satisfaction with current performance.

H10b: An increase in the level of marketing innovations within the past three years is associated with an increase in expectations regarding future performance.

Given the previously discussed literature and the hypotheses presented, we model the firm's product and marketing channel innovations as a function of their awareness of customer needs and competitor actions (i.e. their market orientation) as well as their willingness to be proactive and take calculated risks in response to these needs (i.e. their entrepreneurial orientation). In addition, key farm characteristics, such as farm size and farming experience need to be controlled for, as these have been found to influence the choice of farmers' income diversification activities (Barbieri and Mahoney 2009). From this background, we develop a structural model of managerial orientation, innovation strategy, and firm performance to be applied to data from Ontario farms that market through direct and alternative channels (Fig. 1).

2.3 Survey methods

After developing a questionnaire in consultation with members from Agriculture and Agri-Food Canada, it was administered by Ipsos Agriculture and Animal Health in April and May of 2013. The questionnaire consisted of six sections: demographics, sources of farm income, use of various marketing channels, knowledge of production and marketing-related costs, measurement of the firm's market orientation and entrepreneurial orientation, and satisfaction with financial performance indicators.

To measure the market orientation of the respondents, we modified the original Narver and Slater's (1990) scale to fit the context of the Ontario agri-food industry. This scale has been used in previous studies (Langerak 2003; Farrell et al. 2008; Nasution et al. 2011; Cheng and Krumwiede 2012) and its 14 items relate to three inter-related concepts: a customer orientation, a competitor orientation, and inter-functional coordination. The individual items in the scale measure a respondent's level of agreement with perspectives relating to the firms' collection of information on customer needs and competitor responses (see Appendix 1). After adjusting for the direction of an item statement (agreement with a negative), higher scores indicate greater market orientation.

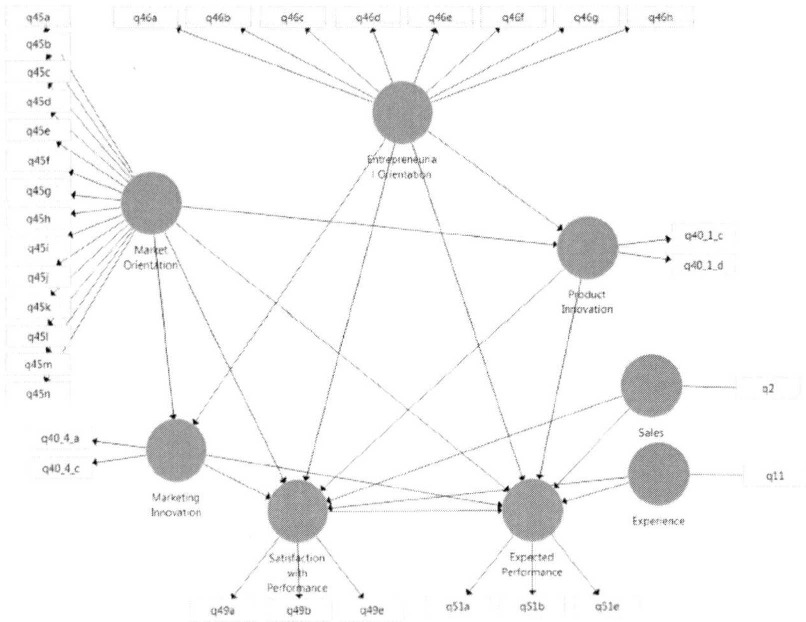

Figure . 1 Path model

The entrepreneurial orientation scale was first developed by Covin and Slevin (1989) and also consists of three inter-related concepts: proactiveness, innovativeness, and risk-taking. It has been primarily used in previous studies that examined innovativeness and firm performance (Keh et al. 2007; Li et al. 2009; Hansen et al. 2011). For the individual items, respondents were given two anchors which attempted to maximize differences in firm posture and strategy for each of the eight items in total (see Appendix 1). Again, after adjusting for direction of a statement, a higher score indicates greater entrepreneurial orientation. As both scales were initially developed for non-agricultural contexts, slight modifications were made to fit the targeted research population.

Innovation can take several forms, ranging from new product development and new packaging and processing to changes in marketing and management practices or organizational structure and processes (Lambrecht et al. 2015). For the purposes of this study, we focus on two areas of change within direct and alternative markets: product innovations (new products and new packaging) and marketing innovations (new channels and changes in marketing communications). We measure these in an incremental/radical innovation framework similar to previous work by Dewar (1986) and Song and Thieme (2009). Therefore, the intensity of innovation was measured for each item on an ordinal scale with three levels: not at all, to some extent, and to a great extent (Table 5).

While studies have examined just how a market orientation and an entrepreneurial orientation influence innovative activity, we were also interested in how changes in these factors were associated with changes in two subjective measures of firm performance. We asked respondents to report their satisfaction with current performance as well as their expectations of future performance (the next 5 years) (Fig. 2). As detailed in Appendix 2, items focused only on financial indicators and included net income, sales, and cash-flow. This method is in line with other recent studies that have asked respondents to rate their satisfaction with certain levels of performance or to compare different aspects of performance to main competitors (Brik et al. 2011; Yannopoulos et al. 2012). Objective measures of performance would not have been obtainable as private family businesses are generally unwilling to share confidential financial data, even in anonymous settings (Shoemaker et al. 2002), and no secondary data exists that also includes measures for market orientation and entrepreneurial orientation. For current performance, the scale anchor points were 'very dissatisfied' (1) and 'very satisfied' (7), and 'will deteriorate significantly' (1) and 'will improve significantly' (7) for expected future performance. Recently, Mahto et al. (2010) examined how family interaction and identification with the family business are significantly related to satisfaction with performance of the family business. As most farm businesses are family businesses, the use of satisfaction with performance (rather than actual performance) may be an adequate workaround, even if satisfaction with similar levels of performance may differ across respondents.

Furthermore, while objective measures of performance such as sales and profit levels may be preferred, studies have shown a strong correlation between subjective and objective measures of performance (Dess and Robinson 1984; Richard et al. 2009; Dawes 1999; Wall et al. 2004). For example, Wall et al. (2004) show that in a sample of UK manufacturing and service sector firms, subjective performance measures demonstrate both convergent and discriminant validity, meaning there is a high correlation between parallel subjective and objective measures and little correlation between nonparallel subjective and objective measures.

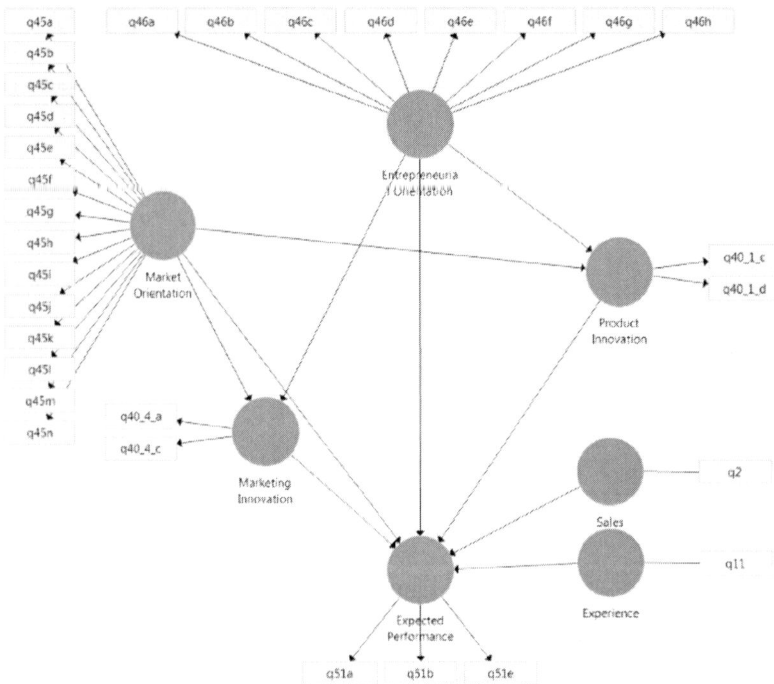

Figure. 2 Path model (expectations of future performance)

Ipsos Agriculture and Animal Health emailed the questionnaires to a proprietary panel of Ontario producers. Specific quotas of farm types (dairy, beef, crops, etc.) were used to limit over and under sampling of specific groups of the diverse Ontario farm population. This panel of farmers has been used in other agricultural research (e.g. Nadella et al. 2014). Respondents who completed the survey received $20 CAD for their time. An email containing a link to the survey was sent to 10,406 unique email addresses that are included in a proprietary database by IPSOS Animal Health. On average, surveys were completed in 34 min and a total of 799 respondents[1] attempted the survey and 405 respondents completed the survey. Based on the 405 completed questionnaires, this amounts to a 3.9% response rate.[2] This study uses a sub-sample of 126 respondents who have been active in farm direct and alternative marketing.

2.3.1 Description of respondents

Table 1 shows the distribution across annual gross farm sales for the full sample, the study sample of 126 farmers involved in direct and alternative marketing, and the 2011 agricultural census for Ontario. While farms with $99,999 CAD in sales or less are underrepresented in the full sample compared to census figures, large farms with annual gross sales of $500,000 or more are similarly over represented. However, it has to be pointed out that Ontario gross farm receipts in 2012—the year of measurement in our survey—were more than 20% above those of 2010—the year of measurement of the census (OMAFRA, 2014). Hence, the bias towards larger operations in our full sample is likely exaggerated. Although the size distribution of the study sample is generally more in line with the census figures, representativeness cannot be inferred. Previous observations that it is mostly small operations that market directly to consumers are, however, confirmed by 56% of study sample having annual gross sales below $100,000, compared to 40% in the full sample.

The respondents in the study sample are key decision makers on their farms (Table 2). A majority of the respondents in the sample have final say in operational (54.8%), financial (52.4%), and marketing (54.8%) decisions, and over 40% of respondents share decision-making in these areas with somebody else on the farm. We also asked respondents about the life cycle stage their farm is at. Ten respondents (8%) self-reported to be in the "getting my farm operation established" category. The vast majority placed their operation in one of two "established" categories: 54 (43%) as "maintaining the operation at a steady level" and 48 (38%) "expanding the operation." The remaining 14 respondents (11%) stated that they were beginning to scale down or planning to sell their operation in the near future.

The respondents in our sample report using a portfolio of methods to market their production (Table 3). A majority of them report to market part of their production wholesale (71%). For these operations, selling wholesale on average accounts for 63% of gross sales and for 51% of net family income.

More specifically, 118 respondents (94%) in the sample report they market direct to consumers through which these operations generate about one third of their gross farm sales and net farm income. Forty-one respondents (33%) report being engaged in other forms of alternative marketing, such as selling to restaurants, independent grocers, or other institutions, thus generating one fifth of sales and net income. On average, the 126 operations in the sample sell through two specific types of direct or alternative sales outlets (Table 4), most commonly through roadside and farm gate stands (45%), followed by farmers' markets, on-farm markets, and online shops (each with 24%).

Table 1 Gross sales of survey respondents in full and study sample compared to census

Annual gross sales brackets	Full sample ($n = 405$)		Study sample ($n = 126$)		Census
	Count	Percent	Count	Percent	Percent[a]
$10,000 to $24,999	62	15.3	35	27.8	17.5
$25,000 to $49,999	53	13.1	18	14.3	12.9
$50,000 to $99,999	44	10.9	17	13.5	11.9
$100,000 to $249,999	74	18.3	20	15.9	13.4
$250,000 to $499,999	66	16.3	19	15.1	9.8
$500,000 to $999,999	52	12.8	4	3.2	6.3
$1 million and over	54	13.3	13	10.3	4.5

[a]2011 federal agricultural census results for the province of Ontario

Table 2 Frequency of decision-making authority among respondents (N = 126)

Decision type	Delegation of decision-making		
	I have the final say in all decisions	I share decision-making with someone else	I have some input but ultimately leave decisions to someone else
Overall operational and production decisions	71 (54.8%)	53 (42.1%)	2 (1.6%)
Financial management decisions	66 (52.4%)	60 (47.6%)	0 (0.0%)
Marketing decisions	69 (54.8%)	54 (42.9%)	3 (2.4%)

Table 3 Marketing methods among respondents (N = 126)

Marketing activity	Yes	No	Share of gross sales[a]	Share of net family income[a]
Marketing farm commodities wholesale	89 (70.6%)	37 (29.4%)	63.3%	50.6%
Marketing farm-fresh products or process products directly to consumers	118 (93.7%)	8 (6.3%)	34.3%	33.2%
Marketing farm-fresh products or processed products directly to restaurants, institutions, or local/independent retailers	41 (32.5%)	85 (67.5%)	21.5%	20.2%
Agri-tourism	14 (11.1%)	112 (88.9%)	13.6%	8.9%

[a] Average shares reported for those respondents engaged in specific marketing activity; hence, reported percentage figures add up to more than 100%

Table 4 Marketing channels used in 2012

Marketing channel	Yes	No
Roadside or farm-gate stand	57	69
On-farm market, café, or restaurant	30	96
Farmers' market	30	96
Online shop	30	96
Events, on- or off-farm	27	99
Restaurant	26	100
Independent grocer	24	102
100 mile markets or similar	12	114
Institutions (schools, hospitals)	11	115
Community-supported agriculture (CSA)	4	122

Table 5 Descriptive statistics of innovation variables and control variables

Variable	Innovation activity reported by respondents		
	Not at all[a]	To some extent	To a large extent
Product innovation (new to farm)			
New or significantly improved processed product	54 (42.9%)	25 (19.8%)	6 (4.8%)
New or significantly improved packaging	54 (42.9%)	25 (19.8%)	11 (8.7%)
Marketing innovation (new to farm)			
Marketing channel	51 (40.5%)	36 (28.6%)	15 (11.9%)
Communications, advertising, or promotion methods	47 (37.3%)	43 (34.1%)	18 (14.3%)
	Median	Minimum	Maximum
Farming experience (years)	23.00	3.00	55.00
Farm gross sales (categorical)[b]	4.00	2.00	8.00

[a]Percentages are based on sample of 126 and may not sum to 100. Residuals are missing data or 'does not apply' responses, which received a score of zero in innovation index generation
[b]Categorical variable where 1 = less than \$10,000; 2 = \$10,000 to \$24,999; 3 = \$25,000 to \$49,999; 4 = \$50,000 to \$99,999; 5 = \$100,000 to \$249,999; 6 = \$250,000 to \$499,999; 7 = \$500,000 to \$999,999; and 8 = \$1 million and over

Overall, it appears that respondents have engaged less often in product and packaging innovation than in marketing channel and communication innovation during the past 3 years. This may indicate that investing in marketing innovations comprises less risk than investing in product innovations relative to the corresponding benefits. In regard to the control variables, Table 1 had already indicated lightly right-skewed distribution across the seven gross farm sales categories so that the median is 1 below the midpoint category. As expected, the study sample exhibits a large range of years of farming experience, i.e. from 3 to 55 years (Table 5).

3. RESULTS AND DISCUSSION

As the key variables in our proposed model are unobservable (i.e. market orientation, entrepreneurial orientation), we must measure these using multiple items from the survey. However, we first must ensure that the measurement items used in our sample are appropriate; therefore, a reliability analysis was conducted to ensure that the measurement scales used in the study exhibited construct validity and were consistent with previous studies. Reliability and validity tests for the scales indicate that they are sufficient as Cronbach alpha (Cronbach 1951) statistics are all relatively high and the average variance extracted (AVE) for most of the scales is above 0.50.[3] Cronbach alpha values for satisfaction with current performance and expectations regarding future performance are all larger than 0.80 and AVE for both are larger than 0.600.

While tests to assess scale reliability, such as Cronbach alpha, show that the scale has adequate internal consistency, it is also important that the degree of overlap among scales is not an issue. Discriminant validity tests assure the researcher that different scales actually measure different factors (Bryant 2000). According to Fornell and Larcker (1981), measurement scales exhibit discriminant validity when the square root of AVE is greater than the correlation between scales. As seen in Table 6, the correlation between measurement scales is always less than the square root of the AVE shown on the diagonal. Given these results, we proceed to the path analysis.

To analyse the relationships between the firm's market orientation, entrepreneurial orientation, and firm performance, we develop a PLS-SEM using SmartPLS 3.0 (Ringle et al. 2015). Management scholars who wish to model interactions between latent constructs when data may not be normal and when sample sizes may be small have used PLS-SEM (Hulland 1999; Hair et al. 2011). PLS-SEM has been used extensively in the literature to examine the relationships between latent constructs and various measures of performance (e.g. Hair et al. 2012; Ashok et al. 2016; Lee et al. 2016). Structural equation modelling (SEM) is a statistical method that enables researchers to measure interactions between unobservable variables (such as a market orientation or entrepreneurial orientation) in complex models when data sets are small. Contrary to covariance-based SEM which minimizes the difference in covariance between the actual data and the values predicted by the model, PLS-SEM attempts to maximize the explained variance within latent constructs iteratively using a series of OLS regressions (Hair et al. 2011; Monecke and Leisch 2012).

In addition to the independent variables, we incorporate two variables that we hypothesize will mediate performance: indices of product innovation intensity and of marketing innovation intensity. Control variables beyond farm management experience and gross farm sales as a size proxy have been considered but have not been included in this analysis for practical reasons.4 In the analysis, we use satisfaction with current performance and expected performance as dependent variables (and also include a control between current performance and expected performance) that are influenced by the independent variables shown in Fig. 1. The results of the models are presented in Table 7.

First, we do find support for hypothesis 1 that market-orientated firms are more likely to engage in market innovations. Our results do not show a significant relationship between a market orientation and product innovation intensity; therefore, hypothesis 2 is not supported. The corresponding coefficients show that an increase in the market orientation scale by one leads to an increased score on the market innovation scale by 0.394. In regard to the role of entrepreneurial orientation, the results show that it is strongly and positively linked with the intensity of both product and marketing innovations, thus providing support for hypotheses 5 and 6. It may be worth noting that the significance is much stronger in both cases when examining the relationship with marketing innovations (1%) compared to product innovations (10%).

Interestingly, an increase in the market orientation of the firm was not shown to have a direct relationship with current performance (hypothesis 3 not supported), and there was similarly no support found for the relationship between an entrepreneurial orientation and expected performance (hypothesis 8 not supported).

In terms of the effect of innovation intensity on satisfaction with current performance or expected future performance, the results from our model do not show any significant effects of product innovation intensity on either performance measure. However, our results do show that the intensity of marketing innovations undertaken over the past 3 years has a positive association with expectations of future performance (H10b supported).

Table 6 Discriminant validity and correlations between factors

	Entrepreneurial orientation	Expected performance	Experience	Market orientation	Marketing innovation	Product innovation	Sales	Satisfaction with performance
Entrepreneurial orientation	0.593							
Expected performance	0.299	0.894						
Experience	0.096	−0.06	1					
Market orientation	0.326	0.410	−0.078	0.719				
Marketing innovation	0.502	0.437	−0.060	0.570	0.766			
Product innovation	0.288	0.049	−0.192	0.233	0.465	0.731		
Sales	0.344	0.118	0.258	0.058	0.127	0.176	1	
Satisfaction with performance	0.323	0.331	0.012	0.193	0.270	0.106	0.184	0.785

Note: Diagonal is square root of AVE; off-diagonal is correlation between variables

Table 7 Results of path analysis

	Path coefficient	*P* values	Indirect effects	*P* values	Total effects	*P* values
Market orientation → marketing innovation	0.394	0.000			0.394	0.000
Market orientation → product innovation	0.136	0.116			0.136	0.116
Market orientation → satisfaction with performance	0.066	0.581			0.104	0.346
Market orientation → expected performance	0.245	0.011	0.084	0.080	0.329	0.000
Entrepreneurial orientation → product innovation	0.182	0.079			0.182	0.079
Entrepreneurial orientation → marketing innovation	0.283	0.000			0.283	0.000
Entrepreneurial orientation → satisfaction with performance	0.190	0.076	0.025	0.570	0.215	0.032
Entrepreneurial orientation → expected performance	0.075	0.444	0.076	0.090	0.151	0.138
Product innovation → satisfaction with performance	−0.027	0.806			−0.027	0.806
Product innovation → expected performance	−0.128	0.166			−0.133	0.177
Marketing innovation → satisfaction with performance	0.104	0.375			0.104	0.375
Marketing innovation → expected performance	0.207	0.026			0.227	0.012
Satisfaction with performance → expected performance	0.188	0.066			0.188	0.066
Sales → expected performance	0.058	0.489			0.078	0.340
Sales → satisfaction with performance	0.106	0.292			0.106	0.292
Experience → satisfaction with performance	−0.027	0.807			−0.027	0.807
Experience → expected performance	−0.073	0.331			−0.078	0.331
	R square	Adjusted *R* square				
Expected performance	0.250	0.207				
Marketing innovation	0.300	0.289				
Product innovation	0.066	0.051				
Satisfaction with performance	0.102	0.058				

Analysis of indirect and total effects within the bootstrapped estimates shows instances where innovation intensity mediates the relationships between the latent constructs. Significant indirect (and total) effects are seen in paths that link market orientation and expected future performance (through product innovation intensity, market innovation intensity, and satisfaction with performance). In addition, there is a significant (albeit small) indirect effect between an entrepreneurial orientation and satisfaction with current performance levels.

The ability of the model to explain satisfaction with current performance is rather limited as the *r*-squared is 0.102, compared to the corresponding value of 0.25 for expected future performance. This difference between the two models could be traced back to the coefficients for testing hypotheses 3, 7, 9a, and 10a, the paths from managerial orientations and innovation intensities to the respective indicator of firm performance. In those four paths, only the path from the firm's entrepreneurial orientation to satisfaction with current performance was significant at the 10% level. However, for expected future performance, positive and highly significant coefficients can be reported. Increases in market orientation lead to higher expected performance ($\beta = 0.245$) supporting hypothesis 4, as does increased engagement in marketing innovation ($\beta = 0.207$), thus supporting hypothesis 10b. These marketing-related paths are further strengthened by the positive correlation between market orientation and marketing innovation intensity (hypothesis 1). Hypothesis 8, which posits that an entrepreneurial orientation is positively correlated with expected performance, is not supported by our analysis.

3.1 Managerial implications

The following key findings of our analysis are particularly relevant for an accurate discussion of managerial implications. First, market orientation and entrepreneurial orientation have been found to be distinct constructs. Both are directly linked to innovation intensity and to different measures of performance. However, without further investigation of the paths to farm performance, these observations would be of mere descriptive value. So building on the first key result, the second one is that both market orientation and entrepreneurial orientation are indirectly—through innovation decisions—related to expected performance. The ability to identify these extended paths of indirect effects is one of the major advantages of the chosen approach of structural equation modelling. Although entrepreneurial orientation is positively related to both product and marketing innovation intensity, its impact is stronger on the latter. Additionally, we find that the effect of a market orientation is more strongly associated with the intensity of marketing innovations.

The stronger relationship between these orientations and market innovation intensity (rather than product innovation intensity) may reflect the findings of Levinthal and March (1993) who suggest that the returns to exploitation may be greater than the returns to exploration for certain firms. One might posit that marketing a current product in a different channel would constitute exploitative search as this requires the firm to continue producing current offerings while searching for new markets in which to sell them. The skills required to be successful in this search may be very different (and more easy to obtain) than the skills required to identify and iterate new product development strategies. This may echo the uncertainty of success that many firms face when developing product-focused innovations (Evanschitzky et al. 2012). Furthermore, as many of the firms used in our study have few employees relative to established food processing firms, the level of human and organizational capital devoted to new product development may be lower, which may lead to a lower success rate and therefore lower expectations of future performance. Another explanation of this result could be that product and packaging innovation may require major investments that present a particular financial burden in the first years after the investment.

But even so, the path with positive and highly significant effects from market orientation to expected performance directly—as well as indirectly through marketing innovation intensity—is striking as it suggests that firms who are aware of customer demands and competitor actions expect significant benefits from marketing-focused innovations. Similarly, our result that increasing market orientation but not entrepreneurial orientation was directly linked to higher expected performance may indicate that more market-oriented firms have more options and flexibility or possibly also more confidence to adapt to market changes to stay profitable. Without these capacities, investments in product-related innovations are also at higher risk of underperforming or even failing. Our survey did not cover how these skills and capacities were or could be obtained, but we would expect that farm operators will increasingly express demand for corresponding learning opportunities and specialized services in these areas.

Finally, although product and marketing innovations are not the only possible paths to increased performance, they are by far the most relevant in the context of direct-to-consumer and alternative marketing approaches. Further, our results corroborate findings of previous research across different industries, as we show that more market-oriented and more entrepreneurial firms have higher levels of expected performance than their rivals (Wiklund and Shepherd 2003; Narver et al. 2004; Verhees and Meulenberg 2004; Beck et al. 2011). The key contribution beyond these previous findings is, however, that a positive impact of entrepreneurial orientation on performance hinges on the firm's market orientation. This is likely a unique characteristic of the relatively small direct-to-consumer and alternative marketing industry in agriculture.

4. CONCLUSIONS

Using a PLS-SEM applied to data from a 2013 survey of direct marketers in Ontario, Canada, this paper examines the importance of a market orientation and entrepreneurial orientation for farm operations. These results show that market awareness coupled with a timely response is an important factor in improving expected future performance.

One of the limitations of this study is that it relies on self-reported and subjective measures of current and future performance. The use of self-reported subjective measures may not have been avoidable since the subjects of inquiry are closely held farm operations. Therefore, the results should be interpreted with caution, especially when it comes to suggesting causal links between entrepreneurial traits, innovation activity, and satisfaction with current performance or expectations of future performance. This also applies to the fact that the data are cross-sectional in nature, as it relies on data from one point in time. Additionally, this study examines the results of marketing and product innovation strategies in Ontario, Canada; therefore, the applicability to other areas may be limited. However, we feel that the results of this research could provide firms in similar political, technological, social, and economic situations a framework from which to build a strategy to compete within direct and alternative markets. Future research in this area could develop new tools and methods to overcome these limitations, either by developing a panel dataset or by establishing the relationship between subjective and objective measures of performance within this sector.

Bearing these limitations in mind, the following general results are highlighted. First, an entrepreneurial orientation has been found to be positively associated with both product and marketing innovation intensity, which suggests that proactive and innovative firms are more likely to engage in new product development or implement changes in how products are marketed. We also find that a market orientation is positively associated with product and marketing innovation. This finding provides broader support for the link between marketing capabilities and innovation among SMEs (Banterle et al. 2011) by finding a similar linkage with respect to a market orientation to product and marketing innovation within another context. Furthermore, by modelling both a market orientation and an entrepreneurial orientation, we are able to observe how increases in these strategic resources may directly and indirectly influence expectations of future performance. For firms operating in highly competitive markets, developing an entrepreneurial orientation alongside a market orientation may enable the firm to become aware of and seize opportunities in new or rapidly evolving markets such as entomology-based protein chains (Pascucci et al. 2015).

Second, since innovation activity was measured as new-to-the-farm adaptations or changes during the most recent 3 years, no significant link between current farm performance and innovation output was found. However, we did find a link between innovations and expected performance. Specifically, we find more market-oriented farms are expecting significant improvements in future performance. This may be a reflection of optimism bias (Meyer 2014), but it also may be the manifestation of a true relationship between awareness of market needs and the firm's ability to meet these needs through changes in their marketing mix.

Fourth and finally, farms with greater entrepreneurial orientation are more likely to develop new products or new packaging for their customers, but there seems to be some uncertainty as to how these innovations are expected to affect performance in the near future. For these farms, focusing on marketing innovations may be a less risky gamble while also making better use of their current capabilities (Carreresi et al. 2015). This may be the result of the necessary skills and capabilities needed to execute the development of a new product line which are less prevalent among small food processing firms relative to the skills and capabilities needed to develop and navigate new marketing channels. Alternatively, it may reflect perceived differences in the risk or real differences in the amortization period of the two types of investment.

As more and more farms utilize direct and alternative markets to sell their products, the development of market-sensing capabilities will become increasingly important. As these markets mature and the novelty of these marketing methods wanes among non-core customers, success within these channels will likely depend on management's ability to identify and react to new and sustainable market trends. In cases such as this, a market orientation would be a valuable resource as it would enable the farm to become aware of opportunities to provide superior value for consumers.

ENDNOTES

[1]Seven hundred ninety-nine respondents attempted the survey. One hundred two respondents were found to be ineligible due to size (less than $10,000 annual gross sales). Two hundred ninety-two respondents started the survey but did not complete it (for unknown reasons), while 405 completed it.

[2]Since the survey excluded farms with less $10,000 annual gross sales, the number of e-mails to eligible recipients must have been considerably smaller than 10,406. Assuming the same share as in those who attempted the survey, i.e. 102 of 799, or 12.8%, the number of eligible recipients would have been 9078, which would yield a response rate of 4.5%. However, this is likely a lower bound as the link was deactivated by IPSOS Animal Health immediately after 405 completed questionnaires had been received.

[3]The only outlier in AVE is the EO scale, which has an AVE of 0.352.

[4]For example, the number of perceived direct competitors was elicited in the survey as a measure of fierceness of competition that could impact innovation behaviour. However, the question did not refer specifically to direct and alternative marketing competitors so that when included in the statistical analysis did not improve model fit and had no significant coefficient estimates. Hence, it was dropped from being reported here. Similarly, the main commodity produced can impact innovation activity due to technical and regulatory reasons. However, including this variable was not feasible either because of small numbers for most product groups and because many direct and alternative marketers produce a variety of commodities that have significant contributions to income generation.

APPENDIX 1

Table 8 Items of managerial orientation scales and scale reliabilities

Measurement scale	Mean	Std. deviation	Item-to-total correlation
Market orientation [Alpha = 0.938; AVE = 0.516]			
The business objectives on our farm operation are driven by customer satisfaction.	5.28	1.598	0.627
We continually monitor our level of commitment to serving customers' needs.	4.79	1.636	0.765
Our strategy for competitive advantage is based on our understanding of customer needs.	5.19	1.484	0.756
Our strategies are driven by our beliefs about how we can create greater value for our customers.	5.15	1.555	0.763
We measure customer satisfaction regularly.	4.98	1.762	0.744
We pay close attention to our customers, even after the sale is made.	5.30	1.646	0.744
We share information with our employees concerning competitors' strategies.	3.98	1.852	0.592
We are quick to respond to competitive actions that threaten us.	4.46	1.657	0.658
We target customers and customer groups where we have, or can develop, a competitive advantage.	4.79	1.603	0.717
We regularly discuss competitors' strengths and strategies.	4.17	1.756	0.619
We regularly visit current customers to see how our products and/or services are meeting their needs.	4.39	1.859	0.741
We discuss reasons for successful and unsuccessful customer experiences on a regular basis.	4.72	1.652	0.817
We coordinate all of our business functions (from buying to producing, selling and accounting) in order to better serve the needs of our target markets.	4.62	1.784	0.770

We understand how everyone in our company can contribute to creating customer value.	5.06	1.599	0.654
Entrepreneurial orientation [Alpha = 0.824; AVE = 0.352]			
A strong emphasis on the use of tried and true products or services for our farm operation.	3.86	1.691	0.431
No new lines of products or services.	3.59	1.786	0.501
Changes in product or service lines have been mostly of a minor nature.	3.41	1.790	0.553
Typically responds to actions which competitors initiate.	3.84	1.286	0.386
Typically seeks to avoid clashes with competitors, preferring a live-and-let-live attitude.	3.52	1.446	0.549
Tend to focus on low-risk investment projects (with normal and certain rates of return).	3.26	1.476	0.658
Tend to focus on low-risk investment projects (with normal and certain rates of return).	3.26	1.476	0.658
Owing to the nature of the business environment, it is best to explore our options gradually via cautious, incremental behaviour.	3.25	1.440	0.664
Typically adopt a cautious wait and see attitude in order to minimize the probability of making costly decision.	3.29	1.469	0.666

Table 9 Items of performance scales and scale reliabilities

	Current performance[a] [Alpha = 0.831; AVE = 0.617]			Expected performance[b] [Alpha = 0.922; AVE = 0.799]		
	Mean	Std. deviation	Loading	Mean	Std. deviation	Loading
Gross farm sales	4.16	1.54	0.620	4.91	1.22	0.961
Net farm income	3.49	1.62	0.742	4.68	1.36	0.849
Cash flow	3.76	1.69	0.956	4.63	1.21	0.867

[a]Scale ranged from 1 (very dissatisfied) to 7 (very satisfied)
[b]Scale ranged from 1 (will deteriorate significantly) to 7 (will improve significantly)

REFERENCES

1. Adams DC, Adams AE (2011) De-placing local at the farmers' market: consumer conceptions of local foods. J Rural Soc Sci 26(2):74–100
2. Ashok M, Narula R, Martinez-Noya A (2016) How do collaboration and investments in knowledge management affect process innovation in services? J Knowl Manag 20:1004–1024. doi:10.1108/JKM-11-2015-0429
3. Banterle A, Cavaliere A, Carraresi L, Stranieri S (2011) Innovativeness in food small business: what is its relationship with marketing? Agric Econ 57:474–483
4. Barbieri C, Mahoney E (2009) Why is diversification an attractive farm adjustment strategy? Insights from Texas farmers and ranchers. J Rural Stud 25:58–66. doi:10.1016/j.jrurstud.2008.06.001

5. Beck L, Janssens W, Debruyne M, Lommelen T (2011) A study of the relationships between generation, market orientation, and innovation in family firms. Fam Bus Rev 24:252–272. doi:10.1177/0894486511409210
6. Beckie MA, Kennedy EH, Wittman H (2012) Scaling up alternative food networks: farmers' markets and the role of clustering in western Canada. Agric Human Values 29:333–345. doi:10.1007/s10460-012-9359-9
7. Boehlje M (2002) Risk in U.S. agriculture: new challenges and new approaches
8. Boehlje M, Bröring S (2011) The increasing multifunctionality of agricultural raw materials: three dilemmas for innovation and adoption. Int Food Agribus Manag Rev 14:1–16
9. Boehlje M, Akridge J, Downey D (1995) Restructuring agribusiness for the 21st century. Agribusiness 11:493–500. doi:10.1002/1520-6297(199511/12)11:6<493::AID-AGR2720110602>3.0.CO;2-G
10. Brik AB, Rettab B, Mellahi K (2011) Market orientation, corporate social responsibility, and business performance. J Bus Ethics 99:307–324. doi:10.1007/s10551-010-0658-z
11. Bryant FB (2000) Assessing the validity of measurement. In: Grimm LG, Yarnold PR (eds) Reading and understanding more multivariate statistics. American Psychological Association, Washington, pp 99–146
12. Cambra-Fierro J, Florin J, Perez L, Whitelock J (2011) Inter-firm market orientation as antecedent of knowledge transfer, innovation and value creation in networks. Manag Decis 49:444–467. doi:10.1108/00251741111120798
13. Camisón C, Villar-López A (2011) Non-technical innovation: organizational memory and learning capabilities as antecedent factors with effects on sustained competitive advantage. Ind Mark Manag 40:1294–1304. doi:10.1016/j.indmarman.2011.10.001
14. Carreresi L, Mamaqi X, Albisu LM, Banterle A (2015) Can strategic capabilities affect performance? Appication of RBV to small food businesses. Agribusiness. doi:10.1002/agr.21451
15. Cheng CC, Krumwiede D (2012) The role of service innovation in the market orientation—new service performance linkage. Technovation 32:487–497. doi:10.1016/j.technovation.2012.03.006
16. Covin JG, Slevin DP (1989) Strategic management of small firms in hostile and benign environments. Strateg Manag J 10:75–87. doi:10.1002/smj.4250100107
17. Cranfield J, Henson S, Blandon J (2012). The effect of attitudinal and sociodemographic factors on the likelihood of buying locally produced food. Agribusiness, 28(2):205–221
18. Cronbach LJ (1951) Coefficient alpha and the internal structure of tests. Psychometrica 16:297–334
19. Cummings H, Kora G, Murray D (1999) Farmers' markets in Ontario and their economic impact
20. Dawes J (1999) The relationship between subjective and objective company performance measures in market orientation research: further empirical evidence. Mark Bull 10:65–75
21. de Lauwere CC (2005) The role of agricultural entrepreneurship in Dutch agriculture of today. Agric Econ 33:229–238
22. Dess GG, Robinson RB Jr (1984) Measuring organizational performance in the absence of objective measures: the case of the privately-held firm and conglomerate business unit. Strateg Manag J 5:265–273
23. Dewar RD, Dutton JE (1986) The adoption of radical and incremental innovations: an empirical analysis. Manage Sci 32:1422–1433
24. Di Stefano G, Gambardella A, Verona G (2012) Technology push and demand pull perspectives in innovation studies: current findings and future research directions. Res Policy 41:1283–1295. doi:10.1016/j.respol.2012.03.021
25. Dodds R, Holmes M, Arunsopha V et al (2013) Consumer choice and farmers' markets. J Agric Environ Ethics 27:397–416. doi:10.1007/s10806-013-9469-4
26. Dukeshire S, Masakure O, Mendoza J et al. (2014) Understanding consumer choices for Ontario produce. Renew Agric Food Syst :1–11. doi:10.1017/S1742170514000234

27. Evanschitzky H, Eisend M, Calantone RJ, Jiang Y (2012) Success factors of product innovation: an updated meta-analysis. J Prod Innov Manag 29:21–37. doi:10.1111/j.1540-5885.2012.00964.x
28. Farrell MA, Oczkowski E, Kharabsheh R (2008) Market orientation, learning orientation and organisational performance in international joint ventures. Asia Pacific J Mark Logist 20:289–308. doi:10.1108/13555850810890066
29. Feldmann C, Hamm U (2015). Consumers' perceptions and preferences for local food: A review. Food Quality and Preference, 40:152–164
30. Fornell C, Larcker DF (1981) Evaluating structural equation models with unobservable variables and measurement error. J Mark Res 18:39–50, doi: http://dx.doi.org/10.2307/3151312
31. Garcia R, Calantone R (2002) A critical look at technological innovation typology and innovativeness terminology: a literature review. J Prod Innov Manag 19:110–132
32. Gellynck X, Banterle A, Kühne B et al (2012) Market orientation and marketing management of traditional food producers in the EU. Br Food J 114:481–499. doi:10.1108/00070701211219513
33. Gracia A (2012) Importance of social influence in consumers' willingness to pay for local food: are there gender differences? Agribusiness 28:361–371. doi:10.1002/agr
34. Grande J, Madsen EL, Borch OJ (2011) The relationship between resources, entrepreneurial orientation and performance in farm-based ventures. Entrep Reg Dev An Int J 23:89–111. doi:10.1080/08985620903183710
35. Gunday G, Ulusoy G, Kilic K, Alpkan L (2011) Effects of innovation types on firm performance. Int J Prod Econ 133:662–676. doi:10.1016/j.ijpe.2011.05.014
36. Guthrie J, Guthrie A, Lawson R, Cameron A (2006) Farmers' markets: the small business counter-revolution in food production and retailing. Br Food J 108:560–573. doi:10.1108/00070700610676370
37. Hair JF, Ringle CM, Sarstedt M (2011) PLS-SEM: indeed a silver bullet. J Mark Theory Pract 19:139–152. doi:10.2753/MTP1069-6679190202
38. Hair JF, Sarstedt M, Pieper TM, Ringle CM (2012) The use of partial least squares structural equation modeling in strategic management research: a review of past practices and recommendations for future applications. Long Range Plann 45:320–340. doi:10.1016/j.lrp.2012.09.008
39. Han JK, Kim N, Srivastava RK (1998) Market orientation and organizational performance: is innovation a missing link? J Mark 62:30–35, doi: http://dx.doi.org/10.2307/1252285
40. Hansen JD, Deitz GD, Tokman M et al (2011) Cross-national invariance of the entrepreneurial orientation scale. J Bus Ventur 26:61–78. doi:10.1016/j.jbusvent.2009.05.003
41. Hong J, Song TH, Yoo S (2013) Paths to success: how do market orientation and entrepreneurship orientation produce new product success? J Prod Innov Manag 30:44–55. doi:10.1111/j.1540-5885.2012.00985.x
42. Hu W, Batte MT, Woods T, Ernst S (2012) Consumer preferences for local production and other value-added label claims for a processed food product. Eur Rev Agric Econ 39:489–510. doi:10.1093/erae/jbr039
43. Hulland J (1999) Use of partial least squares (PLS) in strategic management research: a review of four recent studies. Strateg Manag J 20:195–204. doi:10.2307/3094025
44. Keh HT, Nguyen TTM, Ng HP (2007) The effects of entrepreneurial orientation and marketing information on the performance of SMEs. J Bus Ventur 22:592–611. doi:10.1016/j.jbusvent.2006.05.003
45. Kim M, Curtis KR, Yeager I (2014) An assessment of market strategies for small-scale produce growers. Int Food Agribus Manag Rev 17:187–204
46. Knudson W, Wysocki A, Champagne J, Peterson HC (2004) Entrepreneurship and innovation in the agri-food system. Am J Agric Econ 86:1330–1336
47. Köhler C, Sofka W, Grimpe C (2012) Selective search, sectoral patterns, and the impact on product innovation performance. Res Policy 41:1344–1356. doi:10.1016/ j.respol.2012.03.020

48. Lambrecht E, Taragola N, Kühne B et al (2015) Networking and innovation within the ornamental plant sector. Agric Food Econ 3:1–20. doi:10.1186/s40100-014-0022-1
49. Lamore PR, Berkowitz D, Farrington PA (2013) Proactive/responsive market orientation and marketing—research and development integration. J Prod Innov Manag 30:695–711. doi:10.1111/jpim.12024
50. Langerak F (2003) The effect of market orientation on positional advantage and organizational performance. J Strateg Mark 11:93–115
51. Lee RP, Zhou KZ (2012) Is product imitation good for firm performance? An examination of product imitation types and contingency factors. J Int Mark 20:1–16. doi:10.1509/jim.12.0019
52. Lee H, Smith KG, Grimm CM, Schomburg A (2000) Timing, order and durability of new product advantages with imitation. Strateg Manag J 21:23–30. doi:10.1002/(SICI)1097-0266(200001)21:1<23::AID-SMJ64>3.0.CO;2-0
53. Lee C, Hallak R, Sardeshmukh SR (2016) Innovation, entrepreneurship, and restaurant performance: a higher-order structural model. Tour Manag 53:215–228. doi:10.1016/j.tourman.2015.09.017
54. Levinthal DA, March JG (1993) The myopia of learning. Strateg Manag J 14:95–112. doi:10.1002/smj.4250141009
55. Li Y-H, Huang J-W, Tsai M-T (2009) Entrepreneurial orientation and firm performance: the role of knowledge creation process. Ind Mark Manag 38:440–449. doi:10.1016/j.indmarman.2008.02.004
56. Low SA, Vogel S (2011) Direct and intermediated marketing of local foods in the United States
57. Lumpkin GT, Dess GG (1996) Clarifying the entrepreneurial orientation construct and linking it to performance. Acad Manag Rev 21:135–172
58. Lumpkin GT, Dess GG (2001) Linking two dimensions of entrepreneurial orientation to firm performance: the moderating role of environment and industry life cycle. J Bus Ventur 16:429–451. doi:10.1016/S0883-9026(00)00048-3
59. Mahto RV, Davis PS, Pearce JA, Robinson RB (2010) Satisfaction with firm performance in family businesses. Entrep Theory Pract 34:985–1001. doi:10.1111/j.1540-6520.2010.00393.x
60. Mirzaei OM, Micheels ET, Boecker A (2016). Product and Marketing Innovation in Farm-Based Businesses: The Role of Entrepreneurial Orientation and Market Orientation. International Food and Agribusiness Management Review, 19(2):99–130
61. Meyer WG (2014) The effect of optimism bias on the decision to terminate failing projects. Proj Manag J 45:7–20. doi:10.1002/pmj
62. Monecke A, Leisch F (2012) semPLS: structural equation modeling using partial least squares. J Stat Softw 48:1–32
63. Moreno AM, Casillas JC (2008) Entrepreneurial orientation and growth of SMEs: a causal model. Entrep Theory Pract 32:507–528
64. Moser R, Raffaelli R, Thilmany-McFadden D (2011) Consumer preferences for fruit and vegetables with credence-based attributes: a review. Int Food Agribus Manag Rev 14:121–142
65. Mount P (2012) Growing local food: scale and local food systems governance. Agric Human Values 29:107–121. doi:10.1007/s10460-011-9331-0
66. Nadella K, Deaton B, Lawley C, Weersink A (2014) Do farmers treat rented land differently than the land they own? A fixed effects model of farmer's decision to adopt conservation practices on owned and rented land. In: Agricultural & Applied Economics Association's 2014 AAEA Annual Meeting., pp 1–17
67. Narver JC, Slater SF (1990) The effect of a market orientation on business profitability. J Mark 54:20–35, doi: http://dx.doi.org/10.2307/1251757
68. Narver JC, Slater SF, Maclachlan DL (2004) Responsive and proactive market orientation and new-product success. J Prod Innov Manag 21:334–347. doi:10.1111/j.0737-6782.2004.00086.x
69. Nasution HN, Mavondo FT, Matanda MJ, Ndubisi NO (2011) Entrepreneurship: its relationship with market orientation and learning orientation and as antecedents to

innovation and customer value. Ind Mark Manag 40:336–345. doi:10.1016/j.indmarman.2010.08.002
70. Nelson RR, Winter SG (1982) An evolutionary theory of economic change. Harvard University Press, Boston
71. Ngo LV, O'Cass A (2012) In search of innovation and customer-related performance superiority: the role of market orientation, marketing capability, and innovation capability interactions. J Prod Innov Manag 29:861–877. doi:10.1111/j.1540-5885.2012.00939.
72. OMAFRA (2014). Number of Ontario Census Farms Classified by Total Gross Farm Receipts, 2006 and 2011. http://www.omafra.gov.on.ca/english/stats/census/receipts11.htm
73. Pascucci S, Dentoni D, Mitsopoulos D (2015) The perfect storm of business venturing? The case of entomology-based venture creation. Agric Food Econ. doi:10.1186/s40100-014-0025-y
74. Pérez-Luño A, Wiklund J, Cabrera RV (2011) The dual nature of innovative activity: how entrepreneurial orientation influences innovation generation and adoption. J Bus Ventur 26:555–571. doi:10.1016/j.jbusvent.2010.03.001
75. Rauch A, Wiklund J, Lumpkin GT, Frese M (2009) Entrepreneurial orientation and business performance: an assessment of past research and suggestions for the future. Entrep Theory Pract 33:761–787
76. Regional Analytics and Planscape (2011) Economic impacts of farmers' markets in the province of Ontario
77. Renko M, Carsrud A, Brännback M (2009) The effect of a market orientation, entrepreneurial orientation, and technological capability on innovativeness: a study of young biotechnology ventures in the United States and in Scandinavia. J Small Bus Manag 47:331–369
78. Rhee J, Park T, Lee DH (2010) Drivers of innovativeness and performance for innovative SMEs in South Korea: mediation of learning orientation. Technovation 30:65–75. doi:10.1016/j.technovation.2009.04.008
79. Richard PJ, Devinney TM, Yip GS, Johnson G (2009) Measuring organizational performance: towards methodological best practice. J Manage 35:718–804. doi:10.1177/0149206308330560
80. Ringle CM, Wende S, Becker J-M (2015) SmartPLS 3
81. Rosenbusch N, Brinckmann J, Bausch A (2011) Is innovation always beneficial? A meta-analysis of the relationship between innovation and performance in SMEs. J Bus Ventur 26:441–457. doi:10.1016/j.jbusvent.2009.12.002
82. Roucan-Kane M, Gray AW, Boehlje MD (2011) Approaches for selecting product innovation projects in U.S. food and agribusiness companies. Int Food Agribus Manag Rev 14:51–68
83. Rowley J, Baregheh A, Sambrook S (2011) Towards an innovation-type mapping tool. Manag Decis 49:73–86. doi:10.1108/00251741111094446
84. Schumilas T (2010) The feeders meet the eaters—direct marketing in Ontario's organic sector
85. Shoemaker PJ, Eichholz M, Skewes EA (2002) Item nonresponse: distinguishing between don't know and refuse. Int J Public Opin Res 14:193–201
86. Slater SF, Narver JC (1995) Market orientation and the learning organization. J Mark 59:63–74, doi: http://dx.doi.org/10.2307/1252120
87. Slater SF, Narver JC (2000) Intelligence generation and superior customer value. J Acad Mark Sci 28:120–127
88. Song M, Thieme J (2009) The role of suppliers in market intelligence gathering for radical and incremental innovation. J Prod Innov Manag 26:43–57. doi:10.1111/j.1540-5885.2009.00333.x
89. Sun L, Gómez MI, Chaddad FR, Ross RB (2014) Distribution channel choices of wineries in emerging cool climate regions. Agric Resour Econ Rev 43:87–103

90. Tajeddini K (2010) Effect of customer orientation and entrepreneurial orientation on innovativeness: evidence from the hotel industry in Switzerland. Tour Manag 31:221–231. doi:10.1016/j.tourman.2009.02.013
91. Tajeddini K, Trueman M, Larsen G (2006) Examining the effect of market orientation on innovativeness. J Mark Manag 22:529–551. doi:10.1362/026725706777978640
92. Tan M, Liu Z (2014) Paths to success: an ambidexterity perspective on how responsive and proactive market orientations affect SMEs' business performance. J Strateg Mark :1–22. doi:10.1080/0965254X.2013.876084
93. Thilmany D, Bond CA, Bond JK, (2008). Going local: Exploring consumer behavior and motivations for direct food purchases. American Journal of Agricultural Economics, 90(5):1303–1309
94. Trienekens J, van Uffelen R, Debaire J, Omta O (2008) Assessment of innovation and performance in the fruit chain: the innovation-performance matrix. Br Food J 110:98–127. doi:10.1108/00070700810844812
95. Uematsu H, Mishra AK (2011) Use of direct marketing strategies by farmers and their impact on farm business income. Agric Resour Econ Rev 40:1–19
96. Verhees FJHM, Meulenberg MTG (2004) Market orientation, innovativeness, product innovation, and performance in small firms. J Small Bus Manag 42:134–154, doi: http://dx.doi.org/10.1111/j.1540-627X.2004.00102.x
97. Wall TD, Michie J, Patterson M et al (2004) On the validity of subjective measures of company performance. Pers Psychol 57:95–118. doi:10.1111/j.1744-6570.2004.tb02485.x
98. Wang CL (2008) Entrepreneurial orientation, learning orientation, and firm performance. Entrep Theory Pract 44:635–657
99. Weatherell C, Tregear A, Allinson J (2003) In search of the concerned consumer: UK public perceptions of food, farming and buying local. J Rural Stud 19:233–244
100. Webb JW, Ireland RD, Hitt MA et al (2010) Where is the opportunity without the customer? An integration of marketing activities, the entrepreneurship process, and institutional theory. J Acad Mark Sci 39:537–554. doi:10.1007/s11747-010-0237-y
101. Wiklund J (1999) The sustainability of the entrepreneurial orientation-performance relationship. Entrep Theory Pract 24:37–48
102. Wiklund J, Shepherd D (2003) Knowledge-based resources, entrepreneurial orientation, and the performance of small and medium-sized businesses. Strateg Manag J 24:1307–1314. doi:10.1002/smj.360
103. Wiklund J, Shepherd D (2005) Entrepreneurial orientation and small business performance: a configurational approach. J Bus Ventur 20:71–91. doi:10.1016/j.jbusvent.2004.01.001
104. Yannopoulos P, Auh S, Menguc B (2012) Achieving fit between learning and market orientation: implications for new product performance. J Prod Innov Manag 29:531–545. doi:10.1111/j.1540-5885.2012.00923.x

CHAPTER 2

Determinants of household's consumption preference for processed cocoyam in Enugu State, Nigeria

Adeosun K. P.*, Amaechina E. C. and Nnaji A. P.

Department of Agricultural Economics, University of Nigeria, Nsukka, Enugu State, Nigeria.

ABSTRACT

Cocoyam is a nutritious food crop which is consumed in different parts of Nigeria. Its importance in the food systems of many communities in Nigeria, particularly among the Igbo people is based on its nutritive value and its relative ease in cultivation. Studies on the crop focused more on production and marketing without examining consumer behaviour. Although, an understanding of the determinants of the consumption of various processed forms as well as volume consumed by households is important information for policy makers, this has scarcely been examined in Nigeria. Understanding the socio-economic dynamics that make consumers choose one form of the product and not another is crucial for ensuring the food security of the poor. This study therefore examined the factors that influenced consumption of processed cocoyam and the volume consumed. For the study, multistage sampling technique was used to select the respondents. Data collected was analyzed using multiple linear regression and multinomial logistic regression. The results show that income allocated for cocoyam consumption, hectares of cocoyam cultivated, number of times cocoyam is consumed, distance from home to market and household size are important determinants of volume of cocoyam consumed by households, while, household size, quantity of cocoyam consumed by households, farming experience, age, marital status and income spent on cocoyam are important determinants of consumption of processed cocoyam.

Keywords: Consumer behaviour, household, processed cocoyam, food security.

1. INTRODUCTION

Cocoyam (*Colocasia* spp.*and Xanthosoma* spp.) as a crop has until recently received little attention from international and regional bodies (Agbelemoge, 2013; Onyeka 2014), mainly due to the low value given to it. However, given its potential as an affordable crop for the poor, and the increasing awareness of its health value, there is a growing focus on the crop. It negates fundamentals of increasing demand due to high population pressure as well as urban development. More people now accept the consumption of highly cherished foods within their culture. This threatens the food security of households that consume such crops. According to Onyeka (2014), cocoyam is nutritionally more important than yam and cassava in terms of higher protein, mineral and vitamins content as well as digestible starch. As a tuber crop, it belongs to the class of staple foods that provide most calorific intake by Nigerians (Amusa et al., 2011). Cocoyam remains an underutilized and poorly understood crop in spite of its potential as a food and cash crop and its higher nutritive value (Onyeka, 2014). Okeke et al. (2009) noted that the problem of malnutrition of poor nations will be difficult to solve through food aid from developed countries but rather by effective consumption of indigenous plant foods. The reason is because traditional foods are more likely to meet the household food security needs of the population, particularly the rural households than imported foods.

The crop can be processed into many forms that meet the food needs of households. Amongst the Igbo of Southeastern Nigeria, the cocoyam can be prepared in various forms (Okeke et al., 2009). Cocoyam is available almost all year round (Ndabikunze et al., 2011). Household food inadequacy problem can always be solved when the people can have easy access to food they prefer in the forms they prefer them (Amaza et al., 2009). According to Omotesho et al. (2010), food security may be described as conditions or situations which guarantee easy access to highly nutritious, affordable, socially acceptable and environmentally friendly food within a community. In this case, availability means food can be easily produced by the consumer themselves. Accessibility connotes that food can be transformed into different forms that are preferred by consumers. Affordability suggests the consumers have the purchasing power or ability to obtain food at all times (Omonona and Agoi, 2007). Therefore, if cocoyam is processed into different forms, it creates wider opportunity for its accessibility to consumers, because they have a more diverse spread of options for consuming cocoyam.

Food security can be said to be achieved if people can have access to their preferred food and in the forms they prefer it. The study therefore seeks to determine factors influencing consumption of cocoyam and preference for processed cocoyam.

1.1 Motivation for cocoyam consumption

Cocoyam (*Xanthosoma sagittifolium*) contributes significant portion of the carbohydrate content of the diet in many regions in sub-Sahara Africa and provide edible starchy storage corms or cormels (Sanful and Darko, 2010). According to Opara (2002), cocoyam is perceived to be less important than other tropical roots such as yam, cassava and sweet potato. However, they are still a major staple in some parts of the tropics and sub-tropics, particularly in the rural areas of these regions (Ojinaka et al., 2009). Cocoyam is being put to different uses like other staple foods such as yam, cassava and potatoes. Although, it is not considered as prestigious as yam, its flour has the added advantage that, it is highly digestible and so is used for ingredient in baby foods (Sanful and Darko, 2010). According to Enwelu et al. (2014), consumption of mixture of cocoyam and beans is fairly good and should be encouraged because most people in rural areas eat unbalanced diets usually made up of carbohydrates. Nutritionally, cocoyam is rich in carbohydrates with nutritional value comparable to potato and superior to cassava and yam in the possession of higher protein, mineral and vitamin contents as well as easily digestible starch (Parkinson, 1984; Splittstoesser et al., 1973). It is highly recommended for diabetic patients, the aged, children with allergies and for other persons with intestinal disorders (Plucknett, 1970). These nutritional attributes make it a good base for food preparation for infants, and it has been shown that cocoyam starch can be incorporated in the development of weaning food which is highly digestible and accessible to low-income earners (Oti and Akobundu, 2008).

Cocoyam chips are popular in the local communities in Nigeria and are used in preparing many different local cocoyam delicacies. For instance, consumers of cocoyam believe that it is both energy giving and a light food, a quality that distinguishes it from other energy giving foods like yam and cassava. The corms and cormels of cocoyam are processed by boiling, baking or frying in oil. They are also processed into different products in many parts of Nigeria. All major parts of cocoyam (corm, cormel and leaves) are edible. The young leaves are a nutritious spinach-like vegetable, which provides a lot of minerals, vitamins and thiamine (Ojinnaka et al., 2009). According to Women Group in Kwaso located in the Ashanti region of Ghana, role of cocoyam in the livelihood of rural dwellers is indispensible. When asked if they could do without cocoyam if provided with support in growing alternative crops such as plantain, cassava or yam, the women overwhelmingly exclaimed that doing without cocoyam production is a recipe for hunger which is practically impossible for them to accept. Consumption of cocoyam is seen as part of their culture and therefore cannot be replaced. Cocoyam is more preferred by the aged in the communities, and often used by mothers as weaning food in the absence of commercial baby foods. Cocoyam stores longer even after harvest, and can be left in the ground until needed, thereby providing food all year round (Onyeka, 2014).

2. MATERIALS AND METHODS

2.1 Study area

Enugu State of Nigeria was the study location. Enugu is among the five south eastern states including Imo, Ebonyi, Anambra and Abia. It consists of 36 States. Enugu is located between latitudes 5°61`N and longititudes 6°53`E and 7°55`E (Enugu State Agricultural Development Programme (ENADEP), 2012). The state has a total land mass of about 8,022.96 km^2. It has a population of about 4,185,509 (NPC, 2006). Most of the population lives in rural communities with farming as their major occupation. The major crops grown in the states are yam, cassava, cocoyam, rice, maize as well as variety of fruits and legumes. It boosts the local economy of the state as it is predominantly rural and major occupation is farming (Enwelu et al., 2014).

2.2 Sampling procedure

The units of analysis were households and household heads and were selected to be the target respondents in selected communities of Enugu State. For sampling procedure, a multi stage sampling technique was employed in selecting the households for the study. In the initial stage, one agricultural zone was randomly selected from the list of agricultural zones in Enugu State. The second stage involves the selection of two (2) local government areas randomly from the list of local government areas in Enugu. The third stage involved random selection of ten communities from each of the 2 local government areas, giving a total of 20 communities. Afterwards from each community, nine households were randomly selected giving 180 households.

2.3 Data for the study

Data used for the study were collected from primary sources. The data was obtained using semi-structured questionnaire and group discussion. Personal observations were used to complement the data collected. Data were collected with the use of interview questionnaire. He authors were able to collect information from 170 respondents out of the 180 households selected for the study; this represented 94% of the households sampled for study. The data collected focused on information such as the socioeconomic and institutional characteristics of the cocoyam consumers and factors that facilitate influences, consumer preference for processed cocoyam, perceived attributes of cocoyam and different forms of cocoyam are preferred by consumers.

2.3.1 Model specification

Ordinary linear square (OLS)

Multiple linear regression model was used in the study as stated:

$$Y = \alpha + \beta_1 X_1 + \beta_2 X_{2,} \beta_3 X_3 + \beta_4 X_4 + \beta_5 X_5 + \beta_6 X_6 + \beta_7 X_7 + \beta_8 X_8 + \beta_9 X_9 + \beta_{10} X_{10} + \beta_{11} X_{11} + \beta_{12} X_{12} + \beta_{13} X_{13} + ɥ)$$

where, Y = is dependent variable (volume of cocoyam consumed per month); (X_1 – X_{13}) = explanatory variables (socioeconomic characteristics and institutional factors) ɥ = error term).

Multiple linear regression was employed to determine the volume of cocoyam consumed within households.

Table 1 describes econometric variables that are included in the regression model.

Table 1. Description of variables included in OLS.

Category	Parameters	Coefficient	Unit of measurement
Volume of cocoyam consumed	Y		Kilogram (kg)/month
Constant	-	α	-
Age	X_1	β_1	Years
Gender	X_2	β_2	Discrete
Marital status	X_3	β_3	Discrete
Occupation	X_4	β_4	Discrete
Household size	X_5	β_5	Numbers
Monthly income	X_6	β_6	Naira
Farming experience	X_7	β_7	Years
Farm size	X_8	β_8	Hectares
Distance from home to market	X_9	β_9	Kilogram
Hectare of cocoyam cultivated	X_{10}	β_{10}	Hectares
Income spent on cocoyam consumption	X_{11}	β_{11}	Naira
Price of cocoyam	X_{12}	β_{12}	Naira
number of times cocoyam is consume per week	X_{13}	β_{13}	Numbers

2.3.2 Multinomial logistic regression

A Logistic Regression Model (MLM) was used in the study. Estimating a Multinomial Logistic Regression involves a series of dependent outcome variables in which one is chosen as the comparison variable (Ogundele, 2014). In this case, cooked tuber which is one of the forms cocoyam is consumed was chosen as the comparison group and all other type of processed cocoyam such as *Achicha*, soup thickening, and cocoyam mixed with beans were compared with the comparison outcome. The equation for multinomial logistic model is stated:

$$\Pr(yi = j) = \frac{\exp(Xi\beta j)}{1+\sum_{j=1}^{J}\exp(Xi\beta j)} \qquad (1)$$

And

$$\Pr(yi = 0) = \frac{1}{1+\sum_{j=1}^{J}\exp(Xi\beta j)} \qquad (2)$$

Where ith represents a single consumer, y_i is the comparison outcome (processed cocoyam), while X_I represents a vector of independent variables. The independent variables exploded in this study included: age, gender, marital status, occupation, household size, farming experience, income spent on cocoyam consumed per month, number of times cocoyam is consumed per week within household, and quantity of cocoyam consumed per month.

3. RESULTS AND DISCUSSION

The result of descriptive statistics of the socioeconomic and institutional variables used in the analysis of determinants of consumer preference is presented in Table 2. Some of these variables were included in measuring household characteristics expected to influence choice of processed cocoyam. These factors include age, gender, household size, marital status, occupation (farming or trading), number of years spent in school, farm size, farming experience and distance from home to market. The average age of the respondents was 42 years. The average household size was seven out of which three are males and four are females. The average distance of home to market where the consumer purchases the cocoyam tuber was 10 km. Also, the number of times cocoyam is consumed in a week within the households shows whether the number of times cocoyam is consumed influences different forms in which cocoyam is consumed. Furthermore, farming experience was included as part of variables that influence the preference for processed cocoyam to determine if the number of years spent farming influenced the forms in which they prefer to consume cocoyam. The average number of years of schooling of the household head and farming experience are 9 and 18, respectively. The average amount of money the respondents spent on consumption of cocoyam per month is ₦1861. The gender and marital status was also captured to determine whether it influences preference for processed cocoyam. The average hectare of cocoyam cultivated by respondents is 0.9 hectares, whilst, the average farm size is 2.7 hectares. The average monthly income of the respondents is 18,691 naira.

Table 2. Socioeconomic and institutional characteristics of respondents.

Variable	Mean	Standard deviation	Minimum	Maximum
Age	42.96	15.51	16	90
Gender[a]	0.36	0.48	0	1
Marital status[b]	0.70	0.45	0	1
Occupation (trading)[c]	0.57	0.49	0	1
No of years spent in school	9.67	4.81	0	17
No of male household size	3.2	1.9	1	15
No of female household size	3.8	2.9	1	27
Monthly income	18691.17	15378.85	2000	70000
Farming experience	18.90	12.38	3	75
Farm size	2.70	3.22	0.25	36
Distance from home to market	10.11	16.06	0.50	75
Hectares of cocoyam cultivated	0.87	0.86	0	5
Income spent on cocoyam per month	1861.77	2019.13	0	10000
Number of time cocoyam is consumed Per week	1.33	0.71	0	4

1 if gender[a] is male; 0 otherwise (female), marital status[b] 1 if married; 0 otherwise (divorced), 1 if occupation (trading)[c]; 0 otherwise (farming). Source: Field survey, 2016.

Table 3 shows the result of the multiple regression model. Considering the regression model, it gives a coefficient of multiple determination (R^2) of 0.44. This implies that variations in the explanatory variables explained only 44% of total variation of cocoyam consumed (dependent variable). The result shows that the overall regression equation was significant at 0.05 probability level, since prob > F = 0.000. Factors such as distance from home to market, hectare of cocoyam cultivated, income spent on cocoyam per month and number of times cocoyam is consumed per week are seen to be very important determinants of cocoyam consumed by households. They are significant and have a positive relationship with volume of cocoyam consumed by members of households. For instance, income spent on cocoyam is important because households see cocoyam as a delicacy, therefore the need for cocoyam consumption and the more money spent on cocoyam. This agrees with both the principles of preference and food security, that a consumer will allocate more income to that which has increased utility. Also if food is cheaper that means it is affordable and households can consume it more frequently. This is in accordance with the work of Omotesho et al. (2010) and Omonona and Agoi (2007) which suggest that the consumers have the purchasing power or ability to obtain food at all times. It also agrees with Oti and Akobundu (2008) that cocoyam is accessible by low income earners. The nutritive values of cocoyam may encourage households to cultivate more of it as it can serve as a substitute to other expensive foods that supply the same nutrients, this is supported by Parkinson (1984) which says, cocoyam is rich in carbohydrates with nutritional value comparable to potato, superior to cassava and yam in the possession of higher protein, mineral and vitamin contents. Not only that, cocoyam has been found to be good for infant food for weaning babies and for adult with diabetes (Plucknett, 1970). While, household size is significant and has a negative relationship with the volume of cocoyam consumed. This is against our *a priori* expectation that volume of cocoyam consumed will increase with an increase in the number of household members. This may suggests that different members of the household may prefer other tubers crop such as yam and cassava over cocoyam. However, most aged and children in the household consume cocoyam (Oti and Akobundu, 2008), this is because of softness and nutritive value of cocoyam. Hectare of cocoyam cultivated was significant, this shows that because of high level of importance and functions of cocoyam, households devote more of their land on cultivation of cocoyam. This agrees with interview conducted by Onyeka (2014), among women Group in Kwaso located in the Ashanti region of Ghana that received responses that the role of cocoyam in the livelihood of rural dwellers is indispensible. Also, they see cocoyam as part of their culture and therefore cannot be replaced. The characteristics of cocoyam such as energy giving, carbohydrate content, mineral content, easy to cook, light and can be prepared into different local delicacies has made it's consumption frequent within households. This is supported by result of Sanful and Darko (2010). The distance from home to market is significant and positively related, which is against our *a-priori* expectation. This can still be attributed to the importance of cocoyam to the food security of households. That means member of households can go at any length looking for cocoyam to buy for consumption.

3.1 Processed cocoyam

Cocoyam can be transformed into different forms for consumption and these forms are preferred differently by individual household. Even within households, members have different affinities for forms of processed cocoyam. In Igbo extraction, cocoyam is consumed in different forms such as cooked tuber, achicha, while some prefer it for thickening soup, others mixed it with beans. Table 4 shows the socio economic factors responsible for different forms in which cocoyam is consumed by households.

Table 3. OLS Parameter estimates of determinants of volume of cocoyam consumed.

Variable	Coefficient/standard error	t-value
Gender	3.188(2.141)	1.49
Age	0.097(0.839)	1.16
Marital status	2.617(2.252)	1.16
Number of year spent in school	0.320(0.208)	1.56
Occupation (trading)	0.555(1.976)	0.28
Household size	-0.491(0.254)	-1.98*
Monthly income	-0.000(0.000)	-1.41
Farming experience	0.150(0.097)	1.55
Farm size	0.033(0.299)	0.11
Price of cocoyam	0.001(0.001)	0.88
Distance from home to market	0.204(0.064)	3.16**
Hectare of cocoyam cultivated	2.273(1.601)	2.06**
Income spent on cocoyam per month	0.002(0.000)	3.80***
Number of times cocoyam is consume per week	6.045(1.545)	3.91***
Constant	-9.769(7.059)	-1.38

Source: field survey, 2016; R^2 = 44%; Adjusted R^2 = 39%; F- ratio = 8.75; ***, **, *show significance levels at 1, 5 and 10%, respectively.

The determinants of household's preference for different categories of processed cocoyam were ascertained. Processed cocoyam was categorized into four namely: cooked tuber, *Achicha*, soup thickener and cocoyam mixed with beans. Table 4 presents the result of the findings. The comparison group or base category is cooked tuber. The result shows that some variables significantly influenced preference for processed cocoyam with the χ^2 value of 60.40 at 0.05 level of probability.

First, estimating the category of preference for different processed cocoyam in relation to the comparison group– preference for *Achicha*, the result shows that household size, quantity of cocoyam consumed per month positively and significantly influenced preference for *Achicha* as against cooked tubers. The positive and significant effect of household size indicates that large households are more likely to consume cocoyam in form of *Achicha* than in the form of cooked tuber. Achicha has been seen as a very local delicacy strongly attached to Igbo culture. According to Onyeka (2014), rural households see cocoyam consumption as part of their culture which may be difficult to be replaced by other food types such as yam, cassava and potatoes. This also may be because cocoyam when processed to *Achicha* tends to increase in quantity and may be sufficient for large households. The positive significant effect of quantity of cocoyam suggests that households may eat more cocoyam if it is processed into Achicha, and less when it is consumed as cooked tuber.

Table 4. Parameter estimates of determinants of preference for processed cocoyam.

Preference category	Variables	Multinomial logistic results	Marginal effects
		Coefficients/standard errors	$\frac{dy}{dx}$/standard errors
Achicha	Gender	0.221(0.566)	0.037(0.063)
	Age	-0.018(0.024)	0.002(0.002)
	Marital status	-0.371(0.657)	-0.063(0.064)
	Occupation	0.450(0.516)	0.045(0.061)
	Household size	0.195**(0.0917)	0.013(0.008)
	Farming experience	-0.051**(0.024)	-0.006(0.003)
	Quantity consume	0.031*(0.019)	0.001(0.002)
	Income spent on cocoyam	-0.000***(0.000)	0.000(0.000)
	Constant	2.916(1.095)	
Soup thickening	Gender	-0.096(0.833)	-0.020(0.045)
	Age	-0.070**(0.035)	0.004(0.008)
	Marital status	0.219(0.918)	0.036(0.041)
	Occupation	0.321(0.744)	-0.005(0.042)
	Household size	0.227**(0.115)	0.003(0.005)
	Farming experience	-0.023(0.039)	0.001(0.002)
	Quantity consumed	0.044(0.027)	0.001(0.001)
	Income spent on cocoyam	-0.000***(0.000)	-0.000(0.000)
	Constant	1.819(1.422)	
Cocoyam mixed with beans	Gender	-0.925(1.450)	-4.84e-07(0.000)
	Age	0.321*(0.172)	1.57e-07(0.000)
	Marital status	17.028*(9.066)	0.000(0.000)
	Occupation	-0.556*(1.913)	-4.84e-07(0.000)
	Household size	0.572(0.431)	1.80e-07(0.000)
	Farming experience	-0.435(0.233)	-1.78e-07(0.000)
	Quantity consumed	0.135*(0.108)	4.82e-07(0.000)
	Income spent on cocoyam	-0.002(0.001)	-7.79e-07(0.000)
	Constant	-31.716(0.000)	

Source: Field survey data 2016; Comparison group- cooked tuber; LR χ^2 (24) = 60.40***; ***, **, * show significance levels at 1, 5 and 10%, respectively.

On the other hand, farming experience and income spent on cocoyam consumed per month had a negative and significant effect on consumption of cocoyam in the form of *Achicha* as against cooked tubers. This indicates that households with high number of farming experience may prefer to consume their cocoyam in the form of cooked tuber and households with low number of farming experience prefer to eat theirs in the form of *Achicha*. This suggests that those with high farming experience cultivating cocoyam may think it is easy and take no time to process cocoyam into cooked tuber for fast consumption, while the negative and significant effect of income spent on cocoyam depicts that households spend more on cocoyam if it will be consumed as cooked tuber and less if it will be consumed as *Achicha.* This may be it takes more quantity of tuber to satisfy large household as compared to achicha which transform into large quantity.

Considering using cocoyam for soup thickening as against cooked tuber, the result in the Table 4 shows that some variables, namely: age and income spent on cocoyam had negative and significant effect on eating cocoyam as soup thickening as against cooked tuber, while, household size had positive and significant effect on eating cocoyam as soup thickening as against cooked tuber. The positive and significant effect of household size depicts that large households will prefer to use cocoyam as a soup thickening as against cooked tuber. This is a common culture in Igbo land as most soups prepared are thickened with cocoyam. On the other hand, the negative and significant effect of age of the household suggest that household with elderly household heads will prefer to eat cocoyam as cooked tuber, this support the findings of Pluckett (1970) that cocoyam is highly recommended for diabetic patients and the aged. While the negative and significant effect of income spent on cocoyam indicates that households spend more on cocoyam if it will be consumed as cooked tuber and less if it will be consumed as soup thickener. It also suggests that more quantity will be needed if cocoyam is consumed as cooked tuber as compared to when is used as soup thickener.

Finally, age and marital status have a positive and significant effect and hence control household behavior as regards the consumption of cocoyam in the form of cocoyam mixed with beans as against cooked tuber. For age of the head of the households, findings show that consumption of cocoyam mixed with beans is preferred to cooked tuber if the household head is elderly. This is because beans supply additional nutrient to cocoyam since beans contained protein nutrients. Marital status indicates that household with married people tends to eat cocoyam mixed with beans than against cooked tuber. This may be as a result of having children whose nutrient requirements need to be satisfied with the combinations because cocoyam mixed with beans supply both calories and protein as compared to eating cooked tuber alone.

Farming experience has a negative and significant effect on cocoyam mixed with beans as against cooked tuber. This suggests that households with high farming experience tend to eat less of cocoyam mixed with beans and more of cooked tuber, whilst households with low farming experience prefer cocoyam mixed with beans as against cooked tuber.

4. CONCLUSION

The importance of cocoyam as household's food components cannot be over emphasized as its seen as part of their culture and most importantly cannot be replaced by other food crops. Hence there is an urgent need for cocoyam production to be taken as one of the essential crops focused on in the process of nation's food security attainment. Therefore, households should be encouraged to cultivate more of cocoyam to enable its accessibility and affordability. This study reveals essential household's socioeconomic and institutional characteristics that influenced the volume of cocoyam consumed by households. Results from the study also reveal that households have preference for different forms of processed cocoyam. There is therefore an urgent need to invest into the development of different forms of processed cocoyam to improve their quality and make them available for consumers. This is because evidence from the study shows that households may consume cocoyam often since it is economically affordable and culturally acceptable. Also, there is a need to invest in hybrid cocoyam production to boost the quantity. Generally, cocoyam should be seen as indispensible crop as is well accepted at different forms by households. Therefore, government should include cocoyam as part of arable crops transformation programme.

REFERENCES

1. Agbelemoge A (2013). Utilization of cocoyam in Rural Households in South Western Nigeria. Afr. J. Food Agric. Nutr. Dev. 13:4.
2. Amaza PS, Umeh JC, Hesen J, Adejobi AO (2006). Determinants and Measurement of Food Insecurity in Nigeria: Some Empirical Policy Guide. Paper Presented at the International Association of Agricultural Economists' Conference, Gold Coast, Australia, August 12-26.
3. Amusa TA, Enete AA, Okon UE (2011). Socioeconomic determinants of cocoyam production among small-holder farmers in Ekiti State, Nigeria. Int. J. Agric. Econ. Rural Dev. 4(2):97-109.
4. Enwelu IA, Asogwa NP, Nwalieji HU, Ezeano CI (2014). Assessment of constraints to cocoyam consumption in selected communities of Enugu State, Nigeria. IMPACT: Int. J. Res. Appl. Natl. Social Sci. 2(3):31-40.
5. Ndabikunze BK, Talwana HAL, Mongi RJ, Issa-Zacharia A, Serem AK, Palapala V, Nandi JOM (2011). Proximate and mineral composition of cocoyam (*Colocasia esculenta* L. and *Xanthosoma sagittifoliu* L.) grown along the Lake Victoria Basin in Tanzania and Uganda. Afr. J. Food Sci. 5(4):248-254.
6. NPC (2006). National Population of Nigeria Population figures, NPC, published bulletin, Enugu.
7. Ogundele O (2014). Factor influencing consumers preference for local rice in Nigeria. Afr. J. Mark. Manage. 6(4):49-55.
8. Ojinaka MC, Akobundu ENT, Iwe MO (2009). Cocoyam starch modification effects on functional, sensory and cookies qualities. Pak. J. Nutr. 8:558-567.
9. Okeke EC, Ene-obong HN, Uzuegbunam AI, Ozioko A, Umeh SI, Chukwuone N (2009). The Igbo traditional food system documented in four states in southern, Nigeria. In kuhnklien

IV, Erasmus B & Spigelski D. Indigenous people's food system: the many dimensions of culture, diversity and environment for nutrition and health. FAO Center for Indigenous People's Nutrition and Environment.

10. Omonona BT, Agoi GA (2007). An analysis of food security situation among nigerian urban households: evidence from Lagos state, Nigeria. J. Central Eur. Agric. 8(3):397-406.
11. Omotesho OA, Adewumi MO, Fadimula KS (2010). Food Security and Poverty of the Rural Households in Kwara State, Nigeria. Libyan Agric. Res. Center J. Int. 1(1):56-59.
12. Onyeka J (2014). Status of Cocoyam (*Colocasia esculenta* and *Xanthosoma* spp) in West and Central Africa: production, Household Importance and the threat from Leaf Blight. CGIAR Report of a scoping study Commisioned by the CGIAR Research program on Roots, Tubers and Bananas (RTB). Lima (Peru).
13. Opara LU (2002). Edible Aroids: Postharvest Operation in: AGST/FAO: Danilo, M. (Ed.). Massey University, New Zealand.
14. Oti E, Akobundu ENT (2008). Potentials of cocoyam-soybean-crayfish mixtures in complementary feeding. Niger. Agric. J. 39(1):137-145.
15. Parkinson S (1984). The contribution of aroids in the nutrition of people in the South Pacific. In Chandra S. (ed), Edible Aroids. Clarendon Press, Oxford, UK. pp. 215-224.
16. Plucknett DL (1970). Status and future of the major edible aroid Colocosia, Xanthosoma, Alocasia, Cyrstosperma and Amorphophallus. In Tropical Root Crops Tomorrow: Proceedings of the 2nd International Symposium on Tropical Root Crops, Hawaii pp. 127-135.
17. Sanful RE, Darko S (2010). Production of Cocoyam, Cassava and Wheat Flour Composite Rock Cake. Pak. J. Nutr. 9(8):810-814.
18. Splittstoesser NE, Martin FW, Rhodes AM (1973). The nutritional value of some tropical root crops. Proceed. Trop. Reg. Am. Soc. Hortic. Sci. 17:290-294.

CHAPTER 3

Categorizing households into different food security states in Nigeria: the socio-economic and demographic determinants

Kolawole Ogundari

Department of Applied Economics and Statistics, University of Delaware, Newark, USA

ABSTRACT

The literature reiterates the need to employ indicators that reflect the multidimensional nature of food security. In a quest to capture the multidimensional nature of food security, this study uses a novel ideal that harmonizes two food security indicators (food expenditure ($FOOD_{exp}$) and dietary diversity score (DDS)) to categorize households into different levels of food security states in Nigeria. In addition, the study also examined factors that influence the probability of households being in different levels of food security states in the country. Our estimates show that about 66 and 60% of the households in the sample were food secure based on a single indicator such as $FOOD_{exp}$and DDS, respectively. However, by harmonizing the two indicators, results reveal that about 42% of the households are actually food secure. The implication of this is that about 24 and 18% of the households are transitory food insecure. Results also suggest that the use of a single indicator may wrongly classify households as food secure instead of categorizing them as transitory food insecure. Furthermore, the empirical results show that households that consume only home produced food have high probabilities of being food insecure, while households that consume only market-purchased food are less likely to be food insecure. The implication of this finding is that harmonization of food security indicators helps identify households with different nature of food (in) security problems that require different types of policy interventions most especially in Nigeria. The results of the empirical analysis also imply that market-based intervention policies that facilitate households' access at all time to healthy foods of their choice should be given priority. This gives the household opportunity to benefit from greater varieties of food items offered by the market.

Keywords: Dietary diversity, Food security, Food expenditure, Determinants, Household, Nigeria

1. BACKGROUND

Food is key to sustaining life, as it provides nutrients essential for the maintenance of good health, improvement of wellbeing, and labour productivity. Food insecurity is a global issue that seems to be most severe in Africa and especially sub-Saharan region of Africa. For instance, during the periods 2014 and 2016, the level of undernourishment in sub-Saharan Africa rose to about 220 million when compared to 180 million recorded between 1990 and 1992 (Food and Agricultural organization 2015).

"Food security exists when all people, at all times, have physical and economic access to sufficient, safe, and nutritious food to meet their dietary needs and food preferences for an active healthy life" (Food and Agricultural organization 1996). This definition reinforces the multidimensional nature of food security which includes availability, access, utilization, and sustainability. Availability connotes physical presence of food in adequate amount; accessibility implies efficient purchasing power at all times; utilization reflects the demand for sufficient quantity and quality of food intake; and sustainability refers to the adequacy of food at all times. A number of food security indicators at the micro-level have been identified in the literature (Lokosang et al. 2011; Pangaribowo et al. 2013; Obayelu 2013; Yu and You 2013). Some of these indicators include per capita food consumption (such as per capita nutrient intakes of calorie, protein, and fat), per capita expenditure on food, dietary diversity score/food consumption score, coping strategy index, share of dietary intake, food insecurity access scale (self-report/assessment), and anthropometry measures among others.

However, an issue that has emanated from the literature is whether there exists an indicator of food security that satisfies the multidimensional nature of food security as specified by the definition (Magrini and Vigani 2014). Unfortunately, no single indicator can incorporate all the dimensions of food security (Hoddinot 1999), because a combination of measures and indicators are needed to fully reflect the complex reality of food insecurity problem in any given context (Carletto et al. 2013). Similarly, Habicht and Pelletier (1990) argued that there is no best indicator because the characterization of an indicator in a generic sense as "best" depends ultimately on whether or not it is the most appropriate for the decision to be made.

Despite the variety of food security indicators that exist in the literature, there is lack of consensus on the core household food security indicators that are needed to adequately measure and monitor food security situation around the world both at the micro- and macro-level (Carletto et al. 2013), given that these indicators only focus on one dimension at a time. The absence of an all-encompassing indicator of food security limits the extent to which panoptic food security policies can be made from existing studies. For example, the use of crop yield indicates availability; food consumption and food expenditure denote food accessibility; dietary diversity, anthropometric measures, and food consumption score reflect food utilization; and coping strategy index represents food sustainability among others. The implication of this is that the choice among available food security indicators involves a trade-off, as concept of food security is inherently hierarchical (Barret 2010). Thus, food availability is necessary but not sufficient for food accessibility; food accessibility is necessary but not sufficient for food utilization; and food utilization is necessary but not sufficient for food sustainability.

The absence of more generalized indicators that incorporate more than one dimension of the concept of food security led to the conceptualization of this study. This study harmonizes indicators of food security in order to reflect its multidimensional and also derive clearer policy conclusions from the analysis. This approach could help to categorize households into different levels of food security states. The categorization of households into different levels of food security has a number of policy implications: *First*, it ensures proper identification of households into different states or levels of food security. *Second*, it provides opportunities to target households with appropriate food policy intervention. *Third*, it helps channel policy instruments specifically to households that are most vulnerable.

This study proposes a relatively more encompassing indicator of food security, which incorporates the food accessibility and utilization aspect of food security definition. Due to lack of information on variables representing food availability and sustainability dimensions of food security in the dataset, other aspects of food security are not included in the present study. Hence, we combine (harmonize) two indicators of food security, namely: food expenditure ($FOOD_{exp}$) and the dietary diversity score (DDS) which represent food accessibility and food utilization, respectively. The choice of these indicators for the present study is motivated by two factors. First, these indicators have been widely used in many empirical literatures on food security in the developing countries. Second, these indicators incorporate different dimensions of the concept of food security. Moreover, we believe that limitation in data used for the study may be responsible for the first point, as our data only contains information on expenditure on the variety of food items in the household food basket. Detailed information on nutrient intakes, coping strategy, anthropometric measures, or quantity of food consume by the households among others to construct other dimensions of food security indicators are generally not available in most of the World Bank supported Living Standard Survey (NLSS) data from the developing countries, which we use in the present study.

Harmonization of the two indicators provides the opportunity to stratify (categorize) households in the sample into different states of food security based on acceptable thresholds for food expenditure and dietary diversity score. Households that jointly pass both thresholds are classified into the food secure class; households that fall below both thresholds are categorized as food insecure; while households that pass only one of the thresholds are grouped into different transitory classes of food security. The study further examined the demographic and socio-economic determinants of household dietary diversity (DD) represented by dietary diversity score (DDS) and dietary diversity index (DDI). Also of particular interest in this study is the importance of home-produced and market-purchased food in the reduction of food insecurity. Particularly, we examined the impact of purchased and home-produced food on the probability of being food insecure.

The rest of the paper is organized as follows. The next section describes food security indicators used in the study. "Theoretical framework" focuses on the theoretical framework of the study. "Empirical model" describes the data used in the study. "The data description" presents the empirical model, while "Results and discussion" presents and discusses the results. Concluding remarks are made in the "Conclusions".

2. FOOD SECURITY INDICATORS IN THE STUDY

2.1 Per capita food expenditure

The use of per capita food expenditure ($FOOD_{exp}$) as an indicator of food security has been widespread. It is often used in measuring food accessibility (Smith and Subandoro 2007; Faridi and Wadood 2010). While Hendriks and Msaki (2009) argue that expenditure on food reflects both the concepts of food accessibility and to some extent, the degree of vulnerability to food insecurity, Faridi and Wadood (2010) suggest that higher expenditure proportions are only essential indicators of inter-temporal vulnerability to food insecurity. As a result, food expenditure proxies' food accessibility dimension of food security because food prices and household resources affect the extent to which households have economic access to food. Expenditure on food does not provide any information on the nutritional quantity that household consume, but rather, it reflects the extent to which households' resources affect the amount of food consumed. Similar to the work of Canagarajah and Thomas (2001) and Omonona and Adetokunbo (2007), this study uses the weighted two thirds of the mean of per capita expenditure as a threshold so that a household is referred to as food secure (or food insecure) when the observed per capita food expenditure is greater (or less) than the threshold.

2.2 Dietary diversity score and dietary diversity index

Dietary diversity score (DDS) measures the degree to which the variety of food consumed by households differs in terms of nutrient intakes over a given period of time. The DDS suggest that households gain satisfaction on food consumed not only because "more is better", but because "variety, which DDS represents, is the spice of life". According to USDA-ERS (2012), higher DDS means a household is consuming a diet that is diverse and whose quality is sufficient to supply important micronutrients. The movement of a household from a restricted to diverse diet is often a function of increased income. For instance, it is easy to observe that household meat consumption increases with higher income (Roberts and Schlenker 2010). Thus, DDS represents the food utilization dimension of the food security (Magrini and Vigani 2014). Heady and Ecker (2012) also referred to DDS as an effective indicator of food utilization for two reasons. First, it reflects consumption of both macro- and micronutrients. Second, demand theory suggests that individuals or households that allocate their resources towards higher-quality foods when they have satisfied their basic caloric needs, attain a higher level of utility. However, a major limitation of DDS is its failure to account for the quantity of nutrient consumed by the households. Nevertheless, empirical studies suggest there is a close correlation between DDS and anthropometric measures, which are examples of indicators that take into account the quantities of nutrients consumed in the household diet (Hoddinot and Yohannes 2002). Hence, unlike expenditure on food, DDS accounts for dietary quality both at the individual and household levels (Ruel 2012). Thus, similar to the work of Smith and Subandoro (2007), we consider a household as food insecure (food secure) if their DDS is less (greater) than the average DDS of the households in the upper quartile.

Although traditional demand theory suggests the use of dietary diversity index (DDI) instead of DDS in accounting for consumers' preferences for various food items, DDS has been widely used in many empirical studies. Drescher et al. (2007) also argue that unlike DDI, DDS do not account for the distribution of food groups to reflect the extent of dietary diversification in the household dietary demand. While we acknowledge these limitations of DDS, it is very easy to construct and this may explain why it is widely adopted in food security studies. DDI is an index that ranges between 0 and 1. While DDS is computed from the share of household expenditure on the identified food groups in the household food basket, DDI reflects food utilization dimension of food security.

2.3 Household level food security states

This study harmonizes two indicators representing food accessibility and food utilization dimensions of food security. Weighted two thirds of the mean of food expenditure and average DDS of the households in the upper quartile are used as predetermined thresholds to stratify households into different levels of food security. Thus, this study harmonizes the two indicators in order to classify households into four mutually exclusive levels of food security states. The levels include (1) households that are completely food insecure, (2) transitory food insecure based on $FOOD_{exp}$, although food secure based on DDS, (3) transitory food insecure based on DDS, although food secure based on $FOOD_{exp}$, and (4) completely food secure state as evident from both indicators.

2.4 Theoretical framework

The theoretical framework for modeling the determinants of household food security is, in general, built within the framework of household utility model. Following the work of Singh et al. (1986), we recognize that some households are both consumers and producers of their food and thus, model household utility within the framework of consumer demand and production theories as follows:

$$U_i = u(C_i, l_i | x_i) \tag{1}$$

where U_i is a utility function that is twice differentiable, increasing in its arguments, and strictly quasi-concave; C_i is a vector of the ith household's consumption demand, which include food C_f, and non-food C_{nf}; l_i is the time devoted to leisure and x_i is the vector of household socio-demographic variables that we include, in order to recognize that household utility is derive from the combination of decisions made by household members according to their preferences.

Given the foregoing definition of C_i, it can be specified as:

$$C_i = \left(C_f, C_{nf}\right)' \tag{2}$$

As some households are both consumers and producers of food, C_f can be further considered as a vector of home-produced food f_{hp} and f_{mp} and market-purchased food f_{mp}. Again, within this context, C_f can be stated as follows:

$$C_f = \left(f_{hp}, f_{mp}\right)' \tag{3}$$

Substituting Eqs. 2 and 3 into Eq. 1 gives Becker's (1981) generalized utility function defined as:

$$U_i = u\left[\left(C_f, C_{nf}\right)', \quad l_i | x_i\right] \tag{4a}$$

$$U_i = u\left[\left[\left(f_{hp}, f_{mp}\right)', \quad C_{nf}\right]\right. \tag{4b}$$

For those households that produce food which they also consume and that are consequently subject to certain constraints of production, income and time factors, the optimization of Eq. 4b requires that household's production and consumption decisions are made separately on the assumption that they are all relevant to the market. In this case, production decisions are first made and then subsequently used in allocating the income between consumption of goods and leisure (Strauss 1983). According to Feleke et al. (2005), it is important to make this assumption because food security or food consumption often depends on production variables, but not vice versa.

Similar to the work of Singh et al. (1986), the production, income, and time constraints imposed in the course of optimizing Eq. 4b can be specified as follows:

Production constraint

$$f\left(Q_{hp},\ L,\ A^{0},\ K^{0}\right) = 0 \tag{5}$$

Equation 5 is a typical household production function for food Q_{hp} produced at home and assumed to be twice differentiable, increasing in outputs, decreasing in inputs, and strictly convex; A^0 is the farm size; K^0 is the fixed capital stock; L is total labour used on the farm.

Income constraint

$$P_{hp}\left(Q_{hp}-f_{hp}\right)-P_{mp}f_{mp}-P_{nf}C_{nf}-w\left(L-l_{f}\right)+N=0\ f\left(Q_{hp},\ L,\ A^{0},\ K^{0}\right)=0 \tag{6}$$

From Eq. 6, P_{hp} is the price of food produced, $Q_{hp} - f_{hp}$ is the marketed surplus of food produced; w is the wage rate; l_f is the total family labour supply on the farm; P_{mp} is the price per unit of market-purchased food items; P_{nf} is the price per unit of non-food item; P_{hp} is the price per unit of marketed surplus of food produced; C_{nf} is the demand for non-food items such as education, health, housing, etc.; w is the wage for hired labour; N is the non-farm income adjusted to ensure that Eq. 6 equal to zero.

Time constraint

$$T = l_f + l \tag{7a}$$

$$l_f = T - l \tag{7b}$$

where T is household's time endowment received in each time period, which is allocated between leisure l and time spent working on the farm l_f.

Substituting the right-hand side (RHS) of Eq. 7b into 6 gives:

$$P_{hp}\left(Q_{hp}-f_{hp}\right)-P_{mp}f_{mp}-P_{nf}C_{nf}-w(L-T+1)+N=0 \tag{8}$$

Expanding Eq. 8 gives:

$$P_{hp}Q_{hp}-P_{hp}f_{hp}-P_{mp}f_{mp}-P_{nf}C_{nf}-wL+wT-wl+N=0 \tag{9}$$

Re-arranging Eq. 9 to explicitly account for household income and expenditure gives:

$$\underbrace{P_{hp}Q_{hp} + wT + N - wL}_{\text{HH Income}} = \underbrace{P_{hp}f_{hp} + P_{mp}f_{mp} + P_{nf}C_{nf} + wl}_{\text{HH Expenditure}} \quad (10)$$

Equation 10 shows that the left-hand side (LHS) equals household income (HH income). The household income comprises of the value of farm produce $P_{hp}\,Q_{hp}$, value of HH's time endowment wT, the value of labour used wL, and non-farm income N. Likewise, the RHS is equivalent to household expenditure (HH expenditure). The household expenditure includes the value of home produce food consumed $P_{hp}\,f_{hp}$; value of market purchase food consumed $P_{mp}\,f_{mp}$; value of non-food expenditure $P_{nf}\,C_{nf}$ and purchase of leisure wl. The optimization of Eq. 4b gives rise to income and expenditure equation within the separability assumption, which is necessary to have first order conditions. It is equally possible via optimization of Eq. 10 to yield production and consumption equations separately. This is discussed below.

The demand for inputs and output produced, especially for households that produced their food at home, can be derived by maximizing the first-order condition of the LHS of Eq. 10 with respect to labour (L) and output produced (Q) as:

$$L^* = l^*\left(P_{hp}, w, A^0, K^0\right) \quad (11)$$

$$Q^* = Q^*_{hp}\left(P_{hp}, w, A^0, K^0\right) \quad (12)$$

where L^* is the optimum labour used and Q^* is the optimum output. Substituting Eqs. 11 and 12 into LHS of Eq. 10 gives optimum income/full income Y^* under the assumption of maximized profit π^* as:

$$L^* = l^*\left(P_{hp}, w, A^0, K^0\right) \quad (13a)$$

$$Y^* = wT + \pi^*\left(P_{hp}, w, A^0, K^0\right) + N \quad (13b)$$

where, $\pi^*(P_{hp}, w, A^0, K^0) = P_{hp}\,Q^* - wL$

Household's demand for food C_f can be derived by solving the first-order conditions of the RHS of Eq. 10. However, recall in Eq. 3 that C_f is a vector of f_{hp} and f_{mp} which, in turn, depend on their respective prices. This relationship can be specified as:

$$C_f = c_f\left(P_{hp}, P_{mp}, P_{nf}, w, Y^*\right) \tag{14}$$

Household demand for food also depends on the preferences of its members. These preferences are represented by household demographic characteristics in Eq. 14. Thus, in line with Eq. 13b, we can further specify Y^* in Eq. 14 as:

$$C_f = c_f\left(P_{hp}, P_{mp}, P_{nf}, w, Y^*\left(P_{hp}, w, A^0, K^0, N\right)|x\right) \tag{15}$$

where,

$f = hp, mp$

Equation 15 suggests that household food consumption C_f depends on both food and non-food prices, wages, and household income. Thus, if household demand for food could be referred to as a measure of household food security (FS), then C_f is a reduced form of the utility function in Eq. 1. It allows the evaluation of the effects of household level characteristics as well as economic factors such as income. The relationship can be represented by:

$$C_f \approx FS_i = \left[\mathrm{FOOD}_{\mathrm{exp}}, NT, DDS/DDI, \ldots\mathrm{etc}\right]' \tag{16}$$

where *FS* is taken as a vector of various indicators of household food security, which could be food expenditure $\mathrm{FOOD}_{\mathrm{exp}}$, nutrient intake *NT* such as calorie, protein and other food nutrients, dietary diversity score *DDS* or dietary diversity index *DDI*, and production index (Smith and Subandoro 2007; Lokosang et al. 2011; Obayelu 2013; Pangaribowo et al. 2013).

3. METHODS

3.1 Empirical model

The empirical model is defined within the framework of Eq. 15. Unfortunately, lack of information on prices as a possible determinant of *FS* somewhat limits the scope of the study. Hence, we re-parameterize the relationship between household food security *FS* and its determinants using a reduced form equation in which food and non-food prices are excluded. In this regard, the reduced equation is designed to consider only the socio-economic (income) and demographic (age, education, and proportion of household member by age, etc.) determinants of household food security. This approach of examining the role of demographic determinants other than food prices in explaining food security have been widely adopted in food security literature due to lack of data on food prices in the survey used for the respective studies (see Magana-Lemus et al. 2016; Rose and Chariton 2002; Obayelu 2013).

The empirical specification of the reduced form of Eq. 15 without the food and non-food prices is stated as follows:

$$FS_i = \left(Y_i^*, x_i | \phi, \beta\right) + \eta_i \tag{17}$$

where FS is a vector representing categorical variables denoting household level of food security states (FS_states) derived from the combination of two food security indicators, DDS and DDI. Y^* represents household income; x is a vector of household demographic variables, ϕ and β are the parameters to be estimated, and η is the error term of the regression. The subsequent subsection focuses on the econometric approach employed to examine the determinants of each component of FS, outlined above as well as how DDS and DDI were constructed in the study.

3.2 Determinants of households in different levels of food security states [FS_states]

We assess the determinants of the levels of household food security states using a multinomial logit model. The dependent variable, in this case is the discrete or categorical variable represented by FS_states. According to Rose and Chariton (2002), this type of regression is a generalization of the familiar logistic regression, which is used when there are more than two discrete possibilities for the dependent variable. The multinomial logit model generally captures how households' socio-economic and demographic variables affect the probability that a household in the sample exhibit any of the identified possible levels of food security states in reference to completely food secure households. As noted by Bhat (2003), the use of multinomial logit model is often an important econometric strategy when the choices are unordered. In this case, the test of independence of irrelevant alternatives (IIA) assumption of this model is required. IIA in the context of the present study implies that the odds of choosing completely food secure households as reference households is not affected by the existence of other levels of food security ($m = 1, 2, 3$). In other words, even if households in other levels of food security are taken as a reference point, the outcomes or estimated parameters are expected to be the same. Subsequently, we test the assumption of IIA and the result confirms the suitability of multinomial logit model in the study. We discuss this in detail in "Determinants of households in different levels of food security".

Thus, following Greene (2008), we define the multinomial logit model for the study for M discrete alternatives ($m = 0,1,2,3$) as:[1]

$$P_{im} = P(\text{FS}_{\text{states}_{im}}|x_i)\ \forall m = 0, 1, \ldots\ldots m \tag{18}$$

$$P_{im} = \frac{\exp(\beta_i x_m)}{1 + \sum_{k=1}^{k=3} \exp(\beta_i x_m)}, \text{ if } m = 1, 2, 3 \tag{19}$$

$$P_{im} = \frac{1}{1 + \sum_{k=1}^{k=3} \exp(\beta_i x_m)}, \text{ if } m = 0 \tag{20}$$

where P_{im} is the probability indicator for ith household in the M's FS_states; x is the vector of households' socio-economic and demographic variables (see: Table 1 for the list of the variables); β is the vector of parameters to be estimated; $m = 1$ for food insecure households based on *DDS* indicator only; $m = 2$ for food insecure households based on FOOD_{exp} indicator only; $m = 3$ for food insecure households based on both indicators.

3.3 Determinants of household demand for dietary diversity

Household demand for dietary diversity (DD) has always been modeled in the literature with either DDS, which represents count data, or DDI that represents indices. For example, Hoddinot and Yohannes (2002) and more recently, Woldehanna and Behrman (2013) and Ecker et al. (2013) employed DDS, while Thiele and Weiss (2003), Drescher et al. (2007), and Gaiha et al. (2012) utilized DDI in their respective studies.[2] In order to examine the household social-economic and demographic determinants of DD, we implicitly specify the relationship between DD, household income Y_i and household demographic variables x' as follows:

$$\text{DD}_i = \psi_0 + \phi Y_i + \beta x' + \eta_i \tag{21}$$

where, DD is the vector of DDS and DDI defined as $DD_i = [DDS_i, DDI_i]'$; Y_i and x are variables hypothesized to explain DD; ψ_0, ϕ and β are parameters to be estimated; and η_i is the random error of the fractional response regression.

Table 1 Summary statistics of the variables and food security states

Description	Variable	Mean	Std. dev.
Dietary diversity score	DDS	4.5739	1.2547
Real monthly per capita food expenditure	$FOOD_{exp}$	16,084.9	12,946.2
Food insecurity (FI) indicator 1: identified food security states			
Completely food insecure households from both indicators	FS_states = 3	0.1574	
Transitory food insecure based on $FOOD_{exp}$ & food secure based on DDS	FS_states = 2	0.2394	
Transitory food insecure based on DDS & food secure based on $FOOD_{exp}$	FS_states = 1	0.1788	
Completely food secure households from both indicators	FS_states = 0	0.4244	
Food insecurity (FI) indicator 2: Dietary Diversity Index			
Average Dietary Diversity Index	*DDI*	0.5849	0.1892
Real per capita total expenditure in Naira—a proxy for HH income[a]	HH_INCOME	28,678.08	32,724.82
Household size	HHSIZE	4.8479	2.9067
Average year of education of the Household head (in years)	EDUCATION	7.3680	7.3580
Household head that are male (1/0)	D_GENDER	0.8562	0.3509
Household head with farming as major occupation (1/0)	D_OCCUPATION	0.6257	0.4839
Households that only home-produced food consumed (1/0)	D_OWNPRODUCEONLY	0.0139	0.1172
Household that only purchase food consumed (1/0)	D_PURCHASEONLY	0.3083	0.4618
Households with member within <25 years (1/0)	D_AGE<20	0.0221	0.1470
Households with member within 25–29 years (1/0)	D_AGE25-29	0.0659	0.2481
Households with member within 30–34 years (1/0)	D_AGE30-34	0.1042	0.3056
Households with member within 35–39 years (1/0)	D_AGE35-39	0.1226	0.3279
Households with member within 40–44 years (1/0)	D_AGR40-44	0.1321	0.3386
Households located in the rural areas (1/0)	D_RURAL	0.7613	0.4263
Households located in the south-south SS region (1/0)	D_SOUTH-SOUTH	0.1512	0.3583
Households located in the south-east SE region (1/0)	D_SOUTH-EAST	0.1421	0.3492
Households located in the south-west SW region (1/0)	D_SOUTH-West	0.1697	0.3754
Households located in the north-central NC region (1/0)	D_NORTH-CENTRAL	0.1768	0.3815
Households located in the north-west NW region (1/0)	D_NORTH-WEST	0.2014	0.4010

[a]Note: Expenditure is expressed in Nigerian currency (naira) and 1US$ = 133 naira as at 2003/2004; Variables with D stands for dummies

Similar to the work of Thiele and Weiss (2003) and Drescher et al. (2007), we construct the DDI from household food expenditure, which has been aggregated across food groups using Berry index defined below:[3]

$$\mathrm{DDI}_i = 1-\sum_{j=1}^{J} w_{ji}^2 \ \forall j = 1, 2, \ \ldots\ldots\ldots J \qquad (22)$$

where DDI is as defined earlier and ranges between 0 and 1, with a value towards 1 implying higher dietary diversity and w_{ji} is the expenditure share of jth (in the study, $j = 6$) food group relative to the total expenditure on food by the ith households.

Since the data generating process (DGP) for both DDS and DDI are count and index data respectively, the estimation of Eq. 21 by traditional ordinary least square (OLS) will lead to inefficient results. As a result, we employ appropriate econometric technique applicable to each of the dependent variable. We further discuss about the econometric approaches in the next subsection.

3.4 Zero-truncated Poisson regression model for DDS

As earlier stated, the DDS is a "count" data. Thus, similar to Rashid et al. (2011) we made use of zero-truncated Poisson regression model to estimate Eq. 21 when DDS is used as the dependent variable. In this regard, we assume DDS is random and, in a given time interval, has a Poisson distribution with probability density defined as:

$$pr(y_i = DDS_i) \frac{= e^{-\lambda_i}\lambda^{DDS_i}}{DDS_i!}, \mathrm{DDS}_i = 1, 2, \ \ldots..6 \qquad (23)$$

where DDS is the realized value of a random variable with mean and variance y_i and λ_i respectively. y_i is assumed to be strictly positive ($y_i > 0$). According to Cameron and Trivedi (1998), this is a case of truncation from below that is taken into account when specifying a zero-truncated Poisson model.

To incorporate the set of explanatory variables x and Y stated in Eq. 21 into Eq. 23, and to ensure a non-negative mean y_i, the parameter λ_i is specified as:

$$E\{y_i|Y_i, x_{ik}\} = \lambda_i = \exp(\phi Y_i, x'\beta) = \exp(\psi_0 + \phi Y_i + \beta_1 x_1 + \ldots..\beta_k x_k) \qquad (24)$$

The implicit assumption in the Poisson model is that the variance λ_i is equal to its mean y_i or the data are equally dispersed. Any violation of this assumption has consequences similar to the failure of the assumption of homoscedasticity in the linear regression model (Cameron and Trivedi 1998).

Fractional regression model for DDI$_i$ Since DDI is bounded between 0 and 1 as earlier mentioned, Papke and Wooldridge (1996) argue that such indices is a fractional/proportional data, and it can be better handled by fractional regression model proposed by the same authors. McDonald (2008) shows that the use of OLS, Tobit, and transformed logistic-normal model for fractional dependent variable data would yield inefficient results because the distribution of error terms is likely to be heteroskedastic.[4] This is because, as the conditional variance of the error term approach zero, the conditional mean approaches either of the boundary points of the fractional data.

Thus, Papke and Wooldridge (1996) proposed Quasi-Maximum Likelihood Estimation (QMLE) to estimate the fractional regression model.[5] This model has been widely adopted in similar studies (McDonald 2008; Sauer et al. 2011; Ogundari 2014).

The Bernoulli Log-likelihood function for estimating Eq. 21 when DDI is taken as the dependent variable is specified as:

$$L_i(\beta) \equiv \mathrm{DDI}_i \ell n\left(G\left(x'\right)\right) + \left(1 - \mathrm{DDI}_i\right) \ell n\left(1 - G\left(x'\right)\right) \tag{25}$$

where $0 \leq \mathrm{DDI}_i \leq 1$ denotes the dependent variable while x' refers to the explanatory variables of observation *i*.

Accordingly, Eq. 25 is well defined for $0 < G(y, x_i) < 1$. The QMLE of β is obtained by simply maximizing Eq. 25. Papke and Wooldridge (1996) show that Bernoulli QMLE β is consistent and $\sqrt{N}$—asymptotically normal regardless of the distribution of DDI_i conditional on y and x'. In addition to its consistency and asymptotic normality, this approach does not require any data adjustments for the extreme values and the conditional expectation of DDI $_i$, given the explanatory variables are estimated directly. Asymptotically efficient, unbiased, and consistent estimates are obtained from the QMLE by simply transforming $G(y, x')$ to produce models similar to either logit or probit in the binary choice situation (McDonald 2008). Papke and Wooldridge (1996) used a logistic function for $G(y, x')$ in the framework of generalized linear models (GLM) [that is, $G(y, x') = \frac{\exp(y.x')}{1 + \exp(y.x')}$] which was extensively discussed in their paper. QMLE is estimated by weighted non-linear square allowing for heteroskedasticity and testing procedures, which are asymptotically efficient within a class of estimators (Oberhofer and Pfaffermayr 2009).[6]

In the meantime, QMLE fractional regression employed for the empirical analysis of Eq. 21 when DDI is considered as the dependent variable is specified as:

$$\mathrm{E}(\mathrm{DDI}_i|Y_i, x_k) = G\left(\psi_0 + \phi Y_i + \sum_{k=1}^{K} \beta_k x_{ki} + \eta_i\right) \tag{26}$$

where DDI_i is as earlier defined, x_k and Y_i represent the hypothesized variables to explain DDI_i, ψ_0 and ϕ and β are parameters to be estimated, $G(.)$ is the logistic function while η represents the error term.

3.5 The data description

The study employs data from Nigeria Living Standard Survey (NLSS) conducted from September 2003 to August 2004. The sampling design of the NLSS involves a two-stage stratified random sampling technique. The first stage was a cluster of housing units called Enumeration Area (EA), while the second stage was the random selection of the housing units. There were seven interviewer visits to each selected household at a minimum of 4-day intervals in a cycle of 30 days. Questionnaires were used to obtain information on non-food expenditure, daily food consumption and expenditure, home-produced food, and purchased food items by the households. The NLSS contains information on 19,158 households. We made use of 18,870 households because 288 households had incomplete relevant information.

The information contained in the NLSS includes a detailed value of own-produced food and expenditure on the type of food purchased by the households. For each household, expenditure profile on the following six food groups were included: (1) staples {which include yam, cocoyam, cassava, rice, maize, and millet}, (2) meat and fish, (3) dairy products, (4) fruits and vegetables, (5) fats & oils, (6) sweeteners. Also in the dataset are detailed information on the non-food expenditure, which includes expenditure on education, healthcare, housing (house rent, cost of maintaining the house and the furniture), clothing (clothes, shoes), utilities, house appliance, transportation (transport fares, petrol purchased, maintenance of cars, bicycles), and communication. Others information obtained include household's demographic variables such as: gender, years of education, major occupation of household head, age, and household size. The definition and summary statistics of the variables used in the analysis are presented in Table 1.

4. RESULTS AND DISCUSSION

4.1 Proportion of households in different levels of food security in Nigeria

As mentioned in "Household level food security states", the study combines two food security indicators defined as food expenditure ($FOOD_{exp}$) and dietary diversity score (DDS) to categorize households in the sample into different FS_states. For $FOOD_{exp}$, the threshold for the classification of households as food secure or insecure is taken as 10,723.3 Naira/month, which represent two thirds of mean per capita food expenditure. We consider also the average DDS of households in the upper quintile as the threshold to classify households into food secure/insecure, and this is equivalent to 4.8 food groups out of maximum 6 identified in the sample. Based on the $FOOD_{exp}$threshold, results suggest that about 66% of the households are regarded as food secure. On the other hand, results show that 60% of the households are food secure according to the DDS threshold adopted. Our finding is consistent with previous findings from Nigeria and other SSA countries. In Nigeria, Babatunde et al. (2007) found that about 38% of the households are food insecure while the results of Omonona and Adetokunbo (2007) show that about 49% of households are food insecure. In Ethiopia, Feleke et al. (2005) reported about 40% incidence of food insecurity.

When the two indicators were combined to categorize households in the sample into different levels of food security, the results show that about 42% of the households in the sample are food secure based on both $FOOD_{exp}$ and DDS measures. Results also suggest that 18% of the sample could be classified as transitory food insecure households when $FOOD_{exp}$ was used as an indicator. About 24% of the households are categorized as transitory food insecure households based on DDS. About 16% of the households are classified as completely food insecure when both measures were jointly used as an indicator (see Table 2 for details).

These results show that by combining two indicators to stratify households in the sample into different levels of food security, a comprehensive overview of the nature of food security issues is revealed. Thus, relevant specific policies can be targeted towards vulnerable households in each state or level of food security. For instance, households who are transitory food secure based on DDS indicator can be targeted with policies that could boost household economic access to food. Another importance of these results is that it prevents over characterization of households into food secure categories. Studies that are based on single indicator would have characterized the transitory food secured households as food secure household when in reality, they are not. These might explain why the issue of food security continues to linger despite various national and international interventions because an outcome of single-indicator studies does not tell the complete story of food insecurity. For instance, our estimates show that about 66 and 60% of the households in the sample were food secure based on a single indicator such as $FOOD_{exp}$and DDS, respectively. However, by harmonizing the two indicators our results are able to show that about 42% of the households were actually food secure. The implication of this is that about 24 and 18% of the households regarded as transitory food insecure based on DDS and $FOOD_{exp}$ respectively were wrongly classified as food secure households from a single indicator stand point (see Table 2). Therefore, harmonizing food security indicators is not only consistent with the broad concept of food security but is also necessary to tell the full story of food insecurity. Such approach could help identify households with different nature of food (in) security problems that require different types of policy interventions most especially in developing economies like Nigeria. Likewise, we believe this approach could also help channel limited resources specifically to households that most needed urgent attention in the country.

Table 2 Categorizing households into different food security states

Categorization of households after combining food security indicators	Food security indicators		Proportion of household under different states (%)	Discrete variable representing food security states [FS_states]
	DDS	$FOOD_{exp}$		
Completely food insecure households	Never satisfy	Never satisfy	15.74	FS_states = 3
Transitory food insecure households	Never satisfy	Satisfy	23.84	FS_states = 2
Transitory food insecure households	Satisfy	Never satisfy	17.88	FS_states = 1
Completely food secure households	Satisfy	Satisfy	42.44	FS_states = 0

Note: A transitory food insecure e households are those who are partially food secure based on one indicator; satisfy implies that food security indicators greater than threshold, never satisfy implies that food security indicators less than threshold

4.2 Determinants of households in different levels of food security

Before we go into the socio-economic and demographic determinants of household in the identified levels of food security, it is important to discuss whether the data used in the study fulfills the independence of irrelevant alternative (IIA) assumption for the multinomial logit model. Similar to the work of Hausman and McFadden (1984), the present study uses Hausman-McFadden test under the null hypothesis of no systematic change in the coefficients when one of the outcomes from the model is excluded. We obtain a p value of 0. 8553 which suggest that there is no evidence that the IIA assumption has been violated.

It is also important to mention that we are aware of possible endogeneity of total expenditure used as a proxy for household income in the study. We thus correct for the endogeneity problem using the income of household head as a relevant instrument for the total expenditure within the framework of Durbin-Wu-Hausman test.[7] We regress total expenditure on the income of the household head and other explanatory variables outlined in Table 1 as first stage regression.[8] Thereafter, we conduct a test to confirm the relevance of the instrument and the result was found to be significant with a p value of 0.000 and an F-statistics of 23.1. Thus, the predicted fitted value of the instrumented-total expenditure is then used in the final analysis reported in Tables 3 and 4 as HHINCOME. As noted by Rashid et al. (2011), this type of approach is becoming increasingly popular in the absence of detailed information on household income.

The results of the multinomial model showing the effects of household's socio-economic and demographic characteristics on the probability of households being food insecure, are presented in Table 3. The results show that the probability of households in any of the identified food insecure states decreases significantly and consistently with household size and income, if a household buys all its food, if a household has more members aged below 44 years relative to those with members aged above 44 years, and if a household is located in the south-south, south-east, north- central, south-west, and north-west regions relative to households in the north-east region. The probability under consideration, however, increases consistently for households that consume only home-produced food, and for households with higher education. This may be because of reduced returns to education beyond the basic level and this is evident from the widespread unemployment in Nigeria. These results underscore policy relevance of household financial capabilities or resources, as this can change the lots of food insecure households, enabling them to gain access to adequate food because they are able to participate in food markets of their choice. The finding also draws our attention to the fact that home-produced foods are not enough to guarantee household food security. This supports the findings of Rose and Chariton (2002) that households that depend on own production had a higher probability of being food insecure.

Other results show that farming as occupation of the household head increases the probability of a household being food insecure based on DDS only, but decreases based on $FOOD_{exp}$ only. Also, farming occupation has no significant effect on the probability of being food insecure based on both indicators. Results also suggest that the gender of the household head does not exhibit any significant impact on the level of household food security in the study. Among rural households, the probability of being food insecure increases based on DDS, but decreases based on $FOOD_{exp}$. The probability of being a food insecure household based on both indicators increase among rural households relative to their urban counterpart.

Table 3 Socio-economic and demographic determinants of households being different food-poverty states

Explanatory variables	Probability of food insecure HH based on DDS indicator only		Probability of food insecure HH based on $FOOD_{exp}$ indicator only		Probability of food insecure HH from both indicators	
	Coefficients	Std. error	Coefficients	Std. error	Coefficients	Std. error
LOG_HHINCOME[a]	−0.5658***	0.1331	−4.4065***	0.1115	−3.4135***	0.1213
LOG_HHSIZE	0.4718***	0.0759	−0.7135***	0.0648	−0.4732***	0.0722
LOG_EDUCATION	0.0263**	0.0103	0.1417***	0.0200	0.0824***	0.0112
D_OCCUPATION (farming)	0.2788***	0.0688	−0.6453***	0.0653	0.7501	0.0737
D_GENDER (male)	−0.1029	0.0649	−0.0186	0.0724	0.0273	0.0798
D_HOMEPRODUCEDONLY	1.5369***	0.5955	1.6638***	0.5321	1.6620***	0.5451
D_PURCHASEONLY	−0.2780***	0.0695	−0.0930**	0.0447	−0.4131***	0.0698
D_AGE<25	−0.2322*	0.1340	−0.0628	0.1789	0.1205	0.1764
D_AGE25-29	−0.1940**	0.0849	−0.1075	0.1014	−0.1834*	0.1093
D_AGE30-34	−0.2683***	0.0734	−0.3341***	0.0771	−0.2709***	0.0818
D_AGE35-39	−0.3199***	0.0703	−0.2984***	0.0684	−0.3526***	0.0745
D_AGE40-44	−0.1847***	0.0689	−0.0926	0.0640	−0.1503**	0.0699
D_RURAL	0.2324***	0.0735	−0.5887*	0.0700	0.0204	0.0804
D_SOUTH-SOUTH	−0.7338***	0.0778	−0.1636**	0.0838	−0.5022***	0.0897
D_SOUTH-EAST	−1.0018***	0.0824	−0.3004***	0.0962	−0.7490***	0.1024
D_SOUTH-WEST	−0.3327***	0.0775	−0.8411***	0.0871	−1.0147***	0.0929
D_NORTH-CENTRAL	−0.7022***	0.0833	−0.2828***	0.0856	−0.6943***	0.0930
D_NORTH-WEST	−0.4174***	0.0796	−1.0142***	0.0858	−0.8237***	0.0886
Constant	5.8412***	1.4659	45.3757***	1.2039	34.0435***	1.3198

***, **, *Implies that the estimated parameters are significantly different from zero at 1, 5, and 10% significance level, respectively. Figures in parentheses are standard error

[a]We use income of household as instrument on total expenditure taken as a proxy for household income in the study

In general, these findings show that household socio-economic and demographic characteristics differ in their effects on the probability of households being in the identified states of food security. Furthermore, the results also suggest that categorizing households into different states or levels of food security by combining two-food security indicators reflect nuances that might have been impossible to uncover with just an indicator.

Table 4 Determinants of household demand for dietary diversity

Variables	DDI estimates[a]		DDS estimates[a]	
	Coefficients	Std. error	Coefficients	Std. error
LOG_HHINCOME[b]	0.3222***	0.0106	0.1759***	0.0039
LOG_HHSIZE	0.1739***	0.0108	0.1200***	0.0038
LOG_EDUCATION	−0.0152***	0.0043	−0.0107***	0.0015
D_GENDER	−0.0162	0.0157	0.0043	0.0056
D_OCCUPATION	0.2821***	0.0168	0.0956***	0.0062
D_HOMEPRODUCEDONLY	−2.0441***	0.1416	−1.8111***	0.1236
D_PURCHASEONLY	0.5385***	0.0187	0.1374***	0.0068
D_AGE<25 YEARS	0.1468***	0.0335	0.0475***	0.0121
D_AGE25-29 YEARS	0.1249***	0.0208	0.0496***	0.0076
D_AGE30-34 YEARS	0.0987***	0.0176	0.0504***	0.0063
D_AGE35-39 YEARS	0.0985***	0.0161	0.0509***	0.0057
D_AGE40-44 YEARS	0.0389***	0.0156	0.0238***	0.0056
D_RURAL	0.2239***	0.0167	0.0658***	0.0060
D_SOUTH-SOUTH	0.1775***	0.0175	0.1133***	0.0065
D_SOUTH-EAST	0.1900***	0.0179	0.1127***	0.0064
D_SOUTH-WEST	0.3345***	0.0188	0.1076***	0.0068
D_NORTH-CENTRAL	−0.3412***	0.0234	−0.1003***	0.0094
D_NORTH WEST	0.1261***	0.0186	0.0794***	0.0069
CONSTANT	−3.6708***	0.1248	−0.6249***	0.0462
Model Diagnostics (DDI/DDS)				
Log Pseudo Likelihood / Prob. > F	−8582.43		−33,866.013	
Sample size	18,870		18,870	

***, **,*Implies that the estimated parameters are significantly different from zero at 1, 5, and 10% significance level, respectively. Figures in parentheses are standard error

[a]Both DDI and DDS were estimated with robust option

[b]We use income of household as instrument on total expenditure taken as a proxy for household income in the study

4.3 Determinants of household demand for dietary diversity[9]

Table 4 presents the determinants of household demand for dietary diversity (DD) when DDI and DDS are used as indicators of food security in the study. The Table shows that both the DDI and DDS provide similar significant results. On this basis, as income increases, household diet becomes more diverse. This is consistent with the finding of Langat et al. (2010), and Woldehanna and Behrman (2013) for households in Kenya, and Ethiopia respectively. Our finding, according to Thiele and Weiss (2003), is consistent with the hypothesis that consumption evolves along hierarchical order as income increases.

Other results also show that household dietary diversity increases significantly as household size increase. This is similar to the findings of Rashid et al. (2011), and Woldehanna and Behrman (2013) for households in Bangladesh and Ethiopia, respectively. This result might be arising from the additional sources of income that may be accruing to the household in a situation where most of the members of the household are involved in some income-generating activities.

The education of the household head, which can be taken as a proxy for consumer dietary knowledge and ability to process dietary information has a significant negative relationship with the demand for dietary diversity. This finding is in contrast with previous findings that found households whose head is educated to have higher demand for dietary diversity (Variyam et al. 1998; Rashid et al. 2011, and Woldehanna and Behrman 2013). Also, being a male-headed household has significant negative relationship with the demand for dietary diversity. This result is not surprising because male-headed household dominated the sample used in the study as evident in Table 1. For instance, a typical male-headed household in traditional Africa society is not involved in decision on what to eat and type of food items to purchase in the household food budget.

The results further show that the demand for dietary diversity significantly increases among households headed by farmers in the study, in contrast to what was obtained by Thiele and Weiss (2003) in Germany, where being a farmer means having decreased food diversity. While dietary diversity significantly decreases among households that consume only home-produced food, households that rely only on purchased food have higher dietary diversity. This suggests that dietary diversity of the farmer may be limited by farm size, seasonal and agronomic constraints. Households consuming only purchased food, on the other hand, are not as constrained as the market contains foods sourced from different areas.

The results further show that households whose members are less than 40 years old have a higher demand for higher dietary diversity compared to households whose members are all above 40 years old. This is possible, because those under 40 years have to consume as many nutritional foods as possible to be in the possession of the physical and mental capacities needed to engage in the education and labour activities. Households in rural areas have a higher demand for dietary diversity, relative to households in urban areas. This contradicts Thiele and Weiss' (2003) argument that households in cities tend to have a higher demand for dietary diversity than people living in the rural environment, possibly because foods are less expensive in Nigerian rural areas relative to urban. Relative to households in the north-east region of Nigeria, households in the north-west, south-west, south-east, and south-south regions of the country exhibit higher and significant demand for dietary quality.

CONCLUSIONS

The absence of an all-encompassing indicator of food security limits the extent to which panoptic food security policies can be made from existing indicators. Harmonization of indicators of food security to capture its multidimensional concept in an effort to derive clearer policy conclusions from the empirical analysis has been suggested in the literature. This ensures proper identification of households into different regimes of food security, provides opportunities to target households with different types of policy intervention, help channel policy instruments specifically to households that are most vulnerable, and most importantly provides information on different groups of households with different types of food (in) security problem.

As a result, this study uses a novel idea to combine two popular indicators of food security namely food expenditure ($FOOD_{exp}$) and dietary diversity score (DDS) in the literature to categorize households in Nigeria into four possible states of food security, namely completely food insecure households as revealed by both indicators, transitory food insecure households based only on DDS or $FOOD_{exp}$ indicators, and completely food secure households as revealed by both indicators. Subsequently, we use a multinomial logit model to examine how the socio-economic and demographic characteristics of households influence the probability that a household will be in one of the identified states of food security relative to food secure households. Also, using DDS and dietary diversity index (DDI) as additional sets of indicators of food security and zero-truncated Poisson and fractional regression models, we further investigate the determinants of dietary diversity in Nigeria.

Our estimates show that about 66 and 60% of the households in the sample were food secure based on a single indicator such as $FOOD_{exp}$ and DDS, respectively. However, by harmonizing the two indicators our results reveal that about 42% of the households were actually food secure. The implication of this is that about 24 and 18% of the households regarded as transitory food insecure based on DDS and $FOOD_{exp}$ respectively were wrongly classified as food secure households from a single indicator standpoint.

Also, the empirical results based on the multinomial logit model show that the probability of households being in any of the identified food insecure states relative to food secure households decreases significantly and consistently with income, household size, whether households consume only purchased food, and the geographical region of the household. This probability, however, increases if household consumes only home-produced food.

The results of the determinants of household demand for dietary diversity based on DDS and DDI show that dietary diversity increases with household income and household size, whether the household head is a farmer, whether households consume only purchased food, and whether households reside in the rural areas. We, however, found that educated household heads and those consuming only home-produced food consume less diverse diet. We also found that dietary diversity varies across Nigerian regions.

Therefore, harmonizing food security indicators is not only consistent with the broad concept of food security but is also necessary to tell the full story of food insecurity of households in Nigeria. Conversely, we believe this approach could help identify households with different nature of food (in) security problems that require different types of policy interventions most especially in developing economies like Nigeria. Likewise, we believe this approach could also help channel limited resources specifically to households that most needed urgent attention in the country. Also, understanding these factors that influence household's food security as well as household demand for dietary diversity is crucial in designing food policy programs that will effectively achieve the desired goals. For instance, our findings that household income decreases food insecurity and that home-produced food is not enough to guarantee food security raises the need for Nigeria to focus on market-based intervention policies. Such policies could facilitate households' access at all time to healthy foods of their choice. Such an intervention also gives households opportunity to benefit from greater varieties of food items offered by the market.

We suggest that future research should consider the role of food prices as determinants household food security in Nigeria. This is because we could not incorporate food prices in the analysis due to lack of data on food prices in the survey used for the analysis. In addition, it would be interesting to consider other popular food security indicators that capture effectively other dimensions of food security indicators such as crop yield (availability), calorie intake (accessibility), anthropometric measures (utilization), and coping strategy index (sustainability). Likewise, constructing healthy food diversity can also be explored in future research.

ENDNOTES

[1]The first FS_states denoted by 0 serves as the reference state and it represents households that are food secured.

[2]As noted by Gao et al. (2013), although food diversity is related to diet quality, nevertheless, it may not be as good as measurement as the healthy eat index that is based on dietary guidelines.

[3]There are other frequently used indices such as entropy, Herfindahl, Simpson and among others. The present study employed Berry index because of its simplicity and also it has been used in the previous studies, which provide ease of comparison.

[4]The problem in using OLS on fractional dependent variable is that it is not asymptotically efficient estimator. It is an unbiased but inconsistent estimator.

[5]This is a departure from previous studies such as Thiele and Weiss (2003) and Drescher et al. (2007) that used transformed DDI as dependent variable before estimating with OLS; and from Gaiha et al. (2012) that employed OLS technique directly on the DDI.

[6]QMLE accommodates naturally, non-constant variances and skewness (Oberhofer and Pfaffermayr 2009).

[7]The NLSS contains detail information on income of household head as information on income of other members of households are not readily provided.

[8]For brevity the result of the first stage regression is not presented. This can be requested from the author.

[9]This section focuses on determinants of household demand for dietary diversity based on the identified six food groups used in the study. As noted by Rashid et al. (2011), it is not clear whether diversity between and within the identified food groups is determined by similar factors, which is a serious limitation of this study. Nevertheless, we believe the results obtained are consistent with previous studies, which thus stress further the policy relevance of the findings.

ACKNOWLEDGEMENTS

The author would like to thank the reviewer and editor of this paper for their thoughtful comments and suggestions on the earlier draft of this paper. The author equally thanks Dr. Olufemi Bolarinwa for reading and discussing the earlier draft of the paper. The views in this paper are the author's own and do not necessarily represent those of the University of Delaware.

REFERENCES

1. Babatunde RO, Omotesho OA, Osholotan OS (2007) Economic characteristics and food security status of farming household in Kwara state, North-central Nigeria. Pak J Nutr 6:49–58
2. Barret C (2010) Measuring food insecurity. Science 237:825–828
3. Becker GS (1981) Treatise on the family. Harvard University Press, Cambridge
4. Bhat CR (2003) Random utility-based discrete models for travel demand analysis. In: Goulias KG (ed) Transportation systems planning: methods and applications. CRC Press, Boca Raton, pp 10–30
5. Cameron AC & Trivedi PK (1998) Regression Analysis of Counts Data. Cambridge University Press
6. Canagarajah S, Thomas S (2001) Poverty in wealthy economy: the case of Nigeria. J Afr Econ 10:143 173
7. Carletto C, Zezza A, Banergec R (2013) Towards better measurement of household food security: harmonizing indicator and the role of household surveys. Glob Food Sec 26:30–40
8. Drescher LS, Thiele S, Mensink GB (2007) A new index to measure healthy food diversity better reflects a healthy diet than traditional measures. J Nutr 137:647–651
9. Ecker O, Tan J-FT, Alpuerto V, Diao X (2013) Economic growth and agricultural diversification matters for food and nutrition security in Ghana. Ghana strategy support program, discussion note number 031. IFPRI, Washington DC
10. Faridi R & Wadood SN (2010) An econometric assessment of household food security in Bangladesh. The Bangladesh Development Studies, XXXII: 97-111
11. Feleke ST, Kilmer RL, Gladwin CH (2005) Determinants of food security in Southern Ethiopia at the household level. Agric Econ 33:351–363
12. Food and Agricultural organization (1996) Rome declaration on World Food Security., Worlds food submit. Food and Agricultural organization (FAO), Rome, Accessed Sept 2014
13. Food and Agricultural organization (2015) Regional overview of food insecurity: African food security prospects brighter than ever. FAO, Accra (Accessed July 2016)
14. Gaiha R, Kaieker H, Imai KS, Kulkarni VS, Thapa G (2012) Diet diversification and diet quality in India: an analysis. Discussion paper series RIEN DP 2012-30, Kobe University, Japan
15. Gao Z, Xiaohun Y, Lee J-Y (2013) Consumer demand for diet quality: evidence from the healthy eating index. Aust J Agric Resour Econ 57:301–319
16. Greene WH (2008) Econometric analysis, 6th edn. Prentice-Hall, New Jersey
17. Habicht J, Pelletier J (1990) The importance of context in choosing nutritional indicators. J Nutr 120S:1519–1524
18. Hausman J, McFadden D (1984) Specification test for the multinomial logit model. Econometica 52:1219–1240
19. Heady D & Ecker O (2012) Improving the measurement of food security. IFPRI. Discussion Paper 01225.
20. Hendriks SL, Msaki MM (2009) The impact of smallholder commercialization of organic crops on food consumption patterns, dietary diversity and consumption elasticities. Agrekon 48:184–199
21. Hoddinot J (1999) Choosing outcome indicators of household food security. Technical Guides #7. International Food Policy Research Institute (IFPRI), Washington D.C
22. Hoddinot J, Yohannes Y (2002) Dietary diversity as a food security indicator, FCND Discussion paper No. 136. International Food Policy Research Institute (IFPRI), Washington D.C
23. Langat BK, Salo TK, Nyangiweso PM, Ngeno VK, Korir MK, Kipsat MJ (2010) Household food security in commercialized subsistence economies: factors influencing dietary diversity of smallholder Tea farmers in Nandi South Kenya. Poster presented at the 3rd

African Association of Agricultural Economist (AAAE) and 48th Agricultural Economist Association of South Africa (AEASA) conference, Cape Town
24. Lokosang LB, Ramroop S, Hendriks SL (2011) Establishing a robust technique for monitoring and early warning of food insecurity in post-conflict south-Sudan using ordinal logistic regression. Agrekon 50:101–130
25. Magana-Lemus D, Ishdorj A, Parr Rosson C, Lava-Alvarez J (2016) Determinants of household food insecurity in Mexico. Agric Food Econ 4(10):1–20
26. Magrini E & Vigani M (2014) Technology adoption and multiple dimension of food security: the case of maize in Tanzania. LICOS Centre for Institutions and Economic Reforms. Discussion paper 352/2014.
27. McDonald J (2008) Using Least Squares and Tobit in Second Stage DEA Efficiency Analyses. Flinders Business School Research Paper Series 2008-3 ISSN 1441-3906. Flinders University Adelaide, USA
28. Obayelu AE (2013) Households' food security status and its determinants in the North-Central Nigeria. Food Econ 9:241–256
29. Oberhofer H & Pfaffermayr M (2009) Fractional response models-A replication exercise of Papke and Wooldridge (1996). A working paper in Econometrics and Statistics, University of Innsbruck No.2009-02.
30. Ogundari K (2014) The paradigm of agricultural efficiency and its implication on food security in Africa: what does meta-analysis reveal? World Dev 64:690–702
31. Omonona BT, Adetokunbo AAG (2007) An analysis of food security situation among Nigerian urban households: evidence from Lagos state, Nigeria. J Cent Eur Agric 8:397–466
32. Pangaribowo EH, Gerber N, Torero M (2013) Food and nutrition security indicators: a review. Center for Development Research (ZEF), University of Bonn, Working paper No. 108, Bonn
33. Papke LS, Wooldridge JM (1996) Econometric methods for fractional response variables with an application to 401(K) plan participation rates. J Appl Economet 11:619–632
34. Rashid DA, Smith L, Rahman T (2011) Determinants of dietary quality: evidence from Bangladesh. World Dev 39:2221–2231
35. Roberts MJ & Schlenker W (2010) Why climate change impacts on agriculture could be economically substantial. In: Heal G (ed) Is Economic Growth Sustainable? IEA Conference Volume no. 148. Palgrave Macmillan
36. Rose D, Chariton KE (2002) Quantitative indicators from food expenditure survey can be used to target the food insecure in South Africa. J Nutr 132:3235–3242
37. Ruel MT (2012) Simple dietary diversity indicators: how well do they reflex micronutrient adequacy? Presentation at the FAO symposium
38. Sauer J, Davidora S, Latruffe L (2011) Determinants of smallholders' decisions to leave land fallow. The case of Kosovo. J Agric Econ 63:119–141
39. Singh I, Squire L, Strauss J (1986) Agricultural households models: extensions, application and policy. John Hopkins University Press, Baltimore
40. Smith LC, Subandoro A (2007) Measuring food security using household expenditure surveys. Food security in practice technical guide series. International Food Policy Research Institutes, Washington, D.C
41. Strauss J (1983) Socio-economic determinants of food consumption and production in rural Sierra Leone: application of an agricultural household model with several commodities. MSU International Development papers. Department of Agricultural Economics, Michigan State University, East Lansing
42. Thiele S, Weiss C (2003) Consumer demand for food diversity: evidence for Germany. Food Policy 28:99–115
43. USDA-ER (2012) International food security Assessment, 2012-22/GFA-23. U.S. Department of Agriculture, Economic. Economic Research Service publication, Washington DC
44. Variyam JN, Blaylock J, Smallwood D, Basiotis D (1998) USDA's Healthy Eating Index and nutrition information. Economic Research Service/USDA, Technical Bulletin, 166, Washington D.C

45. Woldehanna T, Behrman JR (2013) What determines Ethiopian children's dietary diversity. Paper presented at center for the study of African Economics (CSAE) held at St. Catherine's college Oxford 17-19 march 2013
46. Yu B, You L (2013) A typology of food security in developing countries. China Agric Econ Rev 5:118–153

CHAPTER 4

Demand relationships in orange exports to Russia: A differential demand system approach focusing on Egypt

Assem Abu Hatab[1,2]

[1]Department of Economics and Rural Development, Faculty ofEnvironmental Agricultural Sciences,Arish University, Al-Arish, Egypt

[2]Department of Economics,Swedish University of AgriculturalSciences, Ulls Väg 27, Box 701375007 Uppsala, Sweden

ABSTRACT

Recent years have witnessed closer diplomatic relations between Egypt and Russia, which have led to significant growth in the countries' bilateral agricultural trade. As a world-leading producer and exporter of oranges, these developments represent an opportunity for Egypt to promote its orange exports to Russia. Another emerging opportunity for Egypt to increase its share in the Russian market for imported oranges has been provided by import embargos imposed by Russia in recent years on agricultural and food commodities from several countries, creating a supply gap of around 25 % in the Russian orange market. To assess the competitiveness of Egyptian oranges and explore the potential export opportunities presented by the Russian market, this paper uses a Rotterdam import allocation model to analyse demand relationships among major orange suppliers to Russia during the period 1996–2014. The results show that in comparison with other orange suppliers, Egypt enjoys a strong comparative advantage in the export of oranges to Russia. The econometric results suggest that both Morocco and Egypt would benefit the most if Russia were to allocate a larger budget to the import of oranges. The expenditure elasticity estimates indicate that an increase in Russia's demand for imported oranges would lead to increases in the quantity of Egypt's orange exports, as well as in its share of the Russian orange market. Furthermore, cross-price elasticity estimates reveal that Egyptian oranges are substitutes for Turkish and South African oranges, implying that Russia has a tendency to switch to these two suppliers when Egyptian oranges become relatively expensive. In light of these results, the adoption of strategies to produce oranges sustainably and cost-effectively, upgrade the orange value chain, acquire processing technologies and enhance the technical and organisational capacity of farmers and exporters could be useful means for promoting exports and boosting the competitiveness of Egyptian oranges on the Russian market.

Keywords: Egyptian orange exports, Russia Import demand, Differential demand system, Rotterdam model

1. BACKGROUND

Fuelled by an increase in real disposable incomes over the past two decades and the growing tendency among domestic consumers to maintain healthier diets, the demand for fresh fruit and vegetables among Russian consumers has grown significantly in recent years (Honkanen and Voldnes 2006; Kolchevnikova 2010; Hamza 2015). With a share of around 7 % of the fruit market, oranges are the second most popular citrus fruit in Russia (Ilyina 2011). The Russian orange market depends entirely on imports since oranges are not grown domestically, with the country having the largest share (around 12 %) of world orange imports (GRC 2012). Between 1996 and 2014, Russia's orange imports have grown by an average of 4.6 % annually, rising from 257.5 thousand tons to around 469 thousand tons (WITS 2016).

Egypt, the world's sixth largest producer and second largest exporter of oranges, has always been one of the main orange suppliers to the Russian market (Hamza 2014). Particularly since 2001, Russia's orange imports from Egypt have increased substantially, skyrocketing from about 8 thousand tons in 2001 to 76 thousand tons in 2003 (Fig. 1). Between 2004 and 2010, the quantity of Russian orange imports from Egypt practically doubled, rising from 110 thousand tons to almost 219 thousand tons. Despite the socio-political unrest in Egypt in 2011, which adversely impacted its foreign trade sector, Egypt's orange exports to Russia have shown relative resilience and have generally continued to grow in the period after 2011 (Fig. 1).

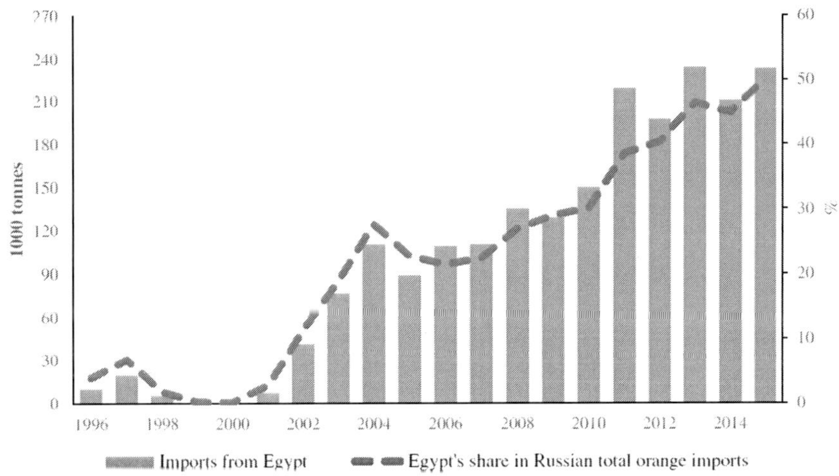

Figure. 1 Russian orange imports from Egypt, 1996–2015. Source: World Bank, WITS (2015)

Generally, this very strong performance of Egyptian orange exports on the Russian market can be attributed to the special attention that Egyptian agricultural policies have paid to the development of the orange supply chain at all levels. Such attention is motivated by the very significant contribution made by oranges to the country's fruit production (one third of total fruit production) and exports (around 10 % of total agricultural exports worldwide) (Hamida 2012; WITS 2016). Furthermore, the promotion of orange exports to Russia has always been an objective of Egypt's agricultural export development strategies aimed at diversifying the map of the country's agricultural exports and reducing its dependency on a limited number of EU and Arab markets. Moreover, the importance of Russia as a target market for Egyptian orange exports has recently increased. Abu Hatab and Nsanimana (2016) indicate that "political proximity" between Egypt and Russia in recent years has revitalised their bilateral agricultural trade relations significantly. Thanks to several trade agreements being signed by the two countries to promote a greater level of trade liberalisation, bilateral agricultural trade has grown from USD 85 million in 1996 to about USD 2.2 billion in 2014, making Russia a major trading partner of Egypt (Parker 2015; WITS 2016). In this context and given Egypt's position in the global orange market, oranges have repeatedly been cited by Egyptian agricultural trade policymakers as an area in which the country could benefit from these improved bilateral agricultural trade relations with Russia and secure a greater foothold in the Russian market for imported oranges.

In addition to these recent improvements in Egypt-Russia bilateral agricultural trade, the recent embargos imposed by Russia on agrifood imports from several destinations, including the EU in 2014 and Turkey in 2015, opened up another opportunity for the promotion of Egyptian orange exports to Russia (Hamza 2014). According to Abu Hatab and Nsanimana (2016), these two embargos have created a supply gap of about 25 % in the Russian market, providing export opportunities to Egypt and other orange suppliers to fill the resulting import void. Although it is too early to assess the impacts of these embargos on Egyptian orange exports to Russia, Fig. 1 shows that there has been an increase of around 11 % in the quantity of Egyptian oranges exported to Russia in 2015 in comparison with the preceding year.

To assess the competitiveness of Egyptian oranges and explore the export opportunities that the Russian market may offer Egyptian oranges, the present study used a source-differentiated approach within a Rotterdam import allocation model to analyse the Russian demand for imported oranges from Egypt and other major suppliers during the period 1996–2014. This paper contributes to the literature in the following ways. First, especially in light of the above developments in Egyptian-Russian bilateral agricultural trade relations as well as the structural changes in the Russian market for imported oranges following the imposed import bans, the subject covered by this paper is very timely and its results could be useful to Egyptian agricultural trade policymakers in the development of informed strategies for the promotion of orange exports to Russia. Second, a closer look at the existing literature on the foreign demand for Egyptian agricultural exports shows that the majority of previous studies have ignored horticultural commodities in favour of other agrifood commodity groups, despite the fact that horticultural commodities are progressively making up an important proportion of Egyptian agricultural exports. Third, previous related empirical studies mainly focus on the EU, USA, and Arab markets, while few studies have so far investigated the demand for Egyptian agricultural exports in Russia and other Eastern European countries. This paper therefore attempts to fill this void in the literature by providing empirical evidence based on the analysis of the Russian demand for imported oranges from major suppliers including Egypt.

The next section in this paper provides a brief assessment of the competitiveness of Egyptian orange exports on the Russian market. This is followed by a presentation of the econometric model and a description of the data and their sources. The econometric results and the estimated demand elasticities are presented thereafter followed by a discussion of the results and concluding remarks.

2. ASSESSING THE COMPETITIVENESS OF EGYPTIAN ORANGE EXPORTS ON THE RUSSIAN MARKET

Several studies have shown that Egyptian oranges enjoy greater competitiveness factors relative to their rivals on the Russian market (Soliman and Bassiony 2012; Soliman 2013; Torayeh 2013; Hamza 2015). A recent USDA report by Hamza (2015) points out that Egyptian oranges traditionally enjoy a set of characteristics that include favourable production conditions, low labour costs and well-established supply chains, which collectively enhance their competitiveness on the Russian market. Specifically in terms of production capacity, recent statistics from MALR (2015) point out that oranges comprise almost half of the total area under fruit cultivation in Egypt. Egyptian orange production increased from 1.6 million tons in 1996 to around 3 million tons in 2014, recording an average annual growth rate of 3.6 % (FAOSTAT 2016). It is worth mentioning that Egypt produces a wide range of orange varieties, including *Baladi* (domestic), *Sukkari* (the sweet orange), *Valencia, blood orange* and *Navel*. However, both *Valencia* and *Navel* varieties represent the major export varieties, whereas other varieties are mainly consumed domestically (MALR 2013).

With respect to product quality and consumer preferences, Rudolf (2015) shows that despite the fact that other competitors (e.g. Turkey and Greece) have a more favourable geographical position with regard to their proximity to the Russian market, the competitiveness of Egyptian oranges on the Russian market relies primarily on their quality and price. Soliman (2013) assessed the competitiveness of Egyptian fresh oranges in a number of European markets, including Russia, using the global value chain analysis approach (GVCA) based on interviews with buyers and key stakeholders in these markets. His results suggest that Egyptian oranges are perceived to be of higher quality than oranges from Turkey and Morocco. Moreover, unlike many other competing countries and due to favourable climatic conditions, Egyptian *Valencia* oranges do not normally experience the "re-greening phenomenon" which breaks down the colour dyes that give oranges their normal colour, causing a greenish tint, i.e. the reappearance of chlorophyll in mature oranges (Abu Hatab and Nsanimana 2016). Increased health concerns among consumers in Russian and other importing markets and fears of the use of chemicals to re-colour oranges therefore result in a high demand for Egyptian oranges on the Russian market.

With regard to export prices, Hassan et al. (2010) compared the export prices of Egyptian oranges with the export prices of other Mediterranean competitors, namely Spain, Israel, Morocco and Turkey. They concluded that Egyptian export prices of oranges are on average approximately 26 % lower than the export prices of these competitors. Furthermore, using annual time series data covering the period 1981–2010, Soliman (2013) found that the average farm prices of Egyptian oranges have usually been lower than the corresponding average world price and that the export price of Egyptian oranges is lowest compared with the prices of traditional competitors such as Morocco, Syria, Tunisia and Turkey. The lower prices of Egyptian oranges further enhance their competitiveness in cost-conscious and heterogeneous markets such as the Russian market. Another important factor that gives Egyptian oranges a competitive edge on the Russian market comes from their early harvest compared to other major producers in the region, thus allowing Egyptian oranges an export season from mid-November to the end of August (Hamza 2015).

To develop a clearer picture of the Russian market for imported oranges, Fig. 2 portrays the distribution of major orange suppliers to Russia during the period 1996–2014. With an average import budget share of about 22.3 %, Morocco tops the list of major orange suppliers to the Russian market. Egypt ranks second, with an average expenditure share of 20.8 %. South Africa comes third (18.5 %) followed by Turkey (15.4 %), while Greece and Argentina lag behind with average budget shares of around 7 and 5 %, respectively. Collectively, these six countries supplied the Russian market on average with nearly 89 % of its total orange imports during the period 1996–2014. The empirical analyses in this paper therefore focused on these orange suppliers.

A closer look at Fig. 3, particularly in the years after 2000, reveals that the budget shares of these major orange suppliers to Russia have experienced dramatic changes over time. For instance, Morocco's budget share declined from about 49 % in 2000 to just 5.4 % in 2014. Albeit from a much lower starting point, Egypt's budget share has, in contrast, increased substantially, from nearly 3 % in 2001 to around 44 % in 2014. In particular, Egypt's share largely surpassed that of other major orange suppliers to Russia during the period 2007–2014, averaging roughly 40 %. With respect to the budget shares of other suppliers, Fig. 3 shows that they have been largely unchanged, with minor increases and decreases over the period 2000–2014.

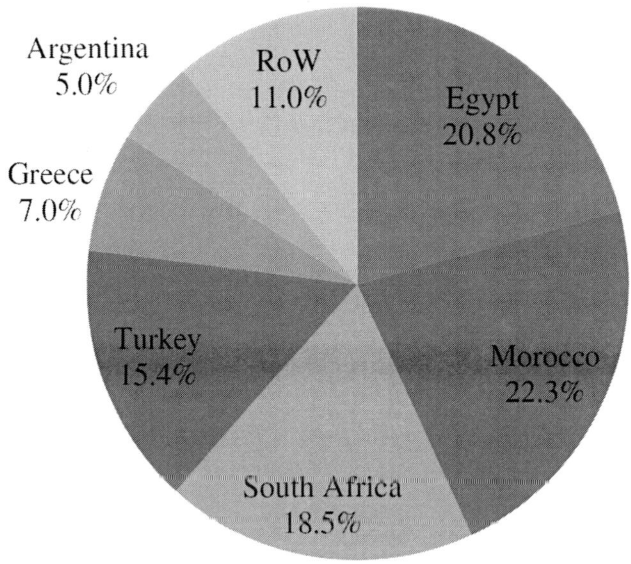

Figure. 2 Russian orange imports by the major country of origin, average 1996 2014. Source: World Bank, WITS (2016)

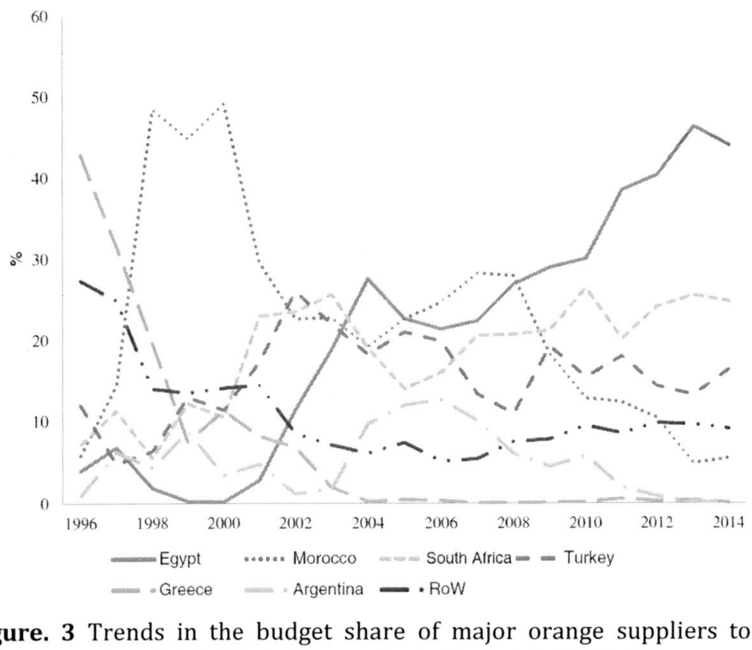

Figure. 3 Trends in the budget share of major orange suppliers to the Russian market, 1996–2014. Source: World Bank, WITS (2015)

To undertake an empirical assessment of the competitiveness of Egyptian oranges, the economic literature suggests that several approaches based on trade data can be used to measure international competitiveness. Originally, the concept of revealed comparative advantage (RCA) was introduced by Balassa (1965). Balassa's RCA index states that a given country has a comparative advantage (disadvantage) in a commodity when the commodity's exports market size in terms of its total exports market size is greater (less) than the commodity's world exports market size in terms of the world total exports market size, i.e. when the value of the index is greater (less) than unity (Abu Hatab and Romstad, 2014). Due to criticisms of Balassa's RCA index (e.g. Laursen 2015 and Leromain and Orefice 2014), Vollrath (1991) introduced three alternative indexes to measure a country's RCA. The first index is the relative trade advantage (RTA_{ad}), which is calculated as the difference between relative export advantage (or Balassa's RCA index) (RXA_{ad}) and relative import advantage (RMA_{ad}):

$$RTA = RXA_{ad} - RMA_{ad} \quad (1)$$

where

$$RXA_{ad} = (X_{ad}/X_{na}) \,/\, (X_{dr}/X_{nr}) = \text{Balassa's RCA index} \quad (2)$$

and

$$RMA_{ad} = (M_{ad}/M_{na}) \,/\, (M_{dr}/M_{nr}) \quad (3)$$

While X_{ad} represents the exports of commodity d by country a, X_{na} represents the exports of all commodities, excluding commodity d, by country a; X_{dr} is the exports of commodity d by the rest of the world, excluding country a; X_{nr} is the exports of all commodities, excluding commodity d, by all countries in the world excluding country a; M_{ad} represents the imports of commodity d by country a; M_{na} represents the imports of all commodities, excluding commodity d, by country a; M_{dr} is the imports of commodity d by the rest of the world; and M_{nr} is the imports of all commodities, excluding commodity d, by all countries in the world excluding country a.

The second index introduced by Vollrath is the natural logarithm of the relative export advantage ($\ln RXA_{ad}$), while the third measure is the revealed competitiveness RC_{ad} defined as follows:

$$RC_{ad} = \mathrm{Ln}(RXAad) - \mathrm{Ln}(RMA_{ad}) \quad (4)$$

Vollrath (1991) points out that positive (negative) values of lnRXA, RTA and RC indicate a revealed competitive advantage (disadvantage). The results of Vollrath's indices of revealed competitive advantage for Egypt and other major suppliers of oranges to Russia are summarised in Table 1. As an average of the period 1996–2014, the results indicated that Egypt had the highest relative export advantage index (Balassa's RCA index) value, implying a greater revealed comparative advantage in the export of oranges in comparison to other competitors on the Russian market. However, the results suggested that Egypt's RXA for oranges has been fluctuating over the years. Morocco's orange RXA has shown a decreasing trend, declining from 34.9 in 1996 to around 5.8 in 2014, implying a reduction in orange competitiveness. With respect to other suppliers, namely South Africa, Turkey and Greece, the results showed that their orange RXAs have improved steadily, although at slower rates and with less vibrations in comparison to Egypt. The values of orange RXA for Argentina and Turkey showed that they significantly lag behind other orange exporters to Russia in terms of revealed comparative advantage.

Table 1 Vollrath's indices of revealed competitive advantage for Egypt and other major orange suppliers to Russia

	Morocco	**Argentina**	**Egypt**	**Greece**	**Turkey**	**South Africa**
Relative export advantage (RXA) – Balassa's RCA index						
1996	34.96	0.72	9.35	7.21	1.62	10.35
2000	19.70	0.38	27.66	7.57	2.47	13.82
2004	15.02	0.67	15.16	7.62	2.29	18.76
2008	18.85	0.61	47.90	8.94	2.82	23.63
2012	7.91	0.24	41.12	8.69	5.42	23.37
2014	5.81	0.25	40.87	10.15	4.16	24.48
Average 1996–2014	16.11	0.58	26.24	9.45	3.38	18.73
Natural logarithm of the relative export advantage (lnRXA)						
1996	3.55	−0.33	2.24	1.98	0.48	2.34
2000	2.98	−0.98	3.32	2.02	0.90	2.63
2004	2.71	−0.40	2.72	2.03	0.83	2.93

2008	2.94	−0.49	3.87	2.19	1.04	3.16
2012	2.07	−1.45	3.72	2.16	1.69	3.15
2014	1.76	−1.38	3.71	2.32	1.42	3.20
Average 1996–2014	2.66	−0.64	3.11	2.23	1.13	2.89
Relative trade advantage (RTA)						
1996	34.96	0.54	9.35	7.08	1.44	10.32
2000	19.70	−0.81	27.66	7.52	2.47	13.77
2004	15.02	0.64	15.16	6.55	2.05	18.74
2008	18.85	0.47	47.90	8.57	2.60	23.52
2012	7.91	0.24	41.11	8.52	5.18	23.32
2014	5.81	0.24	40.87	10.06	4.01	24.40
Average 1996–2014	16.11	0.38	26.24	9.21	3.19	18.68
Revealed competitiveness index (RC)						
1996	17.19	1.41	16.72	4.02	2.18	5.94
2000	16.40	−1.15	17.51	4.98	7.60	5.55
2004	9.03	3.15	9.39	1.96	2.25	7.17
2008	11.74	1.50	18.95	3.19	2.54	5.42
2012	16.52	6.42	8.67	3.96	3.09	6.11
2014	16.19	3.21	10.27	4.71	3.37	5.66
Average 1996–2014	15.04	2.76	12.72	3.98	4.02	6.15

Source: Author's calculations based on data from the World Bank, WITS (2015)

The results of the natural logarithm of the relative export advantage and the revealed competitiveness indices in Table 1 further confirmed Egypt's strong export advantage and competitiveness in comparison to other orange suppliers to Russia. In particular, the results of the revealed competitiveness index revealed that Morocco ranks first with an index value of 15.04, followed by Egypt 12.72, South Africa 6.15, Turkey 4.02, Greece 3.98 and lastly Argentina 2.76. As an overall analysis of the results of Vollrath's indices during the period 1996–2014, it could be concluded that there are competitiveness gains for the majority of orange suppliers to Russia, albeit to varying degrees. Egypt seemed to be the country that has achieved remarkable improvements in the comparative advantage and competitiveness of its orange exports. Alternatively, there was a loss of competitiveness in Morocco and an obvious lack of comparative advantage of Argentinian orange exports in relation to other exporters, while the values of the indices in certain years are below zero.

3. METHODS

3.1 Orange import allocation model

One of the most common approaches to estimating demand relationships is based on a direct differential approximation of the demand function, which generates a set of equations that are local first-order approximations of the underlying relationship between prices, quantities and income (Okrent and Alston 2011). Yu et al. (2004) indicate that the Rotterdam model (Barten 1964; Theil, 1965) has become a prominent vehicle for the econometric analysis of patterns of consumer demand and rigorous testing of the utility maximisation theory. Barnett and Seck (2008) point out that the Rotterdam model gained prominence in food demand analysis, thanks to its locally flexible functional form. Moreover, it does not place a priori restrictions on the possible elasticities at a given point and also features enough parameters to approximate elasticities at a given point. The Rotterdam model has therefore been used quite extensively to analyse agrifood demand from different supply sources (e.g. Fonsah and Muhammad 2008; Feleke and Kilmer, 2009; Seale et al., 2013; Clements and Gao, 2015). Surprisingly, the Rotterdam model has to date received less attention in the literature on foreign demand for Egyptian agrifood commodities, while the few studies featuring differentiation in the source of origin have without exception applied the AIDS model (e.g. Elkady and Abd Elmessih 2005; Abu Hatab and Romstad 2014; Abu Hatab and Surry 2015). In this paper, the Rotterdam model was therefore used to analyse the demand relationships between major orange suppliers to the Russian market during the period 1996–2014. In this context and following Seale et al. (1992) and Theil (1980), total orange demand in the estimated model is determined in the first stage, and conditional on total orange expenditures, the demand for oranges from each source is determined in the second stage.

Therefore, assuming that Russia imports orange from n supplying sources and letting (q_i) denote the imported quantity of orange from supplier i and (p_i) denote the import price of orange from supplier i, the Rotterdam model can be specified as follows:

$$\overline{w}_{it}\ \mathrm{dlog}q_{it} = \theta_i DQ_t + \sum_{j=1}^{n} \pi_{ij}\mathrm{dlog}p_{it} + \varepsilon_{it} \quad (5)$$

where $\overline{w}_{it} = ((w_{it} + w_{i,t-1})/2)$ represents the expenditure share of orange imports from supplier (i) in the year (t). In Eq. (5), $d\log q_{it} = \log(q_{it}/q_{i,t-1})$ and $d\log p_{it} = \log(p_{it}/p_{i,t-1})$. DQ_t is the Divisia volume index which can be expressed by the following formula: $DQ_t = \sum_i \overline{w}_{it} d\log q_{it}$ The random disturbance term (ε_i) is assumed to be normally distributed with a mean of zero. The parameter (θ_i) is the conditional marginal expenditure share on orange imports from supplier i in the total import expenditures of the importing market. The parameter (π_{ij}) is the conditional Slutsky price term or relative price effect. This coefficient measures the impact of the price of jth import on the demand for ith import. In Eq. (5), both the parameters (θ) and (π_{ij}) are assumed to be constant for the estimation. In order to conform to theoretical considerations of the Rotterdam model, these coefficients should satisfy the following parameter restrictions:

Adding-up:

$$\sum_i \theta_i = 1 \text{ and } \sum_i \pi_{ij} = 0 \quad (6a)$$

Homogeneity:

$$\sum_j \pi_{ij} = 0 \quad (6b)$$

Symmetry:

$$\pi_{ij} = \pi_{ji} \quad (6c)$$

In the estimated demand system, the rest of the world (RoW) equation was omitted because the variance-co-variance matrix of (ε_i) is singular and parameter estimates are invariant to the omitted equation. The Russian demand for oranges was estimated based on Eq. (5) by using the iterative, seemingly unrelated regression procedure that is equivalent to the maximum likelihood estimation. A means of likelihood ratio test suggested by Judge et al. (1988) was used to test the validity of the imposition of the theoretical restrictions. The conditional expenditure and the compensated and uncompensated price elasticities are calculated as follows:

Conditional expenditure elasticity:

$$\eta_i = \theta_i / w_i \qquad (7a)$$

Conditional Slutsky compensated price elasticity:

$$\varepsilon_{ij} * = \pi_{ij} / w_i \qquad (7b)$$

Conditional Cournot uncompensated price elasticity:

$$\varepsilon_{ij} = \left(\frac{\pi_{ij}}{w_i} \right) - w_i \eta_i \qquad (7c)$$

3.2 Data and estimation procedures

Annual data covering the period 1996–2014 were used to estimate Russia's import demand for orange imports from Egypt and other major suppliers. As indicated in the second section of this paper, the selection of the suppliers was based on their shares of expenditure. Data on quantities and values of orange imports by Russia were obtained from the World Bank's World Integrated Trade Solution (WITS) database. Values are on cost, insurance and freight (CIF) basis. Due to the unavailability of reliable data on import prices from each supplier, the unit import value, calculated as import expenditure divided by imported quantity, was used as a proxy for import prices. The final import prices are expressed in Russian roubles based on data from the World Bank's World Development Indicators(WDI) database. Table 2 reports descriptive statistics on the variables of the estimated demand model.

Table 2 Summary statistics of orange prices and budget shares in the Russian market, 1996–2014

	Mean	**Standard deviation**	**Coefficient of variation**	**Max.**	**Min.**
Budget shares (measured in percentages)					
Egypt	20.798	15.335	1.356	46.351	0.182
Morocco	22.342	13.583	1.645	49.161	4.864
South Africa	18.488	6.586	2.807	26.409	5.481
Turkey	15.401	5.285	2.914	26.033	4.784
Greece	6.965	12.023	0.579	42.877	0.014
Argentina	4.994	4.011	1.245	12.643	0.098
RoW	16.005	6.277	2.55	31.132	8.81
Prices (measured in Russian roubles/ton)					
Egypt	12,907.735	9955.065	1.297	28,590.117	417.741
Morocco	13,864.427	10,463.912	1.325	30,362.835	412.566
South Africa	14,306.010	10,419.957	1.373	29,296.336	461.672
Turkey	13,057.138	10,052.958	1.293	29,458.259	429.508
Greece	14,052.192	11,239.858	1.250	30,496.363	359.096
Argentina	14,169.481	10,389.449	1.364	30,430.092	443.415
RoW	15,075.326	11,102.323	1.358	32,176.995	510.027

Source: Author’s calculations based on data from the World Bank, WITS (2015) and WDI (2015)

4. RESULTS

4.1 Parameter estimates of orange import demand equations

The estimated parameters for the conditional marginal orange import shares for the Russian market, price coefficients and the corresponding asymptotic standard errors are given in Table 3. Preliminary tests indicated that the calculated likelihood ratio test (LRT) values were lower than the corresponding critical value, suggesting that neither homogeneity nor symmetry conditions could be rejected. The results presented below in Table 3 are homogeneity and symmetry restricted. The goodness-of-fit of the estimated demand system was measured by calculating the squared *R* using the following formula as suggested by (McElroy 1977):

$$R^2 = 1 - \frac{1}{1 + W/[(T-K)(n-1)]}$$

Table 3 Conditional demand estimates for Russian orange imports, 1996–2014

Supplier	**Trend**	**Expenditure**	**Price effects**						
			Turkey	**Egypt**	**Morocco**	**South Africa**	**Argentina**	**Greece**	**RoW**
Turkey	-0.693	0.019	-0.653						
	(-4.371)	(0.007)*	(0.221)*						
Egypt	0.154	0.395	0.061	-0.276					
	(4.75)*	(0.033)*	(0.175)	(0.065)*					
Morocco	0.935	0.244	0.406	0.038	-0.649				
	(4.732)*	(0.038)*	(0.133)*	(0.017)	(0.350)*				
South Africa	0.011	0.153	0.022	0.023	0.024	-0.545			
	(0.190)	(0.035)*	(0.095)	(0.008)*	(0.098)	(0.118)*			
Argentina	-0.347	0.143	0.014	0.059	0.083	0.196	-0.761		
	(-2.712)*	(0.020)*	(0.003)*	(0.030)*	(0.020)*	(0.436)	(0.083)*		

Greece	−0.112	0.046	0.020	0.042	0.020	0.277	0.124	−0.792	
	(0.045)*	(0.013)*	(0.0078)*	(0.029)	(0.015)	(0.257)	(0.055)*	(0.301)*	
RoW	−0.174	0.112	0.131	0.053	0.078	0.003	0.285	0.310	−0.860
	(0.032)*	(0.005)*	(0.032)*	(0.385)	(0.023)*	(0.007)	(0.070)*	(0.078)*	(0.788)

Note: figures within parentheses are asymptotic standard errors

*Significance level ≤0.05

A squared *R* of 0.69 implies that the model can explain the market shares of the total orange imports by Russia. The relative "price effects" are the conditional Slutsky coefficients (π_{ij}), which refer to the change in the market share of supplier *i* for a unit proportionate change in export price. The "expenditure" refers to the relative share received by supplier *i* when an additional Russian rouble was added to the expenditures on imported oranges by Russia. Given that the objective of this paper was to examine the demand for Egyptian oranges on the Russian market, the results and their discussions in the following sections primarily focus on the results related to Egypt's equation, with results related to other suppliers discussed where necessary.

The reported estimates in Table 3 show positive and significant relationships between the Divisia index and Russia's orange imports from all suppliers. This implies a positive and significant relationship between Russian expenditure on imported oranges and the quantity of oranges supplied by major exporters. Specifically, a one-rouble increase in Russian expenditure on orange imports was allocated across the seven suppliers and would be relatively large for Egypt (about 40 cents) and Morocco (23 cents), lower for South Africa (15 cents) and Greece (14 cents), and significantly lower for Argentina (5 cents), Turkey (2 cents), and RoW (1 cent).

The conditional own-price estimates were negative and statistically significant at the 0.05 level for all suppliers, except for RoW where the estimate was positive but insignificant. According to these results, a one per cent increase in the export price of Egyptian oranges would reduce its market share in the Russian market by about 0.28 %. Other suppliers, especially Greece and Argentina, were more sensitive to own-price changes. The conditional cross-price estimates showed that values related to South Africa and Argentina were positive, which implied that a significant competitiveness exists between Egypt and these countries on the Russian market for imported oranges.

4.2 Estimated orange import demand elasticities of the orange import demand elasticities

Abu Hatab and Romstad (2014) illustrated that individual coefficients are likely not to have the usual interpretation or the expected signs because price and income derivatives are non-linear functions of parameters and variables. It is therefore more informative to focus the discussion on the estimated elasticities. Table 4 gives the conditional expenditure elasticities, uncompensated own-price elasticities and their corresponding asymptotic standard errors. The expenditure elasticity estimates for Egypt and Morocco were 1.085 and 1.712, respectively, suggesting that the two countries stand to benefit most from an increase in Russia's expenditure on imported oranges. Specifically, if Russians increase their expenditure on (imported) oranges by 1 %, the market share of Egyptian and Moroccan oranges would increase by 1.1 and 1.7 %, respectively. Expenditure elasticity estimates for South Africa, Greece and Argentina suggested that their total exports would increase but that their market shares would decrease with increases in Russian demand for imported oranges. With respect to Turkey and RoW, the estimated expenditure elasticity coefficients were positive but statistically insignificant, implying that an increase in Russian expenditure on orange imports would not favour these suppliers.

Table 4 Elasticity estimates for Russia's orange imports, 1996–2014

Supplier	**Elasticities**								
	Expenditure	**Own price**	**Cross price**						
			Turkey	**Egypt**	**Morocco**	**South Africa**	**Argentina**	**Greece**	**RoW**
Turkey	0.251	−1.097		−0.322	0.362	0.082	0.216	0.121	0.387
	(1.443)	(0.430)*		(0.142)*	(0.176)*	−0.082	(0.034)*	(0.059)*	(0.524)*
Egypt	1.085	−0.987	0.714		−0.588	0.123	−0.501	−0.131	0.276
	(0.175)*	(0.117)*	(0.107)*		(0.160)*	(0.061)*	(0.422)	(0.062)*	(0.151)*
Morocco	1.712	−0.872	0.914	−0.802		−0.598	−0.473	−0.266	0.385
	(0.269)*	(0.103)*	(0.488)*	(0.191)*		(0.161)*	(0.281)	(0.234)	(0.087)*
South	0.477	−0.86	−0.43	0.395	0.177		0.125	0.255	−0.12

Africa		7	9						3
	(0.225)*	(0.061)*	(0.286)	(0.002)*	(0.025)*		(0.089)	(0.203)	(0.310)
Argentina	0.204	−0.471	0.534	−0.222	0.175	0.125		0.337	−0.682
	(0.110)*	(0.133)*	(0.060)*	(0.288)	(0.104)*	(0.089)		(0.181)*	(0.242)*
Greece	0.316	−0.563	0.113	0.242	0.264	0.143	−0.284		−0.231
	(0.128)*	(0.084)*	(0.44)*	(0.381)	(0.012)*	(0.299)	(0.173)*		(0.038)*
RoW	0.753	−0.241	0.382	−0.563	0.258	−0.457	−0.314	0.181	
	(0.599)	(0.221)	(0.157)*	(0.413)	(0.121)*	(0.345)	(0.306)	(0.382)	

Note: figures within parentheses are the asymptotic standard errors

*Significance level ≤0.05

Own-price elasticity estimates for all orange suppliers to Russia, except for RoW, were negative and statistically significant. Moreover, the estimates for all suppliers are price inelastic with the exception of Turkey (−1.097). Egypt's close-to-unity own-price elasticity implied that if Egypt reduces (increases) the price of its orange exports to Russia by 1 %, its market share increases (decreases) by 0.987 %. The cross-price elasticity estimate for Egypt with Turkey (0.714) indicated that its oranges are a strong substitute for Turkish oranges on the Russian market. Moreover, the positive but low value of the cross-price elasticity (0.123) between Egypt and South Africa indicated limited competition between Egyptian and South African oranges. These findings imply that Russia tends to switch to orange imports from Turkey and South Africa when Egyptian oranges become relatively expensive. In contrast, the negative sign in the cross-price elasticity coefficient for the Egypt-Morocco, Egypt-Greece and Egypt-RoW pairs indicated a complementary relationship between their oranges on the Russian market.

5. DISCUSSION AND CONCLUSIONS

Recent years have seen closer diplomatic relations between Egypt and Russia that have led to a significant growth in their bilateral agricultural trade. As one of the world's leading producers and exporters of oranges, such trade growth represents an opportunity for Egypt to promote its orange exports to Russia. Moreover, in conjunction with Egyptian-Russian bilateral agricultural trade growth in recent years, Russia imposed import embargos on agricultural and food commodities from several countries, creating a supply gap of around 25 % on the Russian orange market, thus representing further export opportunities for Egyptian oranges. Motivated by these developments in the Russian orange market, this paper used a Rotterdam import allocation model to analyse demand relationships among major orange suppliers to Russia during the period 1996–2014.

The empirical results of the Rotterdam model suggested that the orange quantities and market shares of Morocco and Egypt would increase the most if Russia were to allocate a larger budget to the import of oranges. The results also showed that Egypt's export prices had a significant influence on its market share in the Russian market for imported oranges. Previous studies on the export performance of Egyptian agricultural commodities have pointed out that the lack of competitive prices is one of the greatest challenges hindering the access of Egypt's agricultural commodities to foreign importing markets (Soliman and Bassiony 2012; Torayeh 2013, Abu Hatab and Hess 2013). With Egypt's close-to-unity own-price elasticity for oranges, it may be possible to increase Russian demand for Egyptian oranges by implementing policies that help farmers increase their orange production and yield sustainably and cost-effectively. In this respect, it would be necessary to adopt innovative agricultural policies to increase productivity through the implementation of modern and globally agreed production techniques, including irrigation, fertilisation and pest management practices. Moreover, investment in infrastructure and post-harvest services is crucial to connect orange production areas more effectively with their markets and reduce transportation cost and post-harvest losses. Furthermore, improved Egypt-Russia diplomatic and trade relations in recent years and ongoing (agricultural) trade liberalisation negotiations represent a sound opportunity for Egypt to discuss these market access issues in order to enable Egyptian exporters to reduce the prices at which they offer their oranges on the Russian market.

The estimates of cross-price elasticity in particular revealed that Turkey and South Africa are Egypt's main competitors on the Russian market for imported oranges. Russia tends to switch to orange imports from these suppliers when Egyptian oranges become relatively expensive. Given that Turkish oranges are currently excluded from the Russian market, they do not present a competitive threat to Egyptian orange exports to Russia at the moment. A removal of the embargo would likely expose Egyptian oranges to fierce competition with Turkey. This is because Turkey would benefit from its geographical proximity to Russia that gives it a competitive advantage by eliminating the need for refrigerated transportation and therefore allowing it to undercut the Egyptian export price significantly (Gressel 2012). With respect to South Africa, the smaller influence of its competitiveness effect on Egyptian orange exports to Russia might be partially attributed to the fact that it has different production and export seasons ranging from July to September compared to many varieties of Egyptian oranges that are harvested in December. However, Egypt's relative proximity to Russia reduces transportation costs and gives Egypt a competitive advantage that mitigates the competition effect of South Africa (King-Okumu and Aboukheira 2015).

Taking into account the structural changes that the Russian orange market is currently undergoing, competition between traditional exporters and with new entrants is likely to intensify. Although competition may represent a threat to Egyptian orange exports to Russia, it could also act as an efficiency-enhancing factor and a positive driver of innovation that could enhance export competitiveness. In this respect, upgrading the orange value chain, adopting agro-processing technologies, enhancing the technical and organisational capacity of farmers and exporters and strengthening the institutional framework of the orange export sub-sector are important strategies that could enable Egyptian exporters to move up the value chain, offer higher value-added oranges and improve their competitiveness on the Russian market.

Furthermore, despite the calculation results of the revealed comparative advantage in the second section of this paper, indicating that Egypt enjoys a strong comparative advantage in orange exports in comparison to other orange suppliers to Russia, these results should be interpreted with caution since the existence of a comparative advantage alone certainly does not guarantee that a country's exported commodity will be able to "compete" successfully. A country's comparative advantage in a given commodity could alter over time due to changes in any of the other determinants of comparative advantage, such as demand patterns; the level of technology and production systems, government policies, and food safety; and quality standards among others (Cai and Leung 2007). In particular, non-tariff barriers (NTBs) such as sanitary and phytosanitary (SPS) measures are increasingly becoming a major determinant of the performance and market share of Egyptian agricultural exports to Russia and other importing markets (Bassiony 2012; Torayeh 2013). In this context, Trienekens and Zuurbier (2008) illustrated that the growing proliferation and stringency of food safety and quality standards in global importing markets have brought additional challenges to developing countries and raised concerns about their ability to cope with export requirements. Accordingly, the international fresh produce market has increasingly become highly competitive and sophisticated, and compliance with these ever-rising standards has become a critical factor that can adversely affect the competitiveness of exports from developing countries (Martinez and Poole 2004; Okello and Swinton 2007). With respect to the Russian market, Vinokurov et al. (2015) classified and assessed the impact of NTBs in the members states of the Eurasian Economic Commission (including Russia) and concluded that SPS measures represent one of the most important NTBs that increase the cost of exports to Russia on a significantly larger scale than other NTBs and therefore affect the access and competiveness of foreign exporters. Such NTBs could negate the positive effects of ongoing Egypt-Russia trade liberalisation efforts with Egyptian orange exports and should therefore be given priority within the context of these efforts, as well as in the ongoing negotiations aimed at the establishment of a free trade area between Egypt and the Eurasian Customs Union. Moreover, while the primary focus of this paper has been on price relationships between orange suppliers to Russia, more research using different and approaches is needed to provide a quantitative assessment of the trade implications of NTBs on the access and competitiveness of Egyptian orange exports on the Russian market.

REFERENCES

1. Abu Hatab A, Hess S (2013) Opportunities and constraints for small agricultural exporters in Egypt. Int Food Agribus Manag Rev 16:77–100
2. Abu Hatab A, Nsanimana A (2016) A dynamic analysis of Egyptian orange exports to Russia: an ARDL approach. Agricultural Economics Review, the Greek Association of Agricultural Economists. In Press.
3. Abu Hatab A, Romstad E (2014) Competitiveness analysis of Egyptian cotton exports with special focus on the Chinese market. China Agri Econ Rev 6:248–263. doi:10.1108/CAER-02-2013-0026
4. Abu Hatab A, Surry Y (2015) The determinants of import demand for Egyptian potatoes in the European Union. In: Proceedings of the XVIII International Symposium on Horticultural Economics and Management. The International Society for Horticultural Science, Alnarp, Sweden, May 31-June 3, 2015
5. Balassa B (1965) Trade liberalization and "revealed" comparative advantage. Manch Sch Econ Soc Stud 33:99–123Barnett WA, Seck O (2008) Rotterdam model versus almost ideal demand system: will the best specification please stand up? J Appl Economet 23:795–824
6. Barten AP (1964) Consumer demand functions under conditions of almost additive preferences. Econometrica 32:1–38
7. Bassiony H (2012) Impact of non-tariff barriers on Egyptian agricultural trade. In: Proceedings of Minia International Conference for Agriculture and Irrigation in the Nile Basin Countries. El-Minia University, Minia, Egypt, March 29-29, 2012
8. Cai, J, Leung P (2007) A review of comparative advantage assessment approaches in relation to aquaculture development. In: Leung et al. quality standards among others (ed) Species and System Selection for Sustainable Aquaculture, i st edn., Blackwell Publishing, Ames, Iowa, USA. doi: 10.1002/9780470277867.
9. Clements KW, Gao G (2015) The Rotterdam demand model half a century on. Econ Modelling 49:91–103
10. Elkady S, Abd Elmessih E (2005) The competitive ability of the Egyptian potatoes exports in the foreign markets. Egypt J Agri Econ 15:347–368
11. FAOSTAT (2016) Food and agriculture organization of the United Nations' statistical database. FAO, Rome, http://faostat3.fao.org. Accessed 26 July 2016

12. Feleke ST, Kilmer RL (2009) The Japanese market for imported fruit juices. Int Food Agribus Manag Rev 12:1–28
13. Fonsah EG, Muhammad A (2008) The demand for imported apple juice in the United States. JFood Distrib Res 39:57–61
14. GRC, Global Reach Consulting (2012) Russian and world fresh fruit market Russia. GRC, Moscow
15. Gressel J (2012) Egypt-citrus annual. US Department of Agriculture (USDA), USDA Foreign Trade Service. Global Agricultural Information Network, Washington DC
16. Hamida M (2012) Determinants of external demand for Egyptian exports of oranges in the world market. J Appl Sci Res 8:5730–5735
17. Hamza M (2014) Egypt's orange exports benefiting from Russia's ban on EU imports. USDA Foreign Trade Service, Global Agricultural Information Network, Washington DC
18. Hamza M (2015) Egypt-citrus annual. USDA Foreign Trade Service, Global Agricultural Information Network, Washington DC
19. Hassan A, Mohamed A, Elgebaly M, Hassan H (2010) Economic analysis of the current situation for the production, export and the competitiveness of Egyptian oranges in global markets. Aust J Basic Appl Sci 4:5797–5804
20. Honkanen P, Voldnes G (2006) Russian consumers' food habits: results from a qualitative study in Moscow. Norwegian Institute of Fisheries and Aquaculture Research, Tromso
21. Ilyina S (2011) Russians love citrus but U.S. share tiny. USDA Foreign Trade Service, Global Agricultural Information Network, Washington DC
22. Judge GG, Hill RC, Griffiths W, Lutkepohl H, Lee TC (1988) Introduction to the theory and practice of econometrics. John Wiley & Sons, New York
23. King-Okumu C, Aboukheira A (2015) Challenges to achieving sustainable exports of fruits from small-holdings under water scarcity conditions of the Southern Mediterranean. In: Paciello MC (ed) Building sustainable agriculture for food security in the Euro-Mediterranean area. Edizioni Nuova Cultura, Rome
24. Kolchevnikova O (2010) Russian retail food sector nears rebound. USDA foreign agricultural service, Global Agricultural Information network, GAIN report number: RSAT01003, USDA, WashingtonLaursen K (2015) Revealed comparative advantage and the alternatives as measures of international specialization. Eurasian Bus Rev 5:99–115
25. Leromain E, Orefice G (2014) New revealed comparative advantage index: dataset and empirical distribution. Int Econ 139:48–70
26. MALR (2015) Annual agricultural statistics bulletin. MALR, Cairo
27. MALR, Ministry of Agriculture and Land Reclamation (2013) Annual agricultural statistics bulletin. MALR, Cairo
28. Martinez MG, Poole N (2004) The development of private fresh produce safety standards: implications for developing Mediterranean exporting countries. Food Policy 29:229–255
29. McElroy MB (1977) Goodness of fit for seemingly unrelated regressions: a generalization. J Econometrics 6:381–387
30. Okello JJ, Swinton SM (2007) Compliance with international food safety standards in Kenya's green bean industry: comparison of a small-and a large-scale farm producing for export. Appl Econ Perspect Policy 29:269–285
31. Okrent AM, Alston JM (2011) Demand for food in the United States: a review of literature, evaluation of previous estimates, and presentation of new estimates of demand. Giannini Foundation of Agricultural Economics Monograph 48, Berkeley, California
32. Parker JW (2015) Understanding Putin through a Middle Eastern looking glass. Strategic Perspectives issue 19, Center for Strategic Research, Institute for National Strategic Studies, National Defense University, Washington, DC
33. Rudolf M (2015) Egypt's citrus sector is expanding. Online article., Agroberichten Buitenland, http://www.agroberichtenbuitenland.nl/cairo/egyptes-citrus-sector-groeit/. Accessed 28 July 2016
34. Seale JL, Sparks AL, Buxton BM (1992) A Rotterdam application to international trade in fresh apples: a differential approach. Agric Resour Econ Rev 17:138–149
35. Seale JL, Zhang L, Traboulsi MR (2013) US import demand and supply response for fresh tomatoes, cantaloupes, onions, oranges, and spinach. J Agri Appl Econ 45:435–452

36. Soliman I (2013) Oranges sector in Egypt: performance and policies. In: report number D20 (Policy for Competitiveness of the Fresh Orange Export Sector in Mediterranean Partner Countries), USTAINMED Project (Sustainable agri-food systems and rural development in the Mediterranean Partner Countries). https://sustainmed.iamm.fr/index.php/publications/project-reports. Accessed 02 Oct 2016.
37. Soliman I, Bassiony (2012) Egyptian agricultural exports competitiveness. MPRA Paper No. 66820. https://mpra.ub.uni-muenchen.de/66820/ Accessed 30th July 2016.
38. Theil H (1965) The information approach to demand analysis. Econometrica 33:67–87
39. Theil H (1980) The system-wide approach to microeconomics. The University of Chicago Press, Chicago
40. Torayeh NM (2013) The competitiveness of the Egyptian agricultural export in the EU market: should Egypt diversify its trade pattern? Appl Econ Int Dev 13:129–148
41. Trienekens J, Zuurbier P (2008) Quality and safety standards in the food industry, developments and challenges. Int J Prod Econ 113:107–122
42. Vinokurov E, Demidenko M, Pelipas I, Tochitskaya I, Shymanovich G, Lipin A (2015) Assessing the impact of non-tariff barriers in the EEU: results of enterprise surveys. Report 30, EDB Centre for Integration Studies, EDB, Saint Petersburg
43. Vollrath TL (1991) A theoretical evaluation of alternative trade intensity measures of revealed comparative advantage. Weltwirtschaftliches Archiv 127:265–280
44. WDI, World Development Indicators Database (2015). The World Bank. Available at: http://databank.worldbank.org/data/reports.aspx?source=world-development-indicators.
45. WITS, World Integrated Trade Solution, (2015). The World Bank. Available at: http://wits.worldbank.org/wits/.
46. WITS, World Integrated Trade Statistics database (2016) World Bank, Washington DC. http://wits.worldbank.org/. Accessed 23 July 2016.
47. Yu W, Hertel TW, Preckel PV, Eales JS (2004) Projecting world food demand using alternative demand systems. Econ Modelling 21:99–129

CHAPTER 5

Farming household food storage, consumption and sales decision making under price risk in northern Uganda

Charles Owach[1]* Godfrey Bahiigwa[2] Gabriel Elepu[3]

[1]*Food and Agriculture Organization of the United Nations, P. O. Box 521, Kampala, Uganda.*

[2]*International Food Policy Research Institute, Eastern and Southern Africa Regional Office, Addis Ababa, Ethiopia.*

[3]*Department of Agribusiness and Natural Resource Economics, Makerere University, Kampala, Uganda.*

ABSTRACT

This study extended the agricultural household model to explain food storage, consumption and sales behaviors of farming households in northern Uganda using two major staple grains: finger millet and beans. Using secondary data collected by the Uganda Bureau of Statistics from 782 millet and beans producing households (388 households below poverty level and 394 households above poverty level), seemingly unrelated regressions were performed and risk neutrality tests were carried out. It was found that all finger millet and beans producing households in northern Uganda were risk neutral regarding storage and sales decisions with only millet producing households below poverty line being risk averse in their consumption decisions. However, households above poverty line produced and stored more millet and beans implying that they were more food secure than households below poverty level. Therefore, strategies to boost incomes, production and prudent management of millet and beans stocks at the household level are critical for food security alleviation in northern Uganda.

Keywords: Food security, household, precautionary storage, consumption, sales, price risk, Uganda.

1. INTRODUCTION

Numerous studies have shown that precautionary food storage behavior of households differs by income level (Jalan and Ravallion, 2001; Deininger et al., 2007; Carter and Lybbert, 2012; Michler and Balagtas, 2013). Market supply, on the other hand, depends on the volume of grain harvest which is concentrated within a few months of the year in any area, and can fluctuate widely from one year to the next depending on climatic conditions (Park, 2006). According to Ravallion (1987) and Renkow (1990), households engaged in subsistence farming do not often store food for the market, but they store for future consumption. Even when prices are anticipated to rise with a high level of certainty, such households still fail to store food to get arbitrage profits. Deininger et al. (2007) argued that the concern for food security motivates household storage even when arbitrage is unlikely to insure food consumption as opposed for consumption of "all other goods". Meanwhile, better access to credit, for example, increases household food storage for arbitrage and decreases storage for food security purpose implying an overall positive or negative net effect on storage (Park, 2006; Michler and Balagtas, 2013).

Several explanations have been made to understand why many farming households from developing countries store a portion of their output at harvest or invest in storage (Walker and Ryan, 1990; Fafchamps, 1992). Although, storage is used in developed countries to transfer income in the period between two harvests, this is not a realized outcome for most farming households. A study by Stephens and Barrett (2011) found a positive effect of credit access by households on storage (mostly for arbitrage purposes). However, Lee and Sawada (2010) ascertained that increasing access to credit reduced household's reliance on storage. The disparity in the above study results might be because Lee and Sawada's study was based on the premise that grain storage was a form of unproductive savings that households undertook due to credit constraints. Arguments on inadequacy of rural capital markets seem unsatisfactory as households could still transfer income to the planting season in the form of sales cash stored-away, so long as there are functional commodity markets. The scope for the more plausible reason of inter-temporal price arbitrage is limited to farming households with marketable surplus (mostly commercial and not subsistence farmers), and absence of marked seasonal fluctuations in crop commodity prices in some developing countries (Walker and Ryan, 1990).

The importance of food security considerations in explaining household storage inventories in developing countries has been emphasized by a number of researchers. Ravallion (1987) noted that "positive stocks are observed even when expected future price falls short of spot price plus marginal storage cost". The report adds that "...stockholders are also likely to have viewed their stocks as a desirable precaution..." Renkow (1990) expounds on this "food security" reason by attempting to model on-farm storage decisions under price risk. Results showed that even if there is no scope for price arbitrage ($\Delta P_t = 0$), positive stocks may be held due to the "food security" motive (Renkow, 1990). Only in the case of large farming households is there evidence of significant arbitrage motives for holding food stocks (*Ibis*).

In view of these shortfalls, Saha and Stroud (1994) developed an agricultural household model of consumption, storage, savings and labour decisions expanding the reasons for crop storage under price risk beyond speculative behaviour alone, as the commodity arbitrage proponents. Particularly, for small farming households, food storage is usually motivated by their aversion to risk and food security considerations. Despite new insights into the economics of storage, this study used crop production levels to categorize sorghum farmers. Since income plays a critical role in household food security and risk-bearing in many developing countries, it perhaps would have been more meaningful to categorize the farming households using income rather than crop production levels.

In Uganda, most households usually sell off their grain surpluses immediately after harvest, rather than during the off-season when prices are high due to lack of money and proper storage facilities (APS, 1994; Owach, 1998; FAO, 2000). This household behaviour arguably could be the cause for incessant food insecurity and poverty in rural areas among other factors. Of all regions, northern Uganda is the worst hit with chronic poverty levels reaching 26% as compared to the national average of 18% (UBOS, 2015).

Therefore, the motivation of this study was to extend Saha and Stroud's agricultural household model to explain food storage behavior of farming households in northern Uganda using two major staple grains: millet and beans. In this study, households were categorized into two based on income: households above poverty line and households below poverty line.

It was anticipated that findings from this study would have policy implications on alleviation of household food and income security in northern Uganda

2. ANALYTICAL FRAMEWORK

When households anticipate that price increases will be sufficient to cover real costs of storage, they tend to store food. In this case, the typical condition for strictly positive optimal storage would be:

$$\psi E_t [\tilde{p}_{t+1}] > p_t + c \tag{1}$$

Where: ψ = intertemporal discount rate; E_t = expectations; subscript (t) = current period; p_t = current price; $\tilde{p}_{t+1}$ = next period price; c = unit storage cost. In Equation 1, a positive storage level will occur only if the expected discounted future price $\psi E_t [\tilde{p}_{t+1}]$ is greater than the spot price (p_t) plus unit storage cost (c). In Equation 1, commodity storage greater than zero will occur only if the expected discounted future price $\psi E_t [\tilde{p}_{t+1}]$ is greater than the current price (p_t) plus the unit cost of storage (c).

Within agricultural economic cycles, it is common knowledge that optimal choices in agricultural household models are not "separable" in situations of price risk (Sadoulet and de Janvry, 1995; Singh et al., 1986). For the case of subsistence farmers, their optimal choices are not separable because they are both consumers and producers of their own produce. In this study, the model developed by Saha and Stroud (1994) where farm household maximizes a time-wise additively separable and time-invariant utility function over a time horizon of T periods is followed thus:

$$\text{Max } U(.) = \Sigma \psi^t U(C_t, O_t, R_t) \tag{2}$$

Where: U = utility; C_t = food consumption; O_t = consumption of "all other goods"; R_t = consumption of leisure. Household's optimal choices are made in situations of production and price risk. Thus at period (t), the price at (t+1), given by ($\tilde{p}_{t+1}$) is not known to the farmer.

Following Saha and Stroud (1994), it is assumed the stochastic process ($\tilde{p}$) is a stationary Markov process, thus the probability distribution of ($\tilde{p}_{t+1}$) is conditional only on (p_t) and not the whole history of the process. Considering these assumptions, the household's optimisation problem is

$$\text{Max}_{zt}\, H = U[C, p_t M_t + \Delta b_t .... + A_t(L_t) - c(S_t)\, l_t\, (L_t)] + \psi E_t[V^{t+1}(\tilde{p}_{t+1}, \tilde{y}_{t+1})] \tag{3}$$

Where: H = optimal household choices (storage, consumption, sales, labour); C=optimal consumption choice; p_t = current period price; M_t = current period sales; b_t = household savings; A_t = household's net labour income [≡ off-farm labour earnings minus farm labour (hired and family) expenses]; l_t = Leisure; L_t = optimal labour choice [L_t ≡ (hired and family farm labour]; c=inventory cost; S_t = current period optimal storage; $z_t \equiv [C_t, L_t\, b_{t+1}, S_t]$ denotes the vector of decision variables to be optimally chosen by the household; E= expectation operator at period (t). $V^{t+1}=\sum^{T}_{i=t+1}\psi^i\, U[C^*_{i,}, y^*_{i,}, l_i(L_t)]$ denotes the value function, where (*) superscripts indicate optimal choices; $_y$ = Household income. Since $p^{\sim}_{t+1}$ is unknown to the household at period (t), the value of the function is stochastic, denoted by $V^{t+1}[p^{\sim}_{t+1}, y^{\sim}_{t+1}]$.

Saha and Stroud (1994) showed the first order conditions of Equation 3, with optimal choice vector, $z_t^* = [C_t^*, L_t^*, b_{t+1}^*, S_t^*]$ to be:

$$Hc_t \equiv Uc_t - p_t\, Uy_t \equiv 0 \quad (4a)$$

$$H_{Lt} \equiv [p_t\, Q_{tLt}\,(L_t\,)+A_{tLt}\,(L_t\,)+)]U_{yt} + l_{yt}\,(L_t\,)U_t \equiv 0 \quad (4b)$$

$$Hb_{t+1} \equiv -Uy_t + (1+r)\,\psi E_t[Vy^{t+1}] \equiv 0 \quad (4c)$$

$$HS_t \equiv [p_t + C'(S_t)]\, Uy_t + \psi E_t[V_y^{t+1}\, p^{\sim}_{t+1}] \equiv 0 \quad (4d)$$

Where all alphabetic subscripts, except the subscript t, denote partial derivatives and 0 is the null vector of the appropriate dimension.

2.1 Estimation of the model

Following Saha and Stroud (1994), the econometric model used to explore the influence of income and other factors on quantities of millet and beans stored, consumed and sold by households in northern Uganda took the form of a system of three simultaneous equations on storage, consumption and sales. The assumption that household's optimal choices are made in an environment of output price uncertainty implies that at period (t), price at t+1, represented by $p^{\sim}_{t+1}$, is not known to the household. It is assumed that the stochastic process, $p^{\sim}$ follows a stationary Markov process, hence the probability distribution of $p^{\sim}_{t+1}$ is conditional on only p_t, and not on the whole history of the process (Saha and Stroud, 1994). Optimal (Pareto optimality) household choices are therefore defined by:

$$V_t\,(\Phi) \equiv [C_t\,(\Phi), S_t(\Phi), M_t(\Phi)] \quad (5)$$

Where: V_t = optimal household choices; Φ = parameter vector of optimal choice; C_t (Φ), = optimal consumption choice; $S_t(\Phi)$ = optimal storage choice; $M_t(\Phi)$ = optimal sales choice.

Equation 5 gives the dependent variables of the complete system. Equation 6 gives the dynamic relationship of dependent and independent variables. It was assumed that household optimal savings (b_{t+1}), are subsumed in income in each period. From first-order conditions, optimal choices are functions of current crop price, moments of the distribution of random expected future price, price of substitutes and complements, income, education and gender of household head, location and poverty status of household as shown in Equation (6) below:

$$\pi \equiv [p_t, E(p^{\sim}_{t+1}), var(p^{\sim}_{t+1}), Y_t, P^*, P^{**}, Sexhh, Dist, Educ, PovL, Extn, Seas] \quad (6)$$

Where, π = optimal household choice (storage, consumption and sales); $E(p^{\sim}_{t+1})$ = expected future price (measured by the structure $E(p^{\sim}_{t+1}) = \theta p_t$); $var(p^{\sim}_{t+1})$ = variance of future price (as proxy for price volatility); Y_t = current household income (Ushs); P^* = vector of prices of substitutes of millet or beans; P^{**} = vector of prices of complements of millet or beans; Sexhh = sex of household head (male =1; female = 0); Dist = district where household is located (1=Apac; 0=Arua); Educ = number of years spent in school for formal education (years); PovL = poverty line (1=households above poverty line; 0= households below poverty line); Extn = access to extension services (1=household visited by extension personnel; 0 = household not visited by extension personnel); and Seas = production season (1= first season 2009; 0 = second season 2008).

From Equation 5, the set up to estimate a system of seemingly unrelated regression equations would take the form:

$$y = \beta x + \mu \quad (7)$$

Where: y is a vector of the dependent variables (C, S, M), X is a matrix of regressors corresponding to those in Equation 5, β is a vector of parameters to be estimated, and μ is a vector of error terms. Following Saha and Stroud (1994), the structures $E(p^{\sim}_{t+1}) = \theta p_t$; and $var(p^{\sim}_{t+1}) = \theta[p_t - E(p^{\sim}_t)]^2 = \theta[p_t - \theta p_{t-1}]^2$ are then imposed on the moments of the distribution of random price, and these forms are then substituted into Equation 9. The parameter (θ) was jointly estimated from the data for households above and below poverty line. The estimation equation for storage takes the form:

$$S_t = \delta_1 + \delta_2 p_t + \delta_3 \theta p_t + \delta_4 \theta [p_t - \theta p_{t-1}]^2 + \delta_5 Y_t + \delta_6 PovL + \delta_7 P_{cas} + \delta_8 P_{bean} + \delta_9 Dist + \delta_{10} Sexhh + \delta_{11} Educ + \delta_{12} P_{fcas} + \delta_{13} Extn + \delta_{14} Seas + e_1 \tag{8}$$

and this simplifies to:

$$S_t = b_o + b_1 p_t + b_2 p^2_t + b_3 p^2_{t-1} + b_4 p_t p_{t-1} + b_5 Y_t + b_6 PovL + b_7 P_{cas} + b_8 P_{bean} + b_9 Dist + b_{10} Sexhh + b_{11} Educ + b_{12} P_{fcas} + b_{13} Extn + b_{14} Seas + e_1 \tag{9}$$

Where:

(9a) $\beta_o = \delta_1$

(9b) $\beta_1 = \delta_2 + \delta_3 \theta$

(9c) $\beta_2 = \delta_4 \theta$

(9d) $\beta_3 = \delta_4 \theta^3$

(9e) $\beta_4 = -2\delta_4 \theta^2$

and, S_t = current period optimal storage (in kg); β_o to β_{14} = estimation coefficients for the respective variables; p_t = current period millet price (in Ushs per kg); p^2_t = current price squared (in Ushs); p^2_{t-1}= lagged price squared (in Ushs); $p_t p_{t-1}$ = current price times lagged price (in Ushs); P_{cas} = current period cassava price (in Ushs per kg); P_{fcas} = current period fresh cassava price (in shs per kg); P_{bean} = current period beans price (in shs per kg; e_1 = error term. Other variables are as defined in equation (6).

The structure of the optimal consumption equation was

$$C_t = \square_o + \square_1 p_t + \square_2 p^2_t + \square_3 p^2_{t-1} + \square_4 p_t p_{t-1} + \square_5 Y_t + \square_6 PovL + \square_7 P_{cas} + \square_8 P_{bean} + \square_9 Dist + \square_{10} Sexhh + \square_{11} Educ + \square_{12} P_{fcas} + \square_{13} Extn + \square_{14} Seas + e_2 \tag{10}$$

Where C_t= optimal consumption of crop in current period (in kgs); $\square_o$ to $\square_{15}$ = estimation coefficients for the respective variables; e_2 = error term. Other variables in the consumption equation are as defined in Equation 9 above. The optimal sales equation is given by

$$M_t = \square_o + \square_1 p_t + \square_2 p^2_t + \square_3 p^2_{t-1} + \square_4 p_t p_{t-1} + \square_5 Y_t + \square_6 PovL + \square_7 P_{cas} + \square_8 P_{bean} + \square_9 Dist + \square_{10} Sexhh + \square_{11} Educ + \square_{12} P_{fcas} + \square_{13} Extn + \square_{14} Seas + e_3 \tag{11}$$

Where M_t = current period sales (in kg); $\square_o$ to $\square_{15}$ = estimation coefficients for the respective variables; e_3 = error term. Other variables in the sales equation are as defined in Equation 9 above.

Under risk neutrality, household optimal choices of millet or beans storage, consumption and sales would be unaffected by changes in the second moments of random its price. Thus, the coefficients of the quadratic price regressors, $p^2{}_t$, $p^2{}_{t-1}$ and $p_t p_{t-1}$ are tested under the joint null hypothesis for risk neutrality given by,

$$H_o: \square_2 = \square_3 = \square_4 = 0 \qquad (12)$$

Where: $\square_2$, $\square_3$ and $\square_4$ are coefficients of the quadratic price regressors ($\square_2 = \delta_4\theta$; $\square_3 = \delta_4\theta^3$ and $\square_4 = -2\delta_4\theta^2$).

2.2 Data

This study used secondary data collected by Uganda Bureau of Statistics (UBOS) during 2008/2009 agricultural census. UBOS used a stratified two-stage sample design for small and medium-scale households. The first-stage involved selection of enumeration areas (EAs) with probability proportional to size (PPS). The second stage involved selection of households (ultimate sampling units) using systematic sampling, after stratification based on acres of cropland (UBOS 2010). The total UBOS sample size for the two study districts, Arua and Apac, was 1,090 households. Using the UBOS (2010) national poverty line equivalent to Ushs 62,545 (approx. US$34) per month per adult equivalent (in 2005/06 prices), these households were categorized into two: households below and above poverty level. After data was cleaned, only 782 households (388 households below poverty level and 394 households above poverty level) were usable in this study.

The following household data were obtained from UBOS for two crop production seasons (second season 2008 and first season 2009): quantities of finger millet and beans produced, stored, consumed and sold by households, household's income; current and lagged crop prices of finger millet and beans; prices of substitutes and complements; level of education, age and gender of household head; household access to credit, extension services, and membership to farmers' group/association. In addition, household size was computed based on consumption conversion factors of adult-equivalent recommended by World Health Organization (WHO) guidelines (Appleton 2001).

During data analysis, a seemingly unrelated regression (SUR) regression technique was performed to determine factors influencing quantities of millet and beans stored, consumed & sold by households to allow for non separability of household decisions. Then, household response to price risk was tested using coefficients of quadratic price regressors or post-estimation risk neutrality test.

3. RESULTS

3.1 Characteristics of households

Almost three-quarters of sampled households were male headed and, there was no significant difference in age and education level of household head (Table 1). Household access to extension services, credit and membership to farmer groups or associations were generally low in the study districts, although households below poverty line appeared to have had more access to these services than their counterparts. However, as expected, households above the poverty line had higher income (Ushs 70,185 per capita per month) than those below the poverty line (Ushs 38,710 per capita per month) and, this was significant at 1% (Table 1).

3.2 Household production, storage, consumption, and sales of millet and beans

Results in Table 2 show that households above poverty line stored significantly larger quantities of millet than households below the poverty line. The respective per capita per season storage of millet was 56.5 kg (24.4 kg) for households above poverty line (households below poverty line). There was no significant difference in per capita per season millet production, consumption and sales among household groups. For beans, households above the poverty line significantly produced, stored, consumed and sold significantly more beans than those below poverty line. Per capita per season millet production and sales for households above poverty line were 202.9 and 81.4 kg as compared to only 52.6 and 15.5 kg in the case of households below the poverty line. This indicates a better food security situation in households above poverty line than those households below poverty line.

Table 1. Characteristics of millet and beans grow ing households in Apac and Arua Districts.

Characteristic	HHs below poverty line (n=388)	HHs above poverty line (n=394)	Overall sample (N=782)	T-statistic or chi-square
Age of household head (years)	44.7	44.8	44.8	0.0783
Sex of household head: Male (%)	82.7	74.4	78.5	8.1123***
Education of household head (years in school)	5.8	6.2	6.0	1.2452
Household size (adult equivalent)	5.6	3.1	4.3	16.2977***
Household income (Ushs/month per capita)	38,710	70,185	52,253	31.2257***
Household access to extension services (%)	30.4	18.3	24.3	15.6593***
Household access to credit (%)	10.6	4.1	7.3	12.2451***
Membership to farmer group/association (%)	20.1	9.4	14.7	17.8843***

*** = Significant at 1%; Source: UBOS, 2010.

Table 2. Per capita seasonal millet and beans production, storage, consumption and sales in Apac and Arua districts, 2008/09 (in kg)

Crop	HHs below poverty line (n=388)	HHs above poverty line (n=394)	Overall Sample (N=792)	t-statistic
Millet				
Quantity produced	116.1	181.8	152.3	3 1.3454
Quantity stored	24.4	56.5	41.9	2.4246**
Quantity consumed	59.8	86.4	74.4	1.1602
Quantity sold	31.9	38.9	36.0	0.8050
Beans				
Quantity produced	52.6	202.9	116.6	2.7219***
Quantity stored	14.0	45.8	26.0	3.3334***
Quantity consumed	23.1	75.7	45.7	4.0035***
Quantity sold	15.5	81.4	44.9	2.9116***

***, ** = Significant at 1 and 5%, respectively. Source: UBOS, 2010.

3.3 Determinants of household millet storage, consumption and sale

Results in Table 3 indicate that quantities of finger millet stored, consumed and sold varied by type of household and the result was significant at 1% level in each case. Households above the poverty line significantly stored, consumed and sold more finger millet than those below the poverty line by: storage (60%), consumption (65%) and sales (57%). Education of household head was negatively associated with quantities of finger millet stored, consumed and sold. Quantities of millet consumed and sold were positively related to the price of cassava showing that millet and cassava were complementary foods. While households in Apac stored and consumed less millet than those in Arua, household millet consumption and sales in the first season of 2009 were generally lower than in the second season of 2008. As shown in Table 3, the second moments of random price of millet (p^2_t and p^2_{t-1}), price of beans, sex of household head, and access to extension services did not have a significant effect on quantities of finger millet stored, consumed and sold by households.

The hypothesis that under risk neutrality, household optimal choices of storage, consumption and sales would be unaffected by changes in the second moments of random millet price was tested as in Equation 12 above and results are shown in Table 4.

As shown in Table 4, it can be noted that finger millet growing households above poverty line were risk neutral in their storage, consumption and sales decisions. The null hypothesis of risk neutrality for storage, consumption and sales could not also be rejected for both types of households. However, households below the poverty line were risk averse in their consumption decisions.

3.4 Determinants of beans storage, consumption and sale

As in the case of millet, households above the poverty line significantly stored, consumed and sold more beans than those below the poverty line by: storage (83%), consumption (69%) and sales (88%). Quantity of beans sold was negatively related to the price of cassava showing that millet and cassava were complementary enterprises. Also, similar to the millet case, household beans consumption and sales in the first season of 2009 were significantly lower than in the second season of 2008. However, the second moments of random price of beans (p^2_t and p^2_{t-1}), product of current and lagged bean price (p_t p_{t-1}), sex of household head, access to extension services, location of household did not significantly affect quantities of beans stored, consumed and sold (Table 5).

Table 3. Determinants of quantities of f inger millet stored, consumed and sold.

Variable	Parameter estimates (t-values in parentheses)		
	Storage	Consumption	Sales
Current millet price squared (p^2_t)	-0.5252 (1.070)	-0.5176 (1.130)	-0.2382 (0.470)
Lagged millet price squared (p^2_{t-1})	0.1522 (0.370)	-0.2177 (0.560)	-0.6442 (1.480)
Type of household (Above poverty line)	0.5952*** (2.920)	0.6473*** (3.400)	0.5686*** (2.670)
Current bean price (P_{bean})	-0.2626 (0.820)	0.0279 (0.090)	-0.4461 (1.330)
Current cassava price (P_{cas})	0.5621 (1.650)	0.6613** (2.080)	0.7716** (2.170)
Sex of household head (Male)	-0.3920 (1.470)	-0.3455 (1.390)	-0.2312 (0.830)
Extension visit (Yes)	0.5031 (1 900)	-0.0082n (0.030)	0.3751 (1.360)
Education of household head (years)	-0.4342** (2.930)	-0.5000*** (2.650)	-0.7610*** (3.600)
District (Apac)	- 0.7945*** (2.770)	-0.9382*** (3.500)	-0.2784 (0.930)
Season (1st season 2009)	-0.1663 (0.650)	-0.4933** (2.080)	-0.8427*** (3.170)
Constant	7.4375** (2.380)	10.5025*** (3.600)	14.0029*** (4.280)
R^2 = 0.2292 (storage), R^2 = 0.3837 (consumption), R^2 = 0.3372 (sales)			

***, ** = Significant at 1 and 5%, respectively. Source: UBOS 2010.

Table 4. Parameter estimates for f inger millet price expectations and risk neutrality test.

Equation	Coefficient/ statistic	Type of Household		
		Below poverty line (n = 388)	Above poverty line (n = 394)	All households (N = 792)
Storage	(θ)	2.6934	0.5675	0.5478
	(chi^2)	0.8500	9.0400***	3.2800
	(δ_4)	-0.0127	-2.0999	-0.9872
Consumption	(θ)	2.5013	0.6384***	0.6069
	(chi^2)	2.3700	23.7500***	15.3200***
	(δ_4)	0.0270***	-2.7149	-0.9046
Sales	(θ)	1.4108	0.7412***	1.7582
	(chi^2)	14.6100***	24.4100***	21.2000***
	(δ_4)	0.5256	-2.9049	-0.1193

Price expectation parameter, $\theta = \sqrt{B3/B2}$; Price variance parameter, $\delta 4 = B2/\theta$. Chi^2-value is for the joint hypothesis H_0: $\beta_2 = \beta_3 = \beta_4 = 0$; *** = significant at 1%.

Table 5. Determinants of quantity of beans stored, consumed and sold.

Variable	Parameter estimates (t-values in parentheses)		
	Storage	Consumption	Sales
Current bean price squared (p^2_t)	0.1535 (0.220)	-0.5072 (0.760)	0.3882 (0.500)
Lagged bean price squared (p^2_{t-1})	0.7228 (0.960)	-0.9804 (1.360)	0.2385 (0.280)
Current times lagged bean price ($p_t p_{t-1}$)	-0.7064 (0.720)	0.7431 (0.800)	-0.6225 (0.570)
Type of household (Above poverty line)	0.8278*** (3.590)	0.6884*** (3.110)	0.8755*** (3.400)
Current millet price (P_{mill})	-1.5103 (1.650)	-0.1798 (0.200)	-2.4380** (2.390)
Current cassava price (P_{cas})	-0.1229 (0.340)	0.3433 (0.980)	0.0159 (0.040)
Sex of household head (Male)	-0.2334 (0.710)	-0.0379 (0.120)	-0.1375 (0.380)
Extension visit (Yes)	0.4608 (1.740)	0.1621 (0.640)	0.4044 (1.370)
Education of household head (Years)	0.2823 (1.300)	0.2673 (1.280)	0.3768 (1.550)
District (Apac)	0.0065 (0.030)	-0.2274 (1.000)	0.0868 (0.330)
Season (1st season 2009)	-0.0811 (0.370)	-0.6734*** (3.200)	-1.0670*** (4.360)
Constant	9.6890*** (2.870)	10.9809*** (3.400)	17.1925*** (4.580)
R^2 = 0.2160 (storage), R^2 = 0.2780 (consumption), R^2 = 0.3829 (sales)			

***, ** = Significant at 1 and 5%, respectively. Source: UBOS, 2010.

Table 6. Parameter estimates for beans price expectations and risk neutrality test.

Equation	Coefficient/ statistics	Type of Household		
		Below poverty line (n =388)	Above poverty line (n =394)	All households (N =792)
	(θ)	1.7816	1.9607	1.2561
Storage	(Chi^2)	3.1400	7.9500**	2.4200
	(δ_4)	0.2053	-0.2769	0.3703
	(θ)	1.1899	2.2791**	2.2456
Consumption	(Chi^2)	4.6100	12.1200***	6.4300
	(δ_4)	-0.3762	-0.2470	-0.0913
	(θ)	1.5663	1.6925	0.2569
Sales	(Chi^2)	1.5100	28.0500***	2.0900
	(δ_4)	0.2126	0.7841	2.9424

Price expectation parameter, $\theta = \sqrt{B3/B2}$; Price variance parameter, $\delta_4 = B2/\theta$. Chi^2 - value is for the joint hypothesis H_0: $\beta_2 = \beta_3 = \beta_4 = 0$; ***and ** = significant at 1 and 5%, respectively.

Just like it was done for finger millet, the coefficients of the quadratic price regressors, p^2_t, p^2_{t-1} and p_tp_{t-1} under the joint null hypothesis in Equation 12 above was tested for risk neutrality of bean producing households. Results showed that the null hypothesis of risk neutrality could not be rejected for both types of households in the beans storage and sales decisions (Table 6).

The null hypothesis of risk neutrality for beans storage, consumption and sales could not be rejected for both types of households. Thus, all beans producing households irrespective of type were risk neutral when making storage, consumption and sales decisions.

4. DISCUSSION

Finger millet and beans are important staple foods in northern Uganda and, their production is mainly done for subsistence purposes with only unplanned surpluses sold. These crops are grown on a seasonal basis and depending on abiotic and biotic factors, yields fluctuate by season and location. This might explain why production of millet and beans was lower in the first season of 2009 than in the second season of 2008 and, any observed disparities between study districts. Differential household resource endowments and allocation to other competing crops, such as cassava, sorghum, pigeon peas, cow peas, could have also caused variation in production, storage, consumption, and sales of millet and beans.

While this is so, previous studies have revealed that the uncertainty of future food prices and food security concerns causes more food storage as an insurance against high future price if the household has to buy back food for domestic consumption in the future (Michler and Balagtas 2013; Saha and Stroud 1994). Moreover, precautionary food storage has been found to vary with income level of households (Jalan and Ravallion, 2001; Deininger et al., 2007; Carter and Lybbert, 2012; Michler and Balagtas, 2013). Saha and Stroud (1994) found a positive relationship between household income and storage of sorghum in India. Similarly, a study of rice farmers in Bangladesh showed that higher income households stored a large portion (up to 20%) of total rice storage for precautionary purposes (Michler and Balagtas, 2013).

In northern Uganda, precautionary food storage was also significantly higher among finger millet and beans producing households above poverty level than their counterparts. Although, all finger millet and beans producing households in northern Uganda were risk neutral in their storage and sales decisions, households above poverty level seemed to be more food secure than households below poverty level. With the fact that they had more income probably enabled them to produce more millet and beans and, acquire better storage facilities for these grains.

5. CONCLUSION AND POLICY IMPLICATIONS

Results from this study indicate that all finger millet and beans producing households in northern Uganda are risk neutral regarding storage and sales decisions with only millet producing households below poverty line being risk averse in their consumption decisions. However, households above poverty line produced and stored more millet and beans implying that they were more food secure than households below poverty level. Therefore, strategies to boost incomes, production and prudent management of millet and beans stocks at the household level are critical for food security alleviation in northern Uganda. The establishment of minimum levels of millet and beans stocks at the household level would be a more efficient and effective policy response to any price shocks than an outright ban of household sales of these staples. For successful implementation of this policy, it will require massive promotion of use of improved storage facilities among millet and beans producing households. Using the farmer groups/associations approach, households could be mobilized and sensitized on modern millet and beans storage technologies. Moreover, household use of these improved storage technologies could be enhanced by linking them to credit and better produce markets.

REFERENCES

1. APS (Agricultural Policy Secretariat) (1994). Ministry of Finance, Planning & Economic Development. Economics of Crop and Livestock Production. Kampala, Uganda: MFPED.
2. Appleton S (2001). Changes in poverty and inequality in Uganda. In Reinikka R & Collier P (eds.), Uganda's recovery: The role of farms, firms and Government. Washington DC: World Bank.
3. Carter MR, Lybbert TJ (2012). Consumption versus asset smoothing: Testing the implications of poverty trap theory in Burkina Faso. J. Dev. Econ. 99(2):255-64.
4. Deininger K, Jin S, Yu X (2007). Risk coping and starvation in rural China. J. Appl. Econ. 39(11):1341-1352.
5. Fafchamps M (1992). Cash crop production, food price volatility, and rural market integration in the third world. Am. J. Agric. Econ. 74(1):90-99. Food and Agriculture Organization (FAO) (2000). Increasing stockholding capabilities of smallholder farmers in Uganda. Kampala, Uganda: FAO.
6. Jalan J, Ravallion M (2001). Behavioral responses to risk in rural China. J. Dev. Econ. 66(1):23-49.
7. Lee JJ, Sawada Y (2010). Precautionary saving under liquidity constraints: Evidence from rural Pakistan. J. Dev. Econ. 91(1):77-86.

8. Michler JD, Balagtas JV (2013). The determinants of rice storage: Evidence from rice farmers in Bangladesh. Paper prepared for the Agricultural & Applied Economics Association's 2013 AAEA & CAES Joint Annual Meeting, 4-6 August, Washington DC.
9. Owach C (1998). Final report on assessment of agricultural marketing dynamics in Kitgum district.
10. Park A (2006). Risk and household grain management in developing countries. Econ. J. 116(514):1088-1115.
11. Ravallion M (1987). Markets and Famines. Oxford: Clarendon Press.
12. Renkow M (1990). Household inventories and marketed surplus in semi subsistence agriculture. Am. J. Agric. Econ. 72(3):664-675.
13. Sadoulet E, de Janvry A (1995). Quantitative development policy analysis. Baltimore: John Hopkins University Press.
14. Saha A, Stroud J (1994). A household model of on-farm storage under price risk. Am. J. Agric. Econ. 76(3):522-534.
15. Singh I, Squire L, Strauss J (1986). Agricultural household models. Baltimore, John Hopkins University Press.
16. Stephens EC, Barrett CB (2011). Incomplete credit markets and commodity marketing behavior. J. Agric. Econ. 62(1):1-24.
17. UBOS (Uganda Bureau of Statistics) (2010). Uganda census of agriculture 2008/09. Volume II Methodology Report. Ministry of Finance, Planning, and Economic Development, Kampala, Uganda.
18. UBOS (Uganda Bureau of Statistics) (2015). Statistical abstract 2015. Ministry of Finance, Planning, and Economic Development, Kampala, Uganda. Uganda. OXFAM, Kampala, Uganda.
19. Walker TS, Ryan JG (1990). Village and household economies in India's semi-arid tropics. Baltimore, John Hopkins University Press.

CHAPTER 6

Farmgate versus market centre sales: A multi-crop approach

Benjamin Musah Abu, Haruna Issahaku and Paul Kwame Nkegbe*

Department of Economics and Entrepreneurship Development, Faculty of Integrated Development Studies, University for Development, Studies, P O Box 520 Wa, Ghana

AUTHORS' CONTRIBUTIONS

All authors conceived of and planned the study. BMA acquired the data and performed the statistical analysis, with guidance from PKN and HI. BMA drafted the manuscript, and HI and PKN critically reviewed it for important intellectual content. All authors contributed to the interpretation of results and then read and approved the final manuscript.

ABSTRACT

Smallholder farmers have two basic decisions to make regarding selling their surplus produce: selling at farmgate at low prices or travelling to a market centre where higher prices are offered while incurring some transaction costs. Whichever decision is made has implications for poverty alleviation efforts. Previous studies have ignored modelling participation and market choice simultaneously. Taking a multi-crop approach, this paper fills the gap by investigating the key determinants of market participation decisions of smallholder farmers in Ghana using the sample selectivity probit model in order to account for potential endogeneity and selectivity bias and thus obtain unbiased estimates. Household survey data in the Upper West region of Ghana for the 2011 production season are used to achieve the goal of the study. The results reveal that yields of maize and groundnut and market information are the simultaneous determinants of market participation decisions while age of the farmer, yields of the two crops, membership of farmer organisation and prices of the two crops simultaneously determine the choice of market. These imply that policies that enhance productivity of these smallholders and market information are vital in the drive for a commercially oriented agriculture. Also, the incentives to incur transaction costs to market centres to benefit from remunerative prices lie in measures to increase yields.

Keywords: Market participation, Smallholder farmers, Transaction costs, Sample selectivity, probit Ghana

1. BACKGROUND

It has been generally agreed that the way forward for smallholder farmers is increased market participation. Institutions, policymakers and scholars alike continue to emphasise the role of markets in the development of smallholders. For example, Olwande and Mathenge (2012) argue that any pathway that can lift large numbers of the rural poor out of poverty will require some form of transformation of smallholder agriculture into a more commercialised production system. Also, the Ministry of Food and Agriculture's (MoFA) Ghana Commercial Agriculture Project (GCAP) emphasises the importance of graduating from a subsistence-based smallholder system to a sector characterised by a stronger market-based orientation based on a combination of productive smallholders. Consistent with this, Siziba et al. (2011) observe that markets are the pivotal point in the agricultural transformation process.

The increased emphasis on market access relates to its potential to help smallholder farmers break the poverty trap that ensnares them. Several studies (e.g. Al-Hassan et al. 2006; Omiti et al. 2009; Jari and Fraser 2009; Siziba et al. 2011) highlight better incomes and prospects of reducing poverty, sustainable livelihoods, creation of the necessary demand, offering of remunerative prices, expanded production and the attendant adoption of productivity-enhancing technologies and increased economic diversification as some of the benefits of markets to small scale farmers. Muamba (2011) adds that greater market participation will promote the transformation of the farmers' economic status from subsistent or semi-subsistent to specialised farmers where they produce only crops for which they have a comparative advantage. In view of these benefits, the growing realisation of stimulating market participation of poor farmers for the goal of poverty alleviation cannot be overemphasised.

However, market participation decisions affect the extent of poverty alleviation. There are two basic decisions that are open to smallholders in their quest to be market participants: selling at farmgate or selling at a designated market centre. These two decisions are key determinants of the effectiveness of the market system in poverty alleviation. The decision of farmers to participate in the market through the vent of farmgate continues to hamper poverty mitigating efforts. For example, Fafchamps and Hill (2005) indicate that selling at the farmgate is mostly less remunerative. That is, farmgate prices are usually low such that significant gains can be made if sufficient quantities are put out for sale, and given the low productivity of smallholders, there is no doubt that such gains from farmgate would continue to elude them. The problems of the farmgate system of market participation have led to calls for policies to provide price incentives to encourage farmers to sell their harvest at the market rather than at the farmgate (Muamba 2011). Fafchamps and Hill (2005) also note the raising of welfare of farmers through institutional packages such as producer cooperatives as means of avoiding lower prices through the farmgate.

The implication of this is that the orientation of policies towards access to market as a means of fast tracking the development of smallholders should be well defined to exclusively relate to market centre participation rather than either farmgate or market centre participation. In other words, market participation can be beneficial if participants engage in market centre sales where prices are higher than at the farmgate where sale is more of a giveaway. It is worth noting that market centre sale comes with its concomitant transaction costs. That notwithstanding, the wedge between farmgate and market centre prices is usually large enough to cover transaction costs such that market centre is still remunerative than farmgate.

Previous studies have surprisingly ignored modelling participation and market choice simultaneously. This study represents the first attempt and for that matter contributing to the existing literature by implementing a single-stage estimation of the dual problem of participation and market choice of smallholder farmers. We perform the analyses while taking account of endogeneity and sample selection bias. We note also that the literature on market participation in Ghana is nascent. As a result, there is insufficient empirical literature on farmers' dilemma to sell at the farmgate or transport their produce to market centres (Fafchamps and Hill 2005). The few existing studies on the subject in Ghana relate to the factors influencing the probability and the quantity of sale (see Martey et al. 2012;

Abu et al. 2014). By estimating participation and market choice simultaneously, our work differs from the studies of Martey et al. (2012) and Abu et al. (2014). Though the work of Abu et al. (2014) employs a two-step modelling approach, the second stage of the model captures volume of sale rather than choice of market.

The Upper West region is one of the poorest and least developed in Ghana having the least average annual per capita income of GH¢3015.7 as against the national average of GH¢5346.9 (Ghana Statistical Service 2014).[1]Implementing this study in the region is therefore justified. Also, maize and groundnut are chosen for the study due to the fact that they are two of the major crops grown (Ministry of Food and Agriculture 2011) and are of high commercial value in the region.

The rest of the paper is presented as follows. We review related literature on market participation where gaps in these studies are identified. Next, we present the methodology employed, highlighting the study area and data, and methods of data analysis. Estimation results and discussions are presented next. Finally, the conclusion is presented which highlights a summary of the findings and policy implications.

2. LITERATURE REVIEW

Insight into farmgate and market selling dynamics by agricultural households is provided by Furlong and Slotsve (1983) in their exposition of pickup and delivery pricing. The characteristics of pickup and delivery pricing correspond to the farmgate and market centre pricing, respectively, in smallholder marketing. In the smallholder setting, farmers decide to sell at the farmgate or travel to the market and indirectly choose a pickup price or a delivery price. According to Furlong and Slotsve (1983), pickup prices are characterised by discounts. This basically implies that the variant of discounts in farmgate marketing is the lower prices offered. Farmers avoid transaction costs to reach markets but have to offer 'discounts' to buyers who incur such costs to buy at lower prices (pickup prices). However, farmers who decide to incur transaction costs reach markets and then sell at higher prices (delivery prices). The wedge, therefore, between farmgate and market prices is accounted for by transaction costs. The two types of transaction costs, namely fixed and proportional, lead to a low farmgate price at which farmers can sell their crop and a higher price at which they sell that crop at the market (Vance and Geoghegan 2004).

Market participation is viewed as a two-stage phenomenon and hence embodies two decision processes (Goetz 1992; Bellemare and Barrett 2006). First, households decide whether to be net buyers, net sellers or autarchic in the market for that commodity. This is the participation stage. Second, net buyers and net sellers determine the extent or intensity of participation. This is the intensity stage. In the literature, there are two distinct studies: first, those studies that model determinants of participation and intensity and, second, those that model participation and whether households choose to sell at farmgate or market centre. Quite recently, Burke et al. (2015) introduced a third dimension, the triple hurdle approach, which models in the first stage whether a farmer produced a particular crop while the second and third stages maintain the participation and intensity stages.

Three econometric procedures are distinguished in market participation studies. These are the one-step, the two-step and the three-step approaches. The one-step approach is where the focus is on modelling the determinants of the intensity stage. Some econometric models used in the one-step approach include Tobit and truncated regression models.[2] For example, Omiti et al. (2009) use a truncated regression while Holloway et al. (2000, 2001a) and Martey et al. (2012) use Tobit models to model intensity of participation. An underlying assumption of the one-step approach is that the same set of variables determines both the participation and intensity stages (Alene et al. 2008). This assumption is not only restrictive but erroneous since it is possible that a variable could have different effects on participation and intensity of participation (Wooldridge 2002; Ricker-Gilbert et al. 2011; Olwande and Mathenge 2012).

The two-step approach simultaneously or sequentially models participation and intensity, thus helping to resolve the limitation inherent in the single-step approach. Prominent two-step models are the Heckman sample selection and the double hurdle models. Goetz (1992), Makhura et al. (2001), Boughton et al. (2007), Alene et al. (2008) and Siziba et al. (2011) have applied the Heckman model to explain the determinants of participation and intensity. The model estimates a first-stage probit of participation decision, derives an inverse mills ratio (IMR) and then estimates a second-stage ordinary least squares (OLS) with the IMR as a variable to account for selectivity bias. The advantage of the Heckman model is its ability to model the two stages simultaneously and to correct for selectivity bias. However, the appropriateness of the model for participation is questionable. First, according to Ricker-Gilbert et al. (2011), the Heckman model is designed for incidental truncation, where the zeros are unobserved values. However, in market participation studies, a zero value in the data would reflect farmers' optimal choice rather than a missing value (Reyes et al. 2012). It is therefore erroneous to equate these missing observations to zero (Olwande and Mathenge 2012). Second, the second-stage OLS is questionable since it is not known whether the underlying assumptions of OLS are met.

Some other studies (see, for example, Holloway et al. 2005; Olwande and Mathenge 2012; Reyes et al. 2012; Abu et al. 2014) analyse participation and intensity using the double hurdle model. This model estimates a first-stage probit and a second-stage truncated regression. The fundamental assumption underlying the model is that the participation and the intensity stages are independent. Based on this, Smith (2003) argues that assuming dependency between the two equations is not a worthwhile exercise since there is little statistical information available to support dependency in the double hurdle framework. One advantage of the double hurdle model is its ability to nest the Tobit model, thus allowing for the testing of the restrictions implied by the Tobit hypothesis (Olwande and Mathenge 2012). The limitation of the model is inherent in the independence assumption. It is inappropriate to assume that participation and intensity are independent. It is worth noting that though the two-step approaches possess some weaknesses, they are by far better than the one-step approaches. Also, the double hurdle model, despite its weaknesses, is theoretically more plausible to use as compared to its Heckman counterpart.

As mentioned already, the three-step approach is a recent introduction to the literature by Burke et al. (2015) which argues for the introduction of non-producers in the first stage and thus overcomes the weakness of the two-step approach to handle the three stages.

In terms of econometric methodology, frequentist approaches have inundated participation studies and have gained more currency than Bayesian approaches. We note some contributions of Bayesian approaches. Holloway et al. (2001a, b, 2004, 2005, 2008) are notable applications of Bayesian methodology to market participation. Bayesian methodology is usually underpinned by the use of the Markov Chain Monte Carlo (MCMC) method especially the use of the Gibbs sampler (see Gelfand and Smith 1990; Casella and George 1992 for exposition on the Gibbs sampler proposed by Tanner and Wong 1987) to augment data. Another notable contribution is on thresholds where an unknown or random censoring of data is considered instead of the widely used zero censoring threshold. Significant empirical participation studies on this innovative random censoring are provided by Holloway et al. (2000, 2001a, 2004, 2005). In fact, one of the most significant contributions of these non-zero threshold studies is their ability to estimate how far nonparticipating households are from levels of marketable surplus that would encourage and sustain participation. Holloway et al. (2004) find that the zero censoring overstates estimates.

With respect to participation and choice of market, the focus of this study, Fafchamps and Hill (2005) were the first to model whether farmers sell to itinerant traders or directly to markets. They adopted a detailed econometric procedure involving multiple estimations using probit, instrumenting regressions and Tobit models. A number of methodological issues arise from this pioneering study. First, they fail to consider the decision of a farmer to either sell or not. They consider the decision to sell to either market centre or farmgate in the first estimation (ignoring farmers not selling at all) using a probit model. This introduces the problem of truncation and hence resulting in biased and inconsistent estimates. Second, other estimations involve single equations with a number of continuous dependent variables (output, interaction of output and wealth and distance travelled to market). This makes the approach ad hoc, failing to provide a definite procedure for easy replication. Also, the likelihood of getting lost in the estimation procedures is very high. Third, they fail to adopt a two-stage approach simultaneously modelling participation and choice of market.

Muamba (2011) followed up by developing a farm level sequential model taking account of the conditional nature of farmers' decision to participate in the market. While the author adopts an innovative approach to estimating a bivariate probit to account for sample selection and a system of seemingly unrelated regressions, two issues are identified with the approach. First, just like Fafchamps and Hill (2005), the study fails to consider the initial decision of a farmer to either sell or not. Second, the application of the bivariate probit does not account for participation and choice of market. In effect, Muamba (2011) analyses the choice of market and the intensity of participation. A recent introduction of the triple hurdle model by Burke et al. (2015) extends the literature so far. However, it fails to account for choice of market.

To the best of our knowledge, this study is the first attempt at simultaneously estimating participation and choice of market. Thus, our contribution to the existing literature is that, instead of ignoring farmers who do not sell, we model a first-stage equation of factors determining the decision of farmers to sell either maize or groundnut. We then, in a second stage, model the decision to sell either at farmgate or market centre. This implies that what we model in the second stage is what is modelled in the first stages of Fafchamps and Hill (2005) and Muamba (2011). Since the first and second stages are discrete outcomes, none of the two-step approaches explored earlier is applicable. We adopt the sample selectivity probit model to deal with the situation at hand. The sample selectivity probit model assumes that the error terms in the participation and market choice stages are correlated and normally distributed, and this is also its weakness as there is no formal procedure for verifying this. Despite this weakness, it is still useful when modelling two discrete outcomes in a single process.

Two assumptions underpin market participation studies: simultaneous and sequential decision-making. Key et al. (2000) assume that households make simultaneous decisions. This implies that households predetermine either the quantity of sale or choice of market, depending on the study type, before they get to know the information available to them about these choices. The simultaneous decision-making gives traders market power over selling households rendering them vulnerable to exploitation by traders (Bellemare and Barrett 2006). On the other hand, Goetz (1992) and Holloway et al. (2005) assume sequential decision-making, an observation subsequently confirmed empirically by Bellemare and Barrett (2006). The sequential decision-making implies that households make decisions after receiving information from the market. They therefore retain greater flexibility over market decisions, hence reducing traders' capacity to extract much or all the gains from trade (Bellemare and Barrett 2006).

3. METHODS

3.1 Study area and data

This study was conducted in the Upper West region of Ghana involving four agricultural districts: Jirapa-Lambussie, Nadowli, Wa West and Sissala East. Data were collected from 400 farmers (200 each from independent maize and groundnut farmers). For each district, 80 farmers (40 each for the two crops) were selected except for the Jirapa-Lambussie district where 160 farmers were selected because, at the time of data collection, Lambussie-Karni district had not been separated from the Jirapa district.

The drawing of the sample involved the use of a multi-stage sampling procedure in three steps. In the first stage, the four districts were selected based on the fact that they had the highest share in the production of maize and groundnut in the 2011 production season. The second stage narrowed down to specific communities known for the production of the two commodities. The selection was done in consultation with the various selected district offices of MoFA. This was to prevent a random sample of operational areas where maize and groundnut are not intensively cultivated. The third stage involved the selection of respondents. To prevent bias, stratified sampling procedure was employed to create two strata of the population based on gender. The rationale for this stratification was to ensure a proportionate representation of male and female farmers. A semi-structured questionnaire was then administered to the chosen respondents.

3.2 Analytical method

As has been established in the market participation literature, studies typically adopt two-step analytical approach to cater for the unobserved decision to participate and the observed degree or extent of participation. The innovation we table is to model the unobserved decision to participate and the observed choice of participation decision (that is, either selling at farmgate or selling at a market centre). Two problems are anticipated in this analytical framework. First, since nonparticipants are not considered in the second-stage market choice model, proceeding with a univariate probit model would introduce sample selection bias[3] since farmers would self-select into treatment (Faltermeier and Abdulai 2009). Second, the decision of a maize or groundnut farmer to sell either at the farmgate or market centre is conditional on the decision to participate in the market (i.e. the first-stage). Implementing a univariate probit to model the second stage will lead to biased results due to the potential endogenous nature of participating in the market and the choice of market.[4] An appropriate model requires that selectivity and endogeneity are taken into account.

The bivariate probit model is capable of solving the endogeneity problem. However, it is unable to handle the selectivity bias. We therefore rely on sample selectivity probit model to overcome these problems. The sample selectivity probit model is adopted as it has an intrinsic ability to account for selectivity bias and make corrections simultaneously, as well as deal with endogeneity. This model's ability to handle selectivity bias is acclaimed. It simultaneously implements the first and second stages and provides a likelihood ratio test of a rho (ρ) parameter which indicates the relationship between the two equations. The significance of ρ indicates the existence of selectivity bias, and an internal correction of this bias is provided.

Just like other market participation studies, this study is underpinned by the linear random utility assumption (Greene 2008). The rationale for the linear random utility assumption is that an individual faces a choice between two or more alternatives and chooses the alternative which gives the highest utility. This is usually represented as

$$\begin{cases} U_{i0} = X_i'\beta_{i0} + e_{i0} \\ U_{i1} = X_i'\beta_{i1} + e_{i1} \end{cases} \quad (1)$$

where U_{ij} measures utility gained by farmer i from choosing alternative j (with the decision not to participate in the market represented by U_{i0} while participating is represented by U_{i1}), X_i is a vector of characteristics specific to farmer i and attributes associated with alternative j and specific to the ith farmer, β is a vector of unknown parameters, and e_{ij} is random disturbance related to the choice of alternative j by farmer i.

The probability that a farmer chooses an alternative (that is, $Y_i = 1$) as against another (that is, $Y_i = 0$) is associated with the probability distribution of the error differences in the expected utilities from the choices and given by

$$P_i = \text{Prob}(Y_i = 1|X) = \text{Prob}(y_i^* > 0|X) = \text{Prob}\left[e_i > -X_i'\beta|X\right] = F\left(X_i'\beta\right) \quad (2)$$

where F is the cumulative distribution function of $e_i(=e_{i1} - e_{i0})$ evaluated at $X'_i\beta$, and $y^*_i(=u_{i1}-u_{i0})$ is a latent variable, since it is unobservable, and is linked to y_i, the observed binary variable, through the relation

$$\begin{cases} 1 \text{ if } y_i^* > 0 \\ 0 \text{ otherwise} \end{cases} \quad (3)$$

To specify a model that describes how the probability of choosing an alternative and the explanatory variables are related depends on the assumption made about the distribution of the error term. The normal and logistic distributions corresponding to probit and logit models, respectively, are the two mostly assumed distributions in the literature.

Assuming a normal distribution of the error term in the utility model, a probit model can be relied on for analysis. In that case and linking up with Eq. (2), the probability function of choosing an alternative versus another is given by

$$P_i = \text{Prob}[Y_i = 1|X] = \int_{-\infty}^{X_i'\beta} \emptyset(t)dt = \Phi\left(X_i'\beta\right) \quad (4)$$

with $\emptyset(.) = (2\pi)^{0.5}\exp(-t^2/2)$ and $\Phi(.)$ being the density and cumulative distribution functions, respectively, of a standard normal random variable.

In the sample selectivity probit model, just like a bivariate probit model, the assumption of correlated normally distributed error terms in a two-equation system leads to

$$\begin{cases} y_{i1}^* = X_{i1}'\beta_1 + e_{i1}, \ Y_{i1} = 1 \text{ if } y_{i1}^* > 0, \text{ and } 0 \text{ otherwise} \\ y_{i2}^* = X_{i2}'\beta_2 + e_{i2}, \ Y_{i2} = 1 \text{ if } y_{i2}^* > 0, \text{ and } 0 \text{ otherwise} \end{cases} \quad (5)$$

where e_{ij} is the normally distributed error term, $E[e_{i1}|x_{i1}, x_{i2}] = E[e_{i2}|x_{i1}, x_{i2}] = 0$, $\text{Var}[e_{i1}|x_{i1}, x_{i2}] = \text{Var}[e_{i2}|x_{i1}, x_{i2}] = 1$, and $\text{Cov}[e_{i1}, e_{i2}|x_{i1}, x_{i2}] = \rho$. For a typical farmer, Y_{i1} is not observed unless $Y_{i2} = 1$. The appropriate conditional probability, derived by modifying that of bivariate model, is given by Greene (2008) as

$$\text{Prob}[Y_{i1} = 1|Y_{i2} = 1] = \frac{\Phi_2(x_{i1}'\beta_1, x_{i2}'\beta_2, \rho)}{\Phi(x_{i2}'\beta_2)} \tag{6}$$

Following from Eq. (6), the unconditional probabilities for the three kinds of observed outcomes in the model are given as follows:

$$\begin{cases} Y_{i2} = 0: \ \text{Prob}(Y_{i2} = 0 \quad |x_{i1}, x_{i2}) = 1 - \Phi(x_{i2}'\beta_2) \\ Y_{i1} = 0, \ Y_{i2} = 1: \ \text{Prob}(Y_{i1} = 0, \ Y_{i2} = 1\,|x_{i1}, x_{i2}) = \Phi_2(\ x_{i1}'\beta_1, x_{i2}'\beta_2, -\rho) \\ Y_{i1} = 1, \ Y_{i2} = 1: \ \text{Prob}(Y_{i1} = 1, \ Y_{i2} = 1|x_{i1}, x_{i2}) = \Phi_2(x_{i1}'\beta_1, x_{i2}'\beta_2, \rho) \end{cases} \tag{7}$$

where Φ and Φ_2 are, respectively, univariate and bivariate normal cumulative distribution functions.

The probabilities above thus enter the log-likelihood function given by

$$L_i = \sum \ln\left[1 - \Phi(x_{i2}'\beta_2)\right] + \sum \ln\left[\Phi_2(-x_{i1}'\beta_1, x_{i2}'\beta_2, -\rho)\right] + \sum \ln\left[\Phi_2(x_{i1}'\beta_1, x_{i2}'\beta_2, \rho)\right] \tag{8}$$

The maximum likelihood approach is then used to estimate selectivity probit model. The description, measurement and expected signs of variables are presented in Table 1.

As noted by Greene (2008), several 'marginal effects' can be evaluated in a typical bivariate probit, but an interesting step usually involves examining the derivatives of the conditional mean functions. In the light of this, a basic marginal effect function can be stated as

$$\frac{\partial E[Y_1|Y_2 = 1, X]}{\partial X} = \frac{\partial[\text{Prob}(Y_1 = 1|Y_2 = 1, X)]}{\partial X} = \frac{\partial}{\partial X}\left[\frac{\Phi_2(X'\gamma_1, X'\gamma_2, \rho)}{\Phi(X'\gamma_2)}\right] \tag{9}$$

where Φ_2 and Φ are as already defined, γ_j contains all non-zero elements of β_j and likely to also contain some zeros in the positions of variables that appear in only one of the two equations, and $j = 1, 2$. For a dummy explanatory variable, however, the marginal effect is determined by using a modified form of Eq. (9) which will reflect discrete changes in the predicted probabilities.

Table 1 Description, measurements and expected signs of variables

Variable	Description	Measurement	Expected sign
Dependent variables			
PART	Binary variable indicating the decision to participate in the market or not	Dummy: 1 = farmer participates in market (sold maize/groundnut); 0 = otherwise	
POS	Binary variable indicating the decision to sell to the market centre or farmgate	Dummy: 1 = market centre; 0 = farmgate	
Independent variables			
AGE	Age of farmer	Number of years	+/−
GEN	Gender of farmer	Dummy: 1 = if male; 0 = otherwise	+
EDUC	Educational level of farmer	Number of years of schooling	+/−
MARST	Marital status of farmer	Dummy: 1 = if married; 0 = otherwise	+/−
HHSIZE	Household size of farmer	Number of people in the household	+/−
MFBO	Membership of farmer to farmer-based organisation	Dummy: 1 = if member; 0 = otherwise	+
YIELD	Yield of maize or groundnut in the 2011 production season	Ratio of output to farm size	+
ACCRE	Access to credit by farmer	Dummy: 1 = if farmer applied and received credit; 0 = otherwise	+
MKTINFO	Farmer's access to market information	Dummy: 1 = if yes; 0 = otherwise	+
EXTCON	Access to extension services	Dummy: 1 = if yes; 0 = otherwise	+
PRICE	Average price of maize or groundnut sold	Ghana Cedi (GH¢) per kg	+
FOS	Form of sale of groundnut	Dummy: 1 = shelled; 0 = unshelled	+/−

4. RESULTS AND DISCUSSION

Descriptive statistics of households are presented in Table 2. From the table, the mean ages of maize and groundnut farmers are 47 and 42 years, respectively. About 86 and 80 % of maize and groundnut farmers are, respectively, male. Average years spent in school are 3 and 2, respectively. The average farm size cultivated is 1.10 ha for maize-producing farmers and 1.22 ha for groundnut-producing farmers with mean outputs of 550.88 and 520.21 kg, and average yields of 466.58 and 369.91 kg/ha, respectively. With respect to market participation, 48.5 % of maize-producing households sell maize while 83.5 % of groundnut-producing households sell groundnut. About 31.5 and 48.5 % of maize and groundnut are sold at the farmgate. The average prices received by maize and groundnut farmers are GH¢1.37 and GH¢1.67 per kg, respectively.

Table 2 Descriptive statistics of variables

Variable	Maize				Groundnut			
Continuous variables								
	Mean	S.D.	Min.	Max.	Mean	S.D.	Min.	Max.
AGE	46.98	16.09	21	88	42.35	15.44	19	90
HHSIZE	9.81	5.31	2	32	9.84	5.35	2	32
EDUC	2.56	4.49	0	22	2.38	4.09	0	18
FRMSIZE	1.10	0.57	0.40	2	1.22	0.56	0.40	2
OUTPUT	550.88	664.12	50	4150	520.21	497.22	7.5	4000
YIELD	466.58	415.44	41.67	2225	369.91	221.74	18.75	2000
PRICE	1.37	0.37	0.70	2.80	1.67	0.56	0.60	3.9
Categorical variables								
	Percentage				Percentage			
PART	48.5				83.5			
POS	31.5				48.5			
GEN	86.0				80.0			
MARST	84.5				80.0			
ACCRE	22.5				17.0			
MFBO	91.5				87.0			
MKTINFO	63.0				91.5			
EXTCON	41.0				19.5			
FOS	...				65.5			

The results of the determinants of market participation and the choice of market are displayed in Table 3. The Wald chi-square values of both models show that the explanatory variables in the participation and choice equations in both the maize and groundnut models jointly explain participation and choice of market.[5] The significance of the likelihood ratio tests of the rho parameter in both models suggests two things. First, the sample selectivity probit model is preferred to univariate probit models, since the results indicate the existence of sample selection bias if the second-stage equations were estimated alone. Fortunately, the model has an internal system of correcting for this bias. Therefore, the estimates are unbiased and consistent. Second, it shows that market participation and choice of market decisions are made simultaneously, which confirms the work of Key et al. (2000) who assume simultaneous market decisions of households, but contradicts the work of Goetz (1992), Holloway et al. (2005) and Bellemare and Barrett (2006) who assume sequential market decisions of households. This means that households predetermine the choice of farmgate or market centre before realising the information available to them about these choices. According to Bellemare and Barrett (2006), the simultaneous decision-making gives traders market power over selling households rendering them vulnerable to exploitation by traders. This is opposed to the sequential decision assumption where households retain greater flexibility over market decisions, hence reducing traders' capacity to extract much or all the gains from trade (Bellemare and Barrett 2006).

In Table 3, the signs of all the determining variables of market participation and choice of market by maize-producing households, except education and extension services, respectively, are consistent with expectations. For groundnut-producing households, all the variables are consistent with expectations in both the participation and the choice of market models.

We find that as age of the farmer increases, the likelihood of selling maize declines by 1.1 %. This is consistent with the finding of Boughton et al. (2007), Siziba et al. (2011), Olwande and Mathenge (2012) and Burke et al. (2015). Perhaps, older farmers expect to remain food secure and would not want to take the risk of emptying their maize barns as against the younger farmers who expect to enhance their quality of life through incomes from market engagements. The fact that maize is a consumption commodity or staple in Ghana supports this explanation. For choice of market, age is negatively correlated with maize and groundnut market choices. For an additional year of a farmer, the probability of choosing market centre sale decreases by 0.8 and 0.9 % for maize and groundnut producers, respectively. This is consistent with Muamba's (2011) notion that older farmers have less physical ability to carry the crops to the market. It seems plausible that the progressive mindedness and adventurous nature of younger farmers would encourage and stimulate them to incur transaction costs to reach market centres where they can fetch higher prices, consistent with the notion by Enete and Igbokwe (2009) that older farmers are more risk averse in exploring market sales.

Gender is a positive determinant of groundnut market participation, as we find that male farmers are 2.5 % more likely to participate in the groundnut market than female farmers, a result that contradicts that of Burke et al. (2015). This is, however, supported partly by findings in Boughton et al. (2007), Olwande and Mathenge (2012) and Siziba et al. (2011). This observation likely reflects disparities in access to capital (for example, land and finance) between males and females. Also, males often receive the support of females on their farms more than the females do.

Number of years of schooling reduces probability of selling maize by 2.3 %. This observation contradicts the finding of Makhura et al. (2001) and Enete and Igbokwe (2009). The possible explanation for this is that maize is a staple and, hence, farmers with a higher level of education engage in its cultivation for consumption in order to reduce strain on family consumption budgets. Also, it is possible that education could increase the chances of the household head earning non-farm income, hence reducing dependency on income from maize sales (Enete et al. 2004).

Table 3 Sample selectivity probit estimates of determinants of market participation and market choice

Variable	Maize				Groundnut			
	Participation		Market choice		Participation		Market choice	
	Coefficient	Marginal effect	Coefficient	Marginal effect	Coefficient	Marginal effect	Coefficient	Marginal effect
AGE	−0.0286[a] (0.0083)	−0.0112 (0.0033)	−0.0276[a] (0.0100)	−0.0078 (0.0031)	−0.0084 (0.0113)	−0.0001 (0.0002)	−0.0225[b] (0.0118)	−0.0090 (0.0047)
GEN	0.4470 (0.3950)	0.1768 (0.1543)	0.3923 (0.5025)	0.0972 (0.1093)	0.9459[c] (0.4832)	0.0245 (0.0348)	−0.3017 (0.5086)	−0.0953 (0.2041)
EDUC	−0.0598[b] (0.0324)	−0.0234 (0.0127)	−0.0189 (0.0340)	−0.0053 (0.0097)	−0.0525 (0.0465)	−0.0006 (0.0008)	−0.0652 (0.0402)	−0.0263 (0.0159)
MARST	−0.1022 (0.3474)	−0.0397 (0.1342)	−0.3323 (0.4172)	−0.1021 (0.1399)	0.0662 (0.4888)	0.0008 (0.0060)	−0.0961 (0.4390)	−0.0374 (0.1742)
HHSIZE	−0.1072[a] (0.0300)	−0.0420 (0.0119)	0.0050 (0.0374)	0.0014 (0.0107)	−0.0024 (0.0319)	−2.59e-05 (0.0003)	0.0765[c] (0.0335)	0.0302 (0.0131)
YIELD	0.0025[a] (0.0005)	0.0010 (0.0002)	0.0009[c] (0.0004)	0.0003 (0.0001)	0.0130[a] (0.0028)	0.0001 (0.0002)	0.0022[b] (0.0012)	0.0010 (0.0005)
ACCRE	1.5477[a] (0.4811)	0.4811 (0.0927)	0.8197[a] (0.3000)	0.2670 (0.1215)	−0.7232 (0.6565)	−0.0160 (0.0317)	−0.0863 (0.4873)	−0.0500 (0.1865)
MKTINFO	0.7481[a] (0.2730)	0.2905 (0.1024)	−0.6561 (0.4191)	−0.1962 (0.1355)	0.9187[b] (0.4937)	0.0295 (0.0428)	0.0875 (0.8415)	0.0640 (0.3269)
MFBO	1.2993 (0.8678)	0.3849 (0.1503)	1.4231[c] (0.5841)	0.5103 (0.2079)	−0.7104 (0.5089)	−0.0165 (0.0276)	1.7933[a] (0.6487)	0.5489 (0.1148)
EXTCON	0.0881 (0.2552)	0.0345 (0.0997)	−0.4719[b] (0.2773)	−0.1279 (0.0788)	−0.3739 (0.4541)	0.0031 (0.0048)	0.3491 (0.4152)	0.1416 (0.1630)
PRICE			0.7366[b] (0.4331)	0.2075 (0.0822)			2.7987[a] (0.5189)	1.1071 (0.1989)
FOS							1.8253[a] (0.3885)	0.6336 (0.0982)
CONSTANT	0.3104 (0.6962)		−0.2152 (0.9901)		−3.1013[a] (0.9559)		−5.1517[a] (1.5237)	
Rho		1.000 (3.86×10^{-13})				−1.000 (3.40×10^{-10})		
Number of observations	200				200			
Censored observations	103				33			
Wald χ^2	39.28[a]				45.08[a]			
Log likelihood	−110.476				−76.0244			
Likel. ratio test of ρ $\chi^2(1)$	6.37[c]				6.61[c]			

Source: Regression estimates from household survey data, 2012. Note: standard errors in parentheses
[a]At 1 %
[b]Denotes statistical significance at 10 %
[c]At 5 %

Household size is expectedly negatively correlated with maize market participation. For an additional member of a household, the probability of selling maize decreases by 4.2 %. This result confirms the finding of Siziba et al. (2011) and agrees with their argument that households with large family sizes fail to produce marketable surplus beyond their consumption needs. It also confirms the finding of Makhura et al. (2001) who argue that households decide to sell when they cannot consume all they have produced and, hence, the more members the household has, the more likely most of the produce will be consumed, thereby decreasing the possibility for selling. For choice of market, larger households are 3.0 % more likely to sell groundnut to a market centre. This observation is expected since groundnut is a cash crop; hence, it presents the opportunity for larger households to cash in at markets to derive higher income margins to fuel large consumption expenditures. This is easy to achieve since a larger size provides manpower to carry groundnut to market.

The yields of maize and groundnut are significantly associated with higher probability of participating in both markets, which is consistent with expectation, since higher yields are more likely to ensure marketable surplus. For an additional kilogramme yield of maize and groundnut, the probability of market participation increases by 0.1 and 1.3 %, respectively. This finding underscores the importance of increased yield in enhancing the chances of smallholder farmers stepping out of poverty and improving their livelihood through enhanced income from their increased participation in the market. With respect to choice of market, yields of maize and groundnut positively influence the probability of selling at market centres. For every kilogramme yield of maize and groundnut, the probability of selling at market centre increases by 0.03 and 0.1 %, respectively. It is expected that the transaction costs involved in selling larger outputs from larger yields in markets are relatively less than smaller outputs. That is, it is relatively cost-effective to sell smaller outputs at farmgate than at market centre. This finding is consistent with studies (for example, Fafchamps and Hill 2005; Muamba 2011) that use quantity sold in the second stage.

Access to credit has a positive influence on the probability of selling and the choice of maize market. This shows that farmers with access to credit are able to produce enough marketable surpluses and then sell these at market centres. Farmers who have access to credit are 48.1 % more likely to sell maize and 26.7 % more likely to sell at the market. One argument is that access to credit gives the farmer the economic power to cultivate on a large scale and sell at market centres. Also, the obligation of paying back the loan together with the interest provides incentive for producing beyond consumption and to incur transaction costs to reach markets to benefit from remunerative prices. While the influence of credit on participation corroborates the finding of Vance and Geoghegan (2004), it contradicts the negative influence of credit on market centre sale of maize by Muamba (2011).

Access to market information has a positive effect on the probability of selling maize and groundnut. This confirms the finding of Siziba et al. (2011) who argue that access to information reduces risk perceptions. Another possible explanation for this result could be that farmers with access to market information might be easily persuaded to sell than those without such information. Households with access to market information are 29.1 and 3.0 % more likely to sell maize and groundnut, respectively, than those without market information. Access to market information does not influence the choice of market.

Membership in farmer-based organisations (FBOs) has a positive effect on the choice of markets for producers of both crops. This implies that farmers belonging to FBOs are more likely to sell maize and groundnut in market centres than at farmgate. Farmers who belong to FBOs are 51 and 54.9 % more likely to sell maize and groundnut, respectively, at market centres than those who are not members. This is consistent with the finding of Olwande and Mathenge (2012) who argue that farmer group membership increases access to information important to production and marketing decisions. Another plausible reason could be that group membership encourages collective action and strengthens farmers' bargaining and lobbying power, thereby getting them to sell in market centres (Matungul et al. 2001). Muamba (2011) uses a similar argument to explain why farmers who belonged to rotating savings and credit associations (ROSCAs) sell at farmgate—group members are organised as cooperative and have bargaining power over village traders.

Surprisingly, access to extension services is negatively correlated with the choice of maize market. Farmers with access to extension are 47.2 % less likely to sell at market centres. We expected that farmers with access to extension would be better oriented towards market sales since they are better informed not only on production but also on marketing issues. This contradicts some studies (for example, Dinar and Keynan 2001; Holloway and Ehui 2001; Hanson and Just 2001; Frisvold et al. 2001) that have found interesting results for paid extension and its impacts on smallholder agriculture. While we are not certain about this unexpected result, this may be a consequence of the ineffectiveness of extension service in Ghana, or its inability to incorporate market information in its activities. Paid extension is not widespread in Ghana and the general extension services are not effective due to logistical constraints in the midst of dispersed communities. The proportion of the sample (see Table 2) receiving extension services confirms this.

Prices of maize and groundnut positively influence the choice of maize and groundnut markets. Farmers who are faced with higher prices are, respectively, 20.8 and 110.7 % more likely to sell maize and groundnut at market centres. This is consistent with the pickup and delivery pricing arrangements proposed by Furlong and Slotsve (1983). It also confirms the finding of Muamba (2011) regarding cotton sales. He argues that own price is a positive incentive for encouraging farmers to sell to the market. It, however, contradicts his finding on maize sales.

We find a statistically significant positive relationship between form of sale and choice of the groundnut market. Households that shelled groundnut before sale are 63.4 % more likely to sell at a market centre. The possible reason is that shelled groundnut attracts a higher economic value at market centres than at farmgate. Farmers who go through the strenuous activity of shelling would complete the cycle by selling at market centres to fetch higher returns to compensate for their efforts.

CONCLUSIONS

This study examines market participation decisions of smallholder maize and groundnut farmers in the Upper West region of Ghana, employing a procedure that corrects for endogeneity and selectivity bias. Two crops are used to show the differences and similarities in market participation behaviours of farmers cultivating different crops. The results show that though a number of factors determine maize and groundnut producers' market participation, yield of both crops and market information are the simultaneous determinants of market participation, while age of farmer, yield of both crops, membership in FBO and own prices are the simultaneous determinants of choice of market. Only yield of both crops determines participation and choice of market simultaneously. This implies that yield is a very important decision variable in market participation of smallholder farmers, hence a key to unlocking the poverty trap that ensnares most of these farmers. From a policy standpoint, the call by researchers, institutions and governments of especially developing economies for the reorientation of policies towards market access by smallholders can be achieved through policies that enhance productivity of these smallholders. In other words, the quest for commercially oriented smallholders starts with increased yield. From the policy standpoint, the Government of Ghana through the Ministry of Food and Agriculture should make fertiliser and other agro-inputs both physically and financially accessible and also support farmers to use high-yielding crop varieties and appropriate agricultural mechanisation. This should be coupled with institutional capacity building and addressing logistical constraints.

For choice of market, yield, FBOs and price of crops are the important decision variables in encouraging market centre sales. Step one in commercialisation of smallholder agriculture is participation, which is significantly influenced by yield; the second step is selling at remunerative sources (market centres), which is influenced by these three variables (yield, FBOs and price). This implies that incentivising smallholders to incur transaction costs to reach remunerative markets lies in measures that would increase yield (as discussed already), well-functioning farmer associations and offering remunerative prices for produce. A specific policy measure is to create a farmer association unit in MoFA to engage in registration and building capacity of these organisations. Also, price incentives become inevitable in considering alternatives to market centre sales.

Indeed, we acknowledge there could be some cross-equation correlations between maize and groundnut. However, the nature of the data sets used does not allow for this exercise as that might not be very meaningful in this context. This thus constitutes the limitation of the current study, but it at the same time drops hint for further research.

FOOTNOTES

[1] The exchange rate as quoted by www.xe.com as at 11 September 2015 was US$1.00 = GH¢3.955.

[2] A recurrent embrace between empiricism in this particular field arises in the context of censored- versus truncated-regression settings. The theoretical models, for the most part, apply censored regression settings; the technicalities, especially in the case of the Gibbs sampler, require that one important fully conditional distribution is truncated (normal); and thus, while the truncation is necessary, it is brought about, compulsorily, as an intervention required by the censoring of the sales data.

[3] This was tested, and the results confirmed this anticipation.

[4] This potential endogeneity was also tested, and results showed the existence of endogeneity.

[5] A reviewer raised an issue about the existence of cross-equation correlations. That would have been possible if the maize and groundnut were cultivated by same households. In our case, the data were collected from predominantly maize- or groundnut-producing households so that cross-equation correlations might not have very meaningful interpretation.

ACKNOWLEDGEMENTS

The first author is grateful to the International Food Policy Research Institute (IFPRI), Ghana, for providing funds for data collection and allowing for the use of the data. Authors acknowledge, with appreciation, the valuable comments made by the anonymous referees and the editor which helped improve earlier versions of this paper. The usual disclaimer, however, applies.

REFERENCES

1. Abu BM, Osei-Asare YB, Wayo S (2014) Market participation of smallholder maize farmers in the Upper West region of Ghana. Afr J Agric Res 9(31):2427–2435
2. Alene AD, Manyong VM, Omanya G, Mignouna HD, Bokanga M, Odhiambo G (2008) Smallholder market participation under transactions costs: maize supply and fertilizer demand in Kenya. Food Policy 33(4):318–328
3. Al-Hassan RM, Sarpong DB, Mensah-Bonsu A (2006) Linking smallholders to markets. Ghana Stategy Support Programme (GSSP), International Food Policy Research Institute (IFPRI), Background Paper No. GSSP 0001, Accra, Ghana. Available via DIALOG. http://www.pdf.usaid.gov/pdf_docs/Pnadl412.pdf. Accessed 4 Jun 2012
4. Bellemare MF, Barrett CB (2006) An ordered Tobit model of market participation: evidence from Kenya and Ethiopia. Amer J Agr Econ 88(2):324–337
5. Boughton D, Mather D, Barrett CB, Benfica R, Abdula D, Tschirley D, Cunguara B (2007) Market participation by rural households in a low-income country: an asset-based approach applied to Mozambique. Faith Econ 50(Fall 2007):64–101
6. Burke WJ, Myers RJ, Jayne TS (2015) A triple-hurdle model of production and market participation in Kenya's dairy market. Amer J Agr Econ 97(4):1227–1246
7. Casella G, George EI (1992) Explaining the Gibbs sampler. Am Stat 46(3):167–174
8. Dinar A, Keynan G (2001) Economics of paid extension: lessons from experience in Nicaragua Amer. J Agr Econ 83(3):769–776
9. Enete AA, Igbokwe EM (2009) Cassava market participation decisions of producing households in Africa. Tropicultura 27(3):129–136
10. Enete AA, Nweke FI, Tollens E (2004) Gender and cassava processing in Africa. Quarterly J Int Agric 43(1):57–69
11. Fafchamps M, Hill RV (2005) Selling at the farmgate or traveling to market. Amer J Agr Econ 87(3):717–734
12. Faltermeier L, Abdulai A (2009) The impact of water conservation and intensification technologies: empirical evidence for rice farmers in Ghana. Agri Econ 40(3):365–379
13. Frisvold GB, Fernicola K, Langworthy M (2001) Market returns, infrastructure and the supply and demand for extension services. Amer J Agr Econ 83(3):758–763
14. Furlong W, Slotsve G (1983) Will that be pickup or delivery? An alternative spatial pricing strategy. Bell J Econ 14(1):271–274
15. Gelfand A, Smith AFM (1990) Sampling-based approaches to calculating marginal densities. J Am Stat Assoc 85(410):398–409
16. Ghana Statistical Service (2014) Ghana living standards survey round 6 (GLSS6). Accra, Ghana. Available via DIALOG. http://www.statsghana.gov.gh/docfiles/glss6/GLSS6_Main%20Report.pdf. Accessed 4 Oct 2016
17. Goetz SJ (1992) A selectivity model of household food marketing behaviour in Sub-Saharan Africa. Amer J Agr Econ 74(2):444–452
18. Greene WH (2008) Econometric analysis, 6th edn. Prentice Hall, New Jersey

19. Hanson JC, Just RE (2001) The potential for transition to paid extension: some guiding economic principles. Amer J Agr Econ 83(3):777–784
20. Holloway G, Barrett CB, Ehui S (2005) Bayesian estimation of the double-hurdle model in the presence of fixed costs. J Int Agric Trade Dev 1(1):17–28
21. Holloway G, Nicholson C, Delgado C, Ehui S, Staal S (2000) Agro-industrialization through institutional innovation: transactions costs, cooperatives and milk-market development in the East African highlands. Agri Econ 23(2000):279–288
22. Holloway GJ, Ehui SK (2001) Demand, supply and willingness-to-pay for extension services in an emerging-market setting. Amer J Agr Econ 83(3):764–768
23. Holloway GJ, Barrett CB, Ehui S (2001a) Cross-bred cow adoption and milk-market participation in a multivariate, count-data framework. Eurostat special issue: Bayesian methods with applications to science, policy and official statistics., pp 233–242
24. Holloway GJ, Dorfman J, Ehui S (2001b) Tobit estimation with unknown point of censoring with an application to milk-market participation in the Ethiopian highlands. Canadian Journal of Agricultural Economics 49(2001):293–311
25. Holloway GJ, Ehui S, Teklu A (2008) Bayes estimates of distance to market: transactions costs, cooperatives and milk-market development in the Ethiopian highlands. J Appl Econ 23(5):683–696
26. Holloway GJ, Nicholson C, Delgado C, Ehui S, Staal S (2004) A revised Tobit procedure for mitigating bias in the presence of non-zero censoring. Agricultural Economics 31(2004):97–106
27. Jari B, Fraser GCG (2009) An analysis of institutional and technical factors influencing agricultural marketing amongst smallholder farmers in the Kat River Valley, Eastern Cape Province, South Africa. Afr J Agric Res 4(11):1129–1137
28. Key N, Sadoulet E, de Janvry A (2000) Transactions costs and agricultural household supply response. Amer J Agr Econ 82(2):245–59
29. Makhura M-N, Kirsten J, Delgado C (2001) Transaction costs and smallholder participation in the maize market in the Northern Province of South Africa. Paper presented at the 7th Eastern and Southern Africa Regional Maize Conference, Pretoria, pp 11–15
30. Martey E, Al-Hassan RM, Kuwornu JKM (2012) Commercialisation of smallholder agriculture in Ghana: a Tobit regression analysis. Afr J Agric Res 7(14):2131–2141
31. Matungul PM, Lyne MC, Ortmann GF (2001) Transaction costs and crop marketing in the communal areas of Impendle and Swayimana, KwaZulu Natal. Dev Southern Afr 18(3):347–363
32. Ministry of Food and Agriculture (2011) Agriculture in Ghana: facts and figures (2010). Statistics, Research and Information Directorate (SRID), Accra, Ghana. Available via DIALOG. http://mofa.gov.gh/site/wp-content/uploads/2011/10/AGRICULTURE-IN-GHANA-FF-2010.pdf. Accessed 6 Jun 2012
33. Muamba FM (2011) Selling at the farmgate or traveling to the market: a conditional farm-level model. J Dev Areas 44(2):95–107
34. Olwande J, Mathenge M (2012) Market participation among poor rural households in Kenya. Paper presented at the International Association of Agricultural Economists Triennial Conference, Brasil, pp 18–24
35. Omiti JM, Otieno DJ, Nyanamba TO, McCullough E (2009) Factors influencing the intensity of market participation by smallholder farmers: a case study of rural and peri-urban areas of Kenya. Afr J Agric Res Econ 3(1):57–82
36. Reyes B, Donovan C, Bernsten R, Maredia M (2012) Market participation and sale of potatoes by smallholder farmers in the central highlands of Angola: a double hurdle approach. Paper presented at the International Association of Agricultural Economists Triennial Conference, Brazil, pp 18–24
37. Ricker-Gilbert J, Jayne TS, Chirwa E (2011) Subsidies and crowding out: a double-hurdle model of fertilizer demand in Malawi. Amer J Agr Econ 93(1):26–42
38. Siziba S, Kefasi N, Diagne A, Fatunbi AO, Adekunle AA (2011) Determinants of cereal market participation by Sub-Saharan Africa smallholder farmers. Learning Publics J Agric Env Stud 2(1):180–193

39. Smith MD (2003) On dependency in double-hurdle models. Stat Papers 44(4):581–595
40. Tanner MA, Wong W (1987) The calculation of posterior distributions by data augmentation (with discussion). J Am Stat Assoc 82:528–550
41. Vance C, Geoghegan J (2004) Modeling the determinants of semi-subsistent and commercial land uses in an agricultural frontier of Southern Mexico: a switching regression approach. Int Reg Sci Rev 27(3):326–347
42. Wooldridge JM (2002) Econometric analysis of cross section and panel data. Massachusetts Institute of Technology, Cambridge, London

CHAPTER 7

Climate sensitivities and farmland values in Nepal: A spatial panel Ricardian approach

Samrat B. Kunwar* amd Alok K. Bohara

Department of Economics, New Mexico, USA

ABSTRACT

This study presents an application of the Ricardian approach to explore the impact of climate change on farmland values in Nepal. The Ricardian approach is estimated using a panel fixed effects model, and the outcome is compared against two separate models that account for spatial correlation: a spatial autoregressive (SAR) model; and a spatial error model (SEM). The findings suggest that Nepalese farmlands are sensitive to climate change, and this result was consistent in both the non-spatial and the spatial frameworks. The inclusion of the spatial effects, however, revealed the presence of positive spatial autocorrelation and produced conservative estimates of climate change impacts. The net effect of annual increases in average temperature was negative; while the net effect of higher annual average precipitation was a positive outcome on farmland values. In particular, we found that the marginal effect of every degree increase in average annual temperature was Rs.180 /hectare ($1.80) reduction in farmland values. Likewise, for rainfall, it was found that 1 mm increase in average annual rainfall would positively affect farmland value by Rs.225/hectare ($2.25). Finally, the study findings suggested that extreme weather events could also impact the agricultural productivity and the farmland values in Nepal.

Keywords: Climate change, ricardian approach, spatial panel data analysis, Nepalese agriculture, environmental valuation.

1. INTRODUCTION

Climate change is emerging as a significant threat facing the humanity in the 21st century. There is a consensus among researchers that variations in land and water regimes through changes in climate might pose a significant challenge to the natural and human systems (Intergovernmental Panel on Climate Change (IPCC), 2007, 2014).

Agriculture is one area that is highly sensitive to climate due to its reliance on weather patterns and climate cycles for productivity. Agriculture is also the principal use of land globally with approximately 1.2 to 1.5 billion hectares of lands under crops, while another 3.5 billion hectares are used for grazing (Howden et al., 2007).

One country that is predominantly dependent on agriculture is Nepal. Nepal is a tiny developing country located in South Asia between India and China. The Nepalese agricultural sector contributes to more than one-third to the gross domestic product (GDP), and employs more than half of the total labor force (Acharya and Bhatta, 2013). This notable dependence on agriculture makes the Nepalese farming population highly vulnerable to the impacts of climate change.

Past studies suggest that the average annual mean temperature in Nepal has increased at an annual rate of 0.06°C between 1977 and 2000 (Malla, 2009). It has subsequently led to changes in the frequency of temperature extremes with more frequent warmer days and nights; and less frequent colder days and nights (Gum et al., 2009). Precipitation, on the other hand, has not displayed any definitive trends, but evidence indicates an increasing occurrence of intense rainfall events and rising flood days over the years (Gum et al., 2009). Such instances of extreme weather events can result in desirable agricultural land being undesirable as crop yields are restricted.

These changing climatic conditions have led to shifts in cropping patterns and the agricultural sector in Nepal is consequently being severely hurt. Regmi (2007) indicated that the eastern region of Nepal faced rain deficit in 2005/06, and the crop production was reduced by 12.5% on a national basis. Likewise, while Nepal used to be rice exporter in the past, the fluctuating climate conditions has limited the rice yields, and as a result, Nepal has been a rice importer for the past few years.

Nepal's heavy dependence on rain-fed agriculture coupled with the potential distressing effect of climate change, and ultimately on the welfare of the population and the economy of the country itself, necessitates a thorough analysis on the economic impact of climate change on the agricultural sector. An exhaustive assessment of the economic impact would allow for new policy formulation on potential mitigation and adaptation strategies to combat the likely effects of climate change. In this paper, an application of the Ricardian approach is used to evaluate the economic impact of climate change on agricultural productivity in rural Nepal.

The Ricardian approach is a model of climate-land value relationship, which was developed by Mendelsohn et al. (1994) to assess the impact of climate change on farmland values in the United States. The Ricardian Method is, in fact, named after the influential classical economist David Ricardo (1772 to 1823), who argued that in a perfectly competitive market, land values would reflect land profitability.

In their paper, Mendelsohn et al. (1994) evaluated the efficacy of the traditional production function approach in estimating the impacts of climate change with a new method they developed, the '*Ricardian Method*'. The production function approach is based on crop simulation models where the climate change impacts are estimated by varying input variables, including the climate itself. Mendelsohn et al. (1994) suggested that the limitation of the production-function approach in failing to account for the numerous substitutions and adaptations that farmers make could lead to an inherent bias that results in an overestimation of the damages from climate change.

The fundamental idea of the Ricardian approach is that land values and agricultural practices are correlated with an environmental variable, climate. However, some assumptions underlie this framework. The Ricardian model assumes that farmers are rational utility maximizers, and relies on an existence of a competitive economy with perfectly functioning output and input markets. With these assumptions, the Ricardian framework asserts that if the optimal use of farmlands is agricultural production, then the observed market rent on a parcel of land should equal the annual net profits from the production of an agricultural commodity using that land (Mendelsohn et al., 1994). Thus, farmland values are the discounted value of current and future profit. Furthermore, we can observe the relationship between farm values to climate and other variables to infer the optimal use of land. Hence, depending on the positive and negative impact of climate variables, the long-run accumulation of net profit defines land value.

Although the Ricardian method has since garnered widespread attention, there have been some notable criticisms as well because of the strong assumptions it makes (Cline, 1996; Darwin, 1999; Polsky, 2004; Deschenes and Greenstone, 2007). Darwin (1999) maintained irrigation to be an essential variable and omitting it would make the model of Mendelsohn et al. (1994) inconsistent with the Ricardian principle. Cline (1996) argued that the assumption of fixed relative prices in the Ricardian approach makes it a partial-equilibrium analysis. Besides, Cline (1996) also contended that the assumption of infinitely elastic supply of irrigation at current prices is misleading. Polsky (2004) argued that because Ricardian models are strongly aligned with perfect adaptations assumption, the negative impacts are biased to be small. Deschenes and Greenstone (2007) raised doubt on the validity of cross-sectional approaches to Ricardian study and proposed the use of a fixed-effect modeling to get more stable results from the Ricardian function.

To incorporate the limitations in the ergodic assumption of spatially and temporally invariant climate sensitivities, Polsky (2004) modeled a modified regional scale Ricardian analysis by integrating spatial and temporal variations in climate. The author reasoned that ignoring spatial relationship (inter-farmer communications across county borders) to understand climate-land use relationship could not account for climatic effects in different locations or time. Following Polsky (2004) there have been few other studies that have explored the Ricardian approach by explicitly incorporating spatial correlation.

Lippert et al. (2009) accounted for spatial auto-correlation in their analysis of the Ricardian approach in German agriculture by using a spatial lag and a spatial error dependence model. Kumar (2011) studied the impact of climate change on Indian agriculture by addressing the spatial features that could influence the climate sensitivity of agriculture. The paper argued that ignoring the spatial distribution could result in enlarged estimates of climate impacts in Ricardian studies. Their estimates of climate change impacts were more conservative after incorporating spatial correction models, and this finding was consistent in both the spatial lag and the spatial error model specification. Other researchers that have explicitly treated the spatial problem in the Ricardian study are Schlenker et al. (2005) and Chatzopoulos and Lippert (2016).

A separate limitation of numerous Ricardian studies that estimate climate change – land value relationships has been with the use of cross-section data for analysis. Since climate coefficients change over time, analyzing farms' long-term changes using cross-section data may not give reliable estimates. A panel-data approach can be far superior for estimation of any hedonic models, including Ricardian analysis if panel data are available and the time varying and unvarying coefficients are correctly specified (Massetti and Mendelsohn, 2011). A panel data approach also removes year effects and can produce more reliable estimates of the climate coefficients (DeSalvo et al., 2014). Several authors (Massetti and Mendelsohn, 2011; Deschenes and Greenestone, 2011; Massetti et al., 2013), etc. have employed panel data methods to study Ricardian analysis and the trend is rising.

Finally, another issue in many Ricardian studies stems from the use of only historical averages for temperature and precipitation to assess the impact of climate change on agriculture. However, plant physiology literature argues that it is not only the average weather patterns but also the extreme weather events that could have a severe effect on crop yields and agriculture in general (Rosenzweig et al., 2001; Anyamba et al., 2014).

Through this study, we seek to address the research gaps that have been identified in Ricardian analyses, in particular, the concerns mentioned earlier. First, considering the limitations of cross-sectional data approaches in other Ricardian studies, this paper uses panel data approach to enhance estimates reliability. Second, our analysis includes additional climate variables other than seasonal averages that could potentially capture the extremities in climate. Finally, we address the importance of accounting for spatial features and our estimation strategy thereby incorporates spatial methods in the Ricardian approach. Many Ricardian studies ignore the problem of spatial correlation, but when observations are correlated across space, standard approach such as the Ordinary Least Squares (OLS) method can lead to biased and inefficient parameter estimates (LeSage and Pace, 2009). The primary contribution of this paper that separate it from previous Ricardian applications is that we include extreme climate variables, while also explicitly accounting for spatial correlation in a panel data setting to study climate change impacts on agricultural productivity in rural Nepal.

2. MATERIALS AND METHODS

The non-climatic data used in this paper comes from the Nepal Living Standard Survey 2003/04 (NLSS-II[1]) and 2010/2011 (NLSS-III) of the Central Bureau of Statistics (CBS), Nepal. The NLSS survey follows the Living Standards Measurement Survey (LSMS) methodology that has been developed and promoted by the World Bank.

The methodology applied in the NLSS has been used in more than 50 developing countries by the World Bank with the goal to foster increased use of household data as a basis for policymaking. The NLSS survey includes a broad range of topics related to household and community welfare, but the important socio-economic variables necessary for this paper were obtained from the rural community questionnaire of NLSS. As such, this paper focuses on the rural farming communities in Nepal. The rural community questionnaire of the NLSS was developed to interview leaders and knowledgeable persons representing the community of the enumeration areas[2](CBS, 2004).

In this study, the primary sampling units[3](PSUs) are used as the unit of analysis, and not the individual households. The geocoded coordinates were made available only for PSUs, but not for the households, which constrained us to use PSU level analysis to explore the spatial relationship in our study. The total sample of the NLSS-II consists of 4,008 households representing 334 PSUs, from which 100 PSUs were common in NLSS III as well (CBS,2004). The total sample of the NLSS-III was estimated at 7200 households in 600 PSUs (CBS,2010). Among them, the NLSS-III sample is composed of all households visited by the NLSS-II in 100 of its PSU, as mentioned earlier. The final sample selected from NLSS-II and NLSS-III was 155 PSUs for this study[4]. Figure 1 plots the PSU locations used in this study.

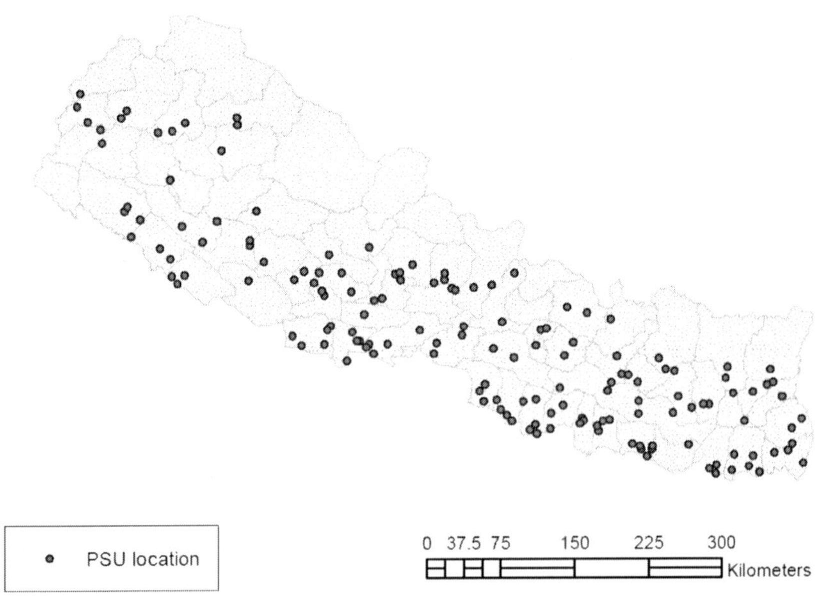

Figure 1. Plot of PSU points (Note: The map shows the location of PSUs used for the study.

In addition to the community welfare data from NLSS, this study used the ground station climate data for daily temperature and precipitation from

1981 to 2010, obtained from the Department of Hydrology and Meteorology, Nepal. The selection of weather stations nearest to each PSU was made in ArcGIS using multiple buffer width of 10 and 25 km radii[5]. Figure 2 shows the graph of the weather stations in Nepal, and Figure 3 presents the graph of buffer analysis undertaken to extract the weather stations nearest to each PSU. Similarly, Table 1 lists the definition of variables used in this study.

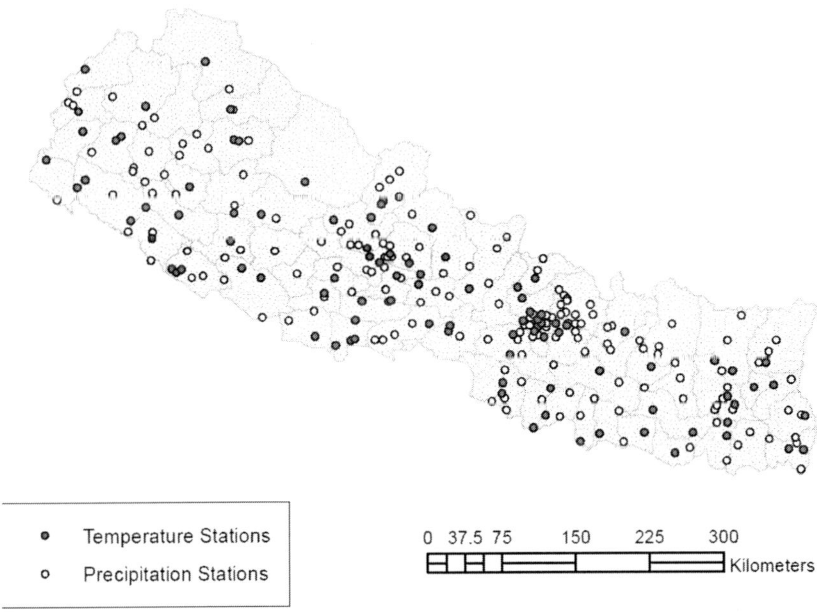

Figure 2. Plot of weather stations in Nepal (Note: The map shows the location of rainfall and temperature stations in Nepal).

2.1 Dependent variable

We used the self-reported average farmland value in each PSU as the measure of net productivity. These values have been converted to a per hectare land value in our analysis. While Ricardian papers often use net revenue or net profits as a proxy for land values, we used the actual farmland value in each PSU, since it was already available in the survey.

A criticism of using net revenue is that it is strongly influenced by the year of analysis (DeSalvo et al., 2014). Land values could be more accurate and an appropriate measure to analyze climate impacts since they reflect the expectations of net revenues across many years (Mendelsohn et al., 2009).

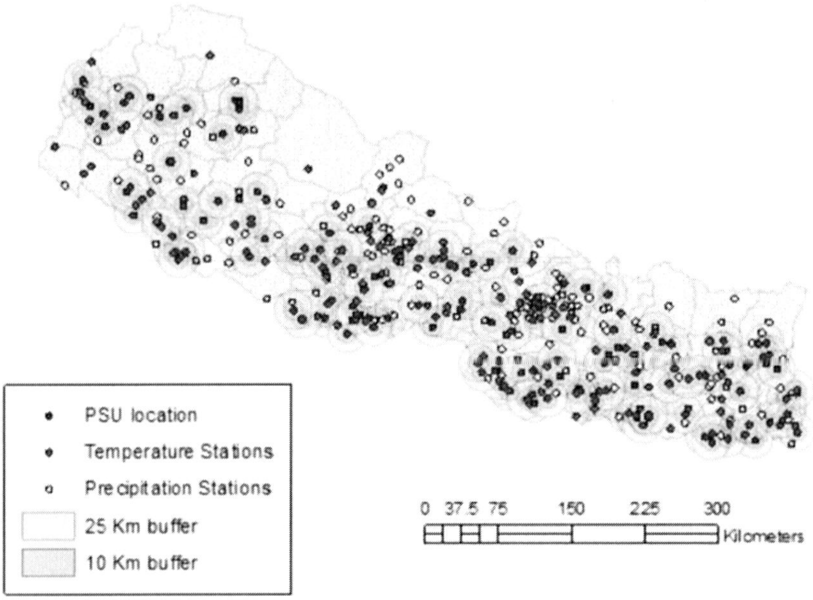

Figure 3. Buffer -10 and 25 km radius (Note: The map shows a buffer of 10 and 25 km radius from each PSU. This method was used to extract the temperature and rainfall stations to each PSU).

2.2 Explanatory variables

2.2.1 Climate variables

In order to construct the climatic variables, the daily temperature, and precipitation data for the time period 1981 to 2010 are used. Temperature and precipitation were classified into four seasons[6]: spring, summer, autumn and winter. We converted the daily average temperature and precipitation data into the four seasonal averages. In order to get the climatic data for 2002, the seasonal average from 1981 to 2002 was taken, and likewise, the seasonal average from 1989 to 2010 was used to capture the 2010 values. The 2002 and 2010 climate data, thus, capture the rolling average for the past 22 years. Using the constructed seasonal averages, the first set of climate variables used were the linear and quadratic measures of seasonal temperature and rainfall. The quadratic variables were introduced to capture the possible nonlinearities in the climate sensitivities.

Table 1. Variable description.

Variable	Definition
Land value	Average market value of farmlands in the PSU (self-reported) (Rs/Ha)
Spring temperature, summer temperature, autumn temperature, winter temperature	Climate-normal annual mean temperature for 29 year preceding each Census year for spring, summer, autumn and winter season (°C)
Spring temperature sq., summer temperature sq., autumn temperature sq., Winter temperature sq.	Square of the climate-normal annual mean temperature for 29 year preceding each Census year for spring, summer, autumn and winter season (°C)
Spring rainfall, summer rainfall, autumn rainfall, winter rainfall	Climate-normal annual mean precipitation for 29 year preceding each census year for spring, summer, autumn and winter season (mm/year)
Spring rainfall sq., summer rainfall sq., autumn rainfall sq., Winter rainfall sq.	Square of the climate-normal annual mean precipitation for 29 year preceding each census year for spring, summer, autumn and winter season (mm/year)
Spring temp. dev, Summer temp. dev, Autumn temp. dev, Winter temp. dev	Deviation of temperature during the spring, summer, autumn and winter season of a Census year from the historical 29 year averages in each of those seasons
Spring rain. dev, summer rain. dev, autumn rain. dev, winter rain. dev	Deviation of precipitation during the spring, summer, autumn and winter season of a Census year from the historical 29 year averages in each of those seasons
WSDI	Warm spell duration index. It represents the annual count of days with at least six consecutive days in which the daily maximum temperature exceeds the 90th percentile of daily maximum temperature for a 5-day running window
SDII	Simple precipitation intensity index. It represents the sum of precipitation in wet days during the day divided by the number of wet days in the year
Population	Total population of a PSU
Road	Access to paved roads in a PSU (yes = 1, no = 0)
Irrigation facilities	Access to irrigation facilities in a PSU (yes=1, no=0)
Electricity	Access to electricity in a PSU (yes=1, no=0)
Market center	Existence of a market center in a PSU (yes=1, no=0)
Farmer's group	Existence of an active user group (farmer's group) in a PSU (yes=1, no=0)

Along with the seasonal averages, we constructed variables to capture climatic deviations, and also climate extremes. The motivation for the inclusion of variables to capture climate extremities comes from the plant physiology literature that argue frequency and intensity of extreme weather events could also have a significant effect on crop yields and agriculture (Rosenzweig et al., 2001; Anyamba et al., 2014). To capture the climatic variation, we constructed the deviation of the seasonal temperature and precipitation for the year 2002 and 2010, from the rolling average of the past 29 years, for each of the four seasons[7].

The other set of climatic variables were employed to capture the extremities in climate, namely, the *warm spell duration index (WSDI)* for temperature, and *simple precipitation intensity index (SDII)* for rainfall[8]. These indices are two of twenty-seven indices that have been recommended to assess extreme weather events by the World Meteorological Commission for Climatology/ World Climate Research Program (CCI/CLIVAR) expert team on climate change detection, monitoring and indices through the CLIMDEX project (*www.climdex.org)*. WSDI represents the annual count of days in each year that is part of a warm spell. More specifically, it represents the annual count of days with at least six consecutive days in which the daily maximum temperature exceeds the 90th percentile of daily maximum temperature for a 5-day running window (Bronaugh, 2015). SDII, on the other hand, represents the sum of precipitation in wet days during the day divided by the number of wet days in the year (Bronaugh, 2015).

2.2.2 Non-climatic variables[9]

The set of control variables used in this paper are access to irrigation facilities, access to a market center, access to a road network, access to electricity; and the presence of farmers group, all within the context of the PSUs. Access to irrigation captures whether the PSU has irrigation facilities available. Access to market center means if the PSU has a market center in that community. Access to road and electricity follow similar explanation as for the case of irrigation and market center. Lastly, farmers group captures the existence of user group in a community (Table 2).

2.3 Conceptual framework

In a Ricardian model, farm performance (land value or net revenue) is regressed on a set of agro-climatic and socio-economic variables to assess the impact of climate change on farm performance. Mendelsohn et al. (1994) argued that the traditional approach to measuring the impacts of climate change on agriculture, the production function approach, was a crop specific analysis and it could overestimate the impacts. To overcome this limitation, the Ricardian approach was developed, and it assumes the following specification (Mendelsohn et al., 1994):

$$V_L = \int_0^{\infty} P_L e^{-\rho t}\, \partial t = \int_0^{\infty} \frac{[P_i Q_i - C_i(Q_i, R, F, Z)]}{L_i} e^{-\rho t}\, \partial t \qquad (1)$$

Table 2. Outlines the descriptive statistics of the dependent and independent variables used in this study.

Variable	N	Mean	St. Dev.
Land value	310	35,290.740	143.103.500
Spring temperature	310	22.719	4.243
Summer temperature	310	28.586	4.190
Autumn temperature	310	19.235	3.543
Winter temperature	310	14.115	3.232
Spring temp. dev	310	1.150	0.994
Summer temp. dev	310	0.630	0.780
Autumn Temp. dev	310	0.808	0.859
Winter Temp. dev	310	0.881	0.950
Spring rainfall	310	40.626	26.231
Summer rainfall	310	252.309	100.455
Autumn rainfall	310	71.807	46.747
Winter rainfall	310	14.827	6.973
Spring rain. dev	310	2.322	2.621
Summer rain. dev	310	3.735	7.967
Autumn rain. dev	310	1.122	0.723
Winter rain. dev	310	0.441	0.446
WSDI	310	40.423	63.001
SDII	310	19.862	6.828
Population	310	1,064.410	1,095.483
Road	310	0.416	0.494
Irrigation facilities	310	0.790	0.408
Electricity	310	0.610	0.489
Market center	310	0.694	0.462
Farmer's group	310	0.097	0.296

Sources: Climate data obtained from Department of Hydrology and Meteorology (DHM, Government of Nepal). PSU sociodemographic data obtained from the Central Bureau of Statistics (CBS), Nepal.

Where, farmland value (V_L) reflects the present value land (L); P_L is the net revenue per hectare;$P_i$and Q_i are the market price and output of the crop i respectively; $C_i(\ )$ is a function of purchased inputs (excluding land); R is a vector of input prices; F is a vector of climatic variables; Z is a vector of socioeconomic variables; t is the time, and is the discount rate.

The Ricardian model assumes that a farmer will maximize his land value (or net revenue) by choosing inputs subject to climate (F) and socio-economic variables (Z). This model relies on a quadratic formulation of climatic variables and is presented as:

$$V = \beta_0 + \beta_1 F + \beta_2 F^2 + \beta_3 Z + u \tag{2}$$

Where, u is an error term. The linear and a quadratic term for temperature and precipitation accounts for the nonlinear shape of the net revenue of the climate response function.

In the study analysis, we regressed the farmland value per hectare in the rural communities of Nepal as a dependent variable against climate and socio-economic variables. The independent climatic variables included the linear and quadratic temperature and precipitation for the four seasons: winter (the average for December, January, and February), spring (March, April, and May), summer (June, July, and August) and autumn (September, October, and November). In addition to the seasonal averages, the study analysis included seasonal temperature and precipitation deviation from the historical 22 years' seasonal average.

Finally, we incorporated WSDI to measure the temperature extreme, and simple precipitation intensity index (SDII) to measure the rainfall extreme. The independent non-climatic variables include the existence of paved road in the PSU, population of the PSU, whether the PSU had irrigation facilities available, having electricity in the PSU, and the existence of market center and farmers group in the PSU.

2.4 Analytical framework

2.4.1 Panel fixed effects (Non-spatial model)

The analytical framework was carried out using a forward specification analysis. First, a panel fixed effects[10] model was run, and the results were compared with a spatial lag and a spatial error model. The general specification of the non-spatial fixed effects model is given by:

$$Y_{it} = \alpha_i + \gamma_t + \beta X_{it}' + \theta {X^2}_{it}' + \delta N'_{it} + \eta Z_{it}' + u_{it} \tag{3}$$

The subscripts i and t in equation (3) denote PSU and time, respectively. The dependent variable Y is the farmland value per hectare, and represents the PSU fixed effects. It is assumed that the PSU fixed effects absorb all the unobserved PSU specific time-invariant factors such as soil and water quality that could influence the crop yields and land values. represents the time fixed effects, and it is presumed to control for time differences in the dependent variable which are common across PSU. The variable X is a vector of climate normals[11]; while N captures the vector of climate deviations and extremes[12]. Finally, Z is a vector of PSU sociodemographic variables[13]; and is an idiosyncratic error term that is assumed to be independent and identically distributed over PSUs and time, with mean zero and variance σ_u^2The fixed effects model concentrates on differences within PSU's. Thus, it explains to what extent farmland values deviate from average PSU farmland value.

While the parameter estimates from equation (3) could provide evidence on the impact of climate change on farmland values in Nepal, the estimates could be biased if land values are spatially correlated. In fact, Tobler (1979) first law of geography states that 'near things' are more related than 'distant things'. This suggests that farmland values could be spatially auto-correlated if there is a dependency between farmland prices. In a developing country like Nepal where farmers may not have sufficient information about their land characteristics, it is likely that land values could depend on interactions across PSUs with other land owners. Patton and McErlean (2003) argue interaction among landlords in order to base their starting price may give rise to spatial relationships.

This provides a motivation to reassess our problem by incorporating the spatial framework. Additionally, spatial modeling might also reduce omitted variable bias and account for spatial heterogeneity from a data-driven perspective. Omitted variable can arise because unobservable factors (for example, location amenities, PSU prestige, water accessibility for irrigation, etc.) could influence the dependent variable (farmland values), and this can be accounted by incorporating a spatial error model (LeSage and Pace, 2009). On the other hand, a spatial autoregressive (lag) model would be crucial if we believe that herd behavior exists in farmland markets, that is, the selling price of farmlands at any particular location acts as a signal that guides the selling price of nearby lands.

2.4.2 Spatial models

Taking into account the potential nature of the spatial relationship in farmland markets, the study analytical framework next incorporated spatial correlation in the Ricardian model. A general specification of the related family of the spatial model takes the following form[14]:

$$y_{it} = \lambda \sum_{j=1}^{N'} w_{ij} y_{jt}' + \alpha_i + \gamma_t + \beta X_{it}' + \theta X^2{}_{it}' + \delta N_{it} + \eta Z_{it}' + u_{it}$$

$$u_{it} = \rho \sum_{j=1}^{N'} w_{ij} u_{it} + v_{it} \qquad (4)$$

Equation (4) presents the specification of the panel spatial model, and it is similar to the non-spatial fixed effects model detailed in equation (3). The interpretation of parameters ($a, y_{,}, \theta, \delta, \eta$) and the variables captured by X, N and Z vectors are same as in equation (3). The additional terms in the spatial model that differentiate it from the non-spatial one are the spatially lagged dependent variable, and the spatial autoregressive process in the error term. The spatial lag coefficient is captured by , while captures the spatial error coefficient. The spatial autoregressive (lag) model (SAR) posits that the dependent variable (farmland value) is influenced by the dependent variables in the adjacent units and on a set of observed local characteristics. The spatial error model (SEM), on the other hand, states that the dependent variable (farmland value) depends on a set of observable characteristics with errors that are correlated across space (Elhorst, 2014).

One of the crucial inputs that spatial models require is the weight matrix W, which summarizes the spatial relations between n spatial units. In particular, the spatial matrix assigns nonzero elements for each observation (row) whose locations (columns) belong to its neighborhood (Anselin and Bera, 1998). A row-standardized weight matrix, where the row of the spatial weight matrix sums to unity, is used in our spatial model. The wy term in equation (4) represents the weighted average of the surrounding observations in the dependent variable; while the wu term represents the weighted average of the surrounding error term. The spatial weight matrix, W, used in this paper is a 5-nearest neighbor weight matrix for the PSUs in our sample. The parameters (λ) and (ρ) measures the extent of the spatial autocorrelation. Furthermore, setting the value of $\rho = 0$ leads to the SAR model that exhibits relationship only in the dependent variable. Similarly, setting $\lambda = 0$ leads to the SEM, resulting in spatial dependence in only the error term. In the case of the spatial models, the parameters to be estimated are the regression coefficients β, θ, δ and η, along with the spatial lag coefficient (λ), and the spatial error coefficient, (ρ).

2.4.3 Marginal impacts

The standard interpretation of estimated parameters as partial derivative is no longer valid in the case of the SAR model. Intuitively, the lag model implies that the farm land values of region i is also influenced by the land values from neighboring regions. The marginal effects in the SAR model take the following form (LeSage and Pace, 2009):

$$\partial y_i / \partial x_{ir} = S_r(W)_{ii}$$
$$\partial y_i / \partial x_{jr} = S_r(W)_{ij} \qquad (5)$$

Where, $Sr\ (W) = (I_n - \phi W)^{-1}\ \beta r$. In equation (5), the subscript i and j represents location i and j respectively, while βr is the coefficient on the rth explanatory variable. $Sr\ (W)$ is a $n \times n$ matrix with the diagonal elements containing the direct impacts and the off-diagonal elements representing the indirect impacts. In the SAR model, the spatial connectivity relationships mean that a change in a single explanatory variable in region i has a 'direct impact' on region i as well as an 'indirect impact' on other regions, $j \neq i$ (LeSage and Fischer, 2008).

The upper quantity in equation (5) captures the impact of a change in an explanatory variable (for example, temperature) at location i on the dependent variable at location i, known as the average direct impact (ADI). For example, the average direct effect shows the impact of climate change on PSU i on the farmland values of PSU i. The lower quantity, on the other hand, captures the effect of a change in the explanatory variable at location j on the dependent variable at location i, with $j \neq i$ and is known as the average indirect effect (AII). For example, the average indirect effect, also known as the neighboring effect, measures the impact of an increase in climate at PSU i on the farmland value of neighboring PSUs, averaged over all neighboring PSUs. The average direct effect can be interpreted as the own derivative, while the average indirect effect captures the cross derivative. Lastly, the average total effect (ATI) is the sum of average direct and indirect effect, and it measures the total cumulative impact. In this paper, the average total effects estimate how changes in climate affect total farmland values, taking into account both own-PSU and spillover effects.

2.4.4 Marginal climate impacts

Since the study climate variables contain linear and quadratic coefficients in raw form, we evaluated the marginal climate impacts (MCI) for rainfall and temperature to ease interpretation. Recalling equation (3), the MCI[15] of average annual changes in climate (temperature or precipitation) on the mean farmland value per hectare can be expressed as:

$$E\left(\frac{\partial LANDVALUE}{\partial CLIMATE}\right) = \beta + 2\theta * E(CLIMATE) \qquad (6)$$

In this study, the MCI represents the change in Rs./ha of farmland value per ^{0}C or mm/year, evaluated at the mean annual climate for farmlands in Nepal.

3. RESULTS

The data summary (Table 2) shows that most of the variables have low standard deviation, indicating data homogeneity. The regression results[16] from the non-spatial fixed effects model (Table 3, Column 1) suggested that the spring and summer temperature; and the spring, autumn and winter rainfall impacted the farmland values. Likewise, the spring temperature deviation and the winter rainfall deviation, as well as the extreme indices in WSDI and SDII also affected the farmland values. The non-climatic variables that were found to be significant were irrigation facilities, population and the existence of market center in the PSU, all of which had a positive impact on the farmland values. The significance of these variables was almost identical in the spatial lag and the spatial error model as well (Table 3, Column 2 and 3).

While the significance of most variables were comparable in all three models, the magnitudes of the estimated climate coefficients were smaller in spatial models than the non-spatial fixed effect model. Furthermore, the spatial correlation parameter (ρ) in the spatial error model and the spatial autoregressive parameter (λ) in the spatial lag model were both significant, suggesting the presence of spatial correlation. The positive coefficient on the spatial lag parameter (λ) indicates that farmland values are positively affected by land values in the neighboring PSU's. This finding substantiates the need to incorporate spatial effects in our modeling, and ignoring these effects to explore the impact of climate change on farmland values can lead to biased estimates. In order to choose the best spatial model, we relied on the lowest AIC and the BIC values (Table 3). While the estimates[17] of the AIC and the BIC values largely favored the spatial models over the non-spatial one, the SAR model was only slightly preferred over the SEM model based on the goodness of fit estimates (Table 3).

We also found evidence that extremities in weather could affect farmland values too. This was revealed by the fact that areas with more warmer days throughout the year had higher farmland values; while areas with more intense precipitation had lower farmland values. Additionally, the climate change impact estimates displayed non-linear relationship with farmland values in certain seasons, and this result is consistent with the Ricardian hypothesis proposed by Mendelsohn et al. (1994).

Finally, we looked at the marginal effect of average annual temperature and rainfall changes in farmland values. The net effect of increasing annual temperature was negative; while the net effect of higher annual precipitation was a positive outcome on farmland values (Table 4). In particular, we found that the marginal effect of every degree increase in average annual temperature was Rs.180 /hectare ($1.80) reduction in farmland values. In contrast, every mm increase in annual average rainfall led to an increase in farmland values by Rs.225/hectare ($2.25) (US$1 = Nepali Rs.100 conversion rate of 22 April, 2015 used throughout).

Table 3. Spatial and Non-spatial regression result

Variable	Non-spatial model	Spatial models	
	(1)	(2)	(3)
	Fixed effects	Spatial error	Spatial lag
Climate Normals			
Spring temperature	-4.705.098*** (1.499.624)	-3.770.218* (2.138.728)	-3.956.385* (2.145.844)
Spring temperature sq.	101.512*** (28.603)	83.418* (45.813)	82.806* (45.999)
Summer temperature	5.991.240** (2.465.669)	4.007.011* (2.653.571)	6.610.327** (2.731.009)
Summer temperature sq.	-104.679* (58.102)	-74.317 (53.995)	-95.566* (55.429)
Autumn temperature	-2.353.482* (1.328.044)	-1.305.782 (1.845.721)	-1.641.706 (1.852.688)
Autumn temperature sq.	45.178 (36.104)	19.373 (48.755)	28.268 (49.354)
Winter temperature	1.806.967 (1.656.178)	1.312.581 (1.649.897)	1.085.378 (1.653.009)
Winter temperature sq.	-68.647 (53.326)	-60.404 (55.602)	-46.325 (55.485)
Spring rainfall	34.195** (16.741)	31.688* (17.358)	34.674* (17.907)
Spring rainfall sq.	-0.161** (0.081)	-0.152 (0.110)	-0.169 (0.114)
Summer rainfall	10.342 (8.554)	16.377 (10.488)	15.076 (10.507)
Summer rainfall sq.	-0.011 (0.008)	-0.017 (0.016)	-0.017 (0.016)
Autumn rainfall	55.846*** (19.832)	49.805*** (-13.656)	51.647*** (13.674)
Autumn rainfall sq.	-0.393*** (0.130)	-0.337*** (0.055)	-0.354*** (0.054)
Winter rainfall	279.972*** (77.915)	275.596*** (64.144)	264.975*** (65.576)

Winter rainfall sq.	-5.282***	-5.065***	-5.185***
	(1.417)	(1.373)	(1.421)
Deviation from climate normals			
Spring temp. dev	104.573**	58.676	68.320
	(52.130)	(62.889)	(64.283)
Summer temp. dev	98.275	57.341	84.828
	(90.009)	(72.634)	(72.766)
Autumn temp. dev	-113.153	-150.053**	-141.447**
	(92.014)	(69.786)	(71.460)
Winter temp. dev	8.259	36.376	23.136
	(67.114)	(71.332)	(71.545)
Spring rain dev	-17.951	-12.812	-12.410
	(16.312)	(20.012)	(20.101)
Summer rain dev	4.772	5.824	6.346
	(4.563)	(6.904)	(6.887)
Autumn rain dev	-86.028	-81.182	-79.463
	(107.969)	(65.157)	(64.607)
Winter rain dev	252.092*	274.139**	255.970**
	(135.636)	(118.769)	(119.601)
Climate extremes			
WSDI	3.372***	3.324**	3.699**
	(1.286)	(1.558)	(1.535)
SDII	-27.106**	-29.827*	-27.633*
	(11.518)	(16.475)	(16.439)
Controls			
Irrigation facilities	253.018***	212.951*	250.600**
	(78.588)	(116.922)	(119.621)
Electricity	138.826	131.655	122.340
	(87.675)	(117.869)	(118.015)
Road	46.480	-45.574	-0.484
	(100.604)	(99.349)	(99.096)
Population	0.140**	0.165***	0.145***
	(0.069)	(0.044)	(0.044)
Farmer's group	94.422	160.621	94.629
	(69.942)	(143.045)	(137.966)
Market center	222.409***	232.431**	234.405**
	(80.611)	(91.972)	(92.028)

Rho	-	0.357***	-
	-	(0.072)	-
lambda	-	-	0.351***
	-	-	(0.065)
Log Likelihood	-3.048.864	-2.777.865	-2.776.498
AIC	6,161.728	5.621.730	5.618.996
BIC	6281.299	5.745.037	5.742.303
Observations	310	310	310
Number of PSU	155	155	155
PSU FE	YES	YES	YES
Year FE	YES	YES	YES

*** p<0.01, ** p<.5, * p<0.1. Dependent variable is the farmland value per hectare (Rs./Ha). The values in the parenthesis are the standard errors. Column 1 lists the output of the non-spatial fixed effects model. Column 2 and 3 are the output of the spatial fixed effects model. Column 2 is the spatial error model (SEM), while column 3 is the spatial lag model (SAR).

Variable	Mean land value	95% confidence interval	
Temperature (°C)	-180.692	-187.257	-174.127
	(3.349)	-	-
Precipitation (mm)	-225.850	192.060	259.641
	(17.24)	-	-

Notes: Annual marginal at the average Nepalese climate measured as a change in average land value per hectare. These values are calculated from the ATI values of the SAR model using equation (6). Standard errors are derived using the delta method.

4. DISCUSSION

The findings from both, the non-spatial and the spatial models, suggested that climate does seem to have an impact on the value of farms in Nepal. For instance, the non-spatial fixed effect model showed evidence that the average temperature during the spring and summer season; and the average rainfall during the spring, winter and the autumn season affected the farmland values.

Additionally, the presence of a nonlinear relationship between climate and land values, although present only in certain seasons, is consistent with the findings from other Ricardian studies (Mendelsohn et al., 1994; Deressa et al., 2005; Seo and Mendelsohn, 2008). The effect of temperature on farmland value was more pronounced than that of rainfall, and this suggests higher sensitivities of crop growth to temperature changes (Lobell and Burke, 2008). Regarding precipitation, the result indicated that although higher rainfall is conducive for crop development, excess rainfall could hurt the crops, and, thereby the farmland values.

In essence, the significant quadratic variables imply that climate and farmland values have a nonlinear relationship, and it is consistent with the hypothesis of Ricardian approach (Mendelsohn et al., 1994). The positive coefficient in the quadratic terms for temperature (rainfall) suggests a minimally productive level of temperature (rainfall) and either more or less temperature (rainfall) would increase land values. The negative quadratic coefficients for temperature (rainfall) indicate that there is an optimal level of climate variable from which the value function decreases in both directions (Mendelsohn et al., 1994).

The findings from Table 3 shows that the significance of the variables for both the non-spatial (Column 1) and spatial models (Column 2 and 3) was almost alike. While the sign and significance of most coefficients in the three models were comparable, the magnitudes of the climate normal variables in the non-spatial model were larger in absolute value compared to the spatial models. This finding is consistent with other papers that have examined spatial and non-spatial modeling in the context of Ricardian framework (Kumar, 2011; Baylis et al., 2011).

The linear and the quadratic temperature variable for the spring season was significant across all models, while for the summer, the variables were significant in the non-spatial (column 1) and the SAR model (column 3). Winter temperature did not have any effect on farmland values across all three models, while the linear term for the autumn temperature was significant only in the non-spatial model. Looking at column (3)[18], the turning point for spring temperature occurred at 23.88°C. This indicates that average spring temperature above 23.88°C is associated with higher crop yields, which results in higher farmland values as well.

Similarly, the turning point for summer temperature in column (3) is 28.98°C, indicating that farmland values decline when average summer temperature exceeds 29°C. This result makes sense when we consider the major agricultural outputs of Nepal. The major crops grown in Nepal are paddy, wheat, and maize (Malla, 2009); and the optimal temperature range estimated in this paper is consistent with literature that have explored these crop's life cycle. Karn (2014) found that the critical temperature threshold for rice yield in Nepal to be 29.9°C, and temperature beyond that would lead to a decline in rice yields. Bhatt et al. (2014) found that the critical maximum threshold for maize production to be 27°C in Eastern Nepal. Bannayan et al. (2004) also suggest the optimum temperature for maize growth, in general, should be between 22 to 25°C. These findings could potentially explain the results found in this paper for the decline in farmland values at temperatures below 23.88°C and beyond 28.98°C. Since the suitable temperature for both maize and rice in Nepal lies between about 22 to 30°C, it seems plausible that farmland values increase in that temperature range.

The other significant climatic variables across all three models were the winter rainfall deviation, WSDI, and SDII. In particular, higher rainfall deviation during the winter season; and areas with higher annual warmer spells, both had a positive impact on farmland values. On the other hand, areas with more intense and excessive rainfall, in general, were associated with lower farmland values. Although the winter temperature was not significant in any of the models, we believe that is due to the growth requirement of winter crops in Nepal. The main winter crops in Nepal are wheat and barley, which have been found to be highly sensitive to winter rainfall, moreso than temperature (Krishnamurthy et al., 2013). In fact, Krishnamurthy et al. (2013) state that the winter crops in Nepal are extremely sensitive to small changes in rainfall patterns while the impact of temperature on these crops is low.

The coefficients on the linear and the quadratic precipitation variables suggest that autumn and winter average rainfall affect farmland values and this result is consistent across all three models. Similarly, higher spring precipitation had a significant and positive consequence on farmland values across all three models. This result also seems plausible when we consider the harvesting period of the major crops in Nepal. The harvesting period in Nepal for rice starts from mid-October to December, while wheat is harvested in winter period. Thus, the positive coefficients in the autumn and winter rainfall imply the presence of suitable environment for these crops during harvest time, which is positively reflected on the land values. Likewise, the negative coefficients on the quadratic terms for winter and autumn precipitation imply that excessive rainfall during these seasons could potentially damage the harvest, thereby negatively affecting the land values.

The other findings were that PSU's with access to irrigation, market center, and with higher population have a positive impact on farmland values. These results also seem reasonable since PSUs with irrigation facilities would not need to rely on rain-fed agriculture for crop growth and thus, these areas have higher land values. Similarly, presence of market center provides opportunities to easily purchase different agricultural products to improve yield; and higher population implies location with better amenities that could be driving population growth, both of which would result in higher farmland values as suggested by the results in this paper. In fact, the positive impact of irrigation facilities and market center on Nepalese farmland valuation has also been confirmed by Joshi et al. (2017).

The findings from the spatial analysis revealed the need to incorporate spatial models to enhance estimation reliability. With regards to the choice between the two spatial models, we looked at the model performance parameters, the AIC, and the BIC values, which suggested SAR as a slightly better model. While the SAR model was preferred from an econometric perspective, this lag model seems probable from an intuitive viewpoint as well. It is reasonable to assume that since farmers may not know the inherent value of their land due to insufficient information about land characteristics, especially in developing countries like Nepal, land prices could thereby depend on landowner interactions across communities.

Similarly, one can also argue that farmlands surrounded by expensive lands could potentially be worth more than those surrounded by inexpensive lands. Additionally, agricultural land markets are highly localized with many buyers being farmers looking to add fields near to their existing operation (Baylis et al., 2011), which further strengthens the argument for the use of the lag model. Anselin et al. (2008) state that for an equilibrium outcome of a spatial or social interaction process where the value of a dependent variable for one agent is jointly determined with that of neighboring agents, a SAR is considered to be ideal.

We then looked at the marginal impacts of the SAR model (Appendix Table 1)[19]. The direct effect showed that for every degree increase in the spring temperature beyond the threshold value of 23.88°C, farmland values increased by Rs.85/hectare ($0.85). However, beyond 28.98°C temperature in the summer season, every degree increase in temperature reduced farmland values by Rs.98/hectare ($0.98). The average indirect impact for the spring and summer temperature were also significant, indicating that the temperature at a particular PSU also affects the land values of neighboring PSUs (as defined by the *W* matrix). The indirect effect implied that for every degree increase in spring temperature beyond 23.88°C at a particular PSU, the land values in the neighboring PSUs increased by Rs.42/hectare ($0.42).

Likewise, beyond 28.98^{0} C in the summer, farmland values in the neighboring PSUs declined by Rs.49/hectare ($0.49). The significance of the indirect effect does not seem implausible since climate is not vastly dissimilar in a small spatial scale. Therefore, if spring temperature affects the land values at a particular PSU, it is likely to affect the neighboring PSUs as well. Finally, the average total impact of an increase in the spring temperature beyond 23.88°C suggested that for every degree increase, the total farmland value increased by Rs.127/hectare ($1.27), when taking into account the own-PSU effect and the spillover effect of a change in spring temperature. The impact of summer temperature on farmland values can be interpreted in a similar fashion. Taking into account the findings from changes in temperature, the overall net marginal effect of a degree increase in average annual temperature was a reduction in farmland values by Rs.180/hectare ($1.80) (Table 4).

Regarding rainfall, the findings suggested that the average direct impact of a mm increase in rainfall during the spring season increased farmland values by Rs.35/hectare ($0.35). Precipitation during autumn and winter season also had an impact on the farmland values. Similar to the case of spring temperature, the average indirect impact indicates that precipitation during these seasons was not only affecting farmland values at that PSU, but also the land values in the neighboring PSUs. The average direct impact of an increase in precipitation during autumn and winter season was Rs.52/hectare ($0.52) and Rs.271/hectare ($2.71) respectively. However, excessive rainfall destroys crops, and it is reflected in the lower land value captured by the negative quadratic terms. Finally, the overall net marginal impact of a mm increase in the annual mean rainfall was an increase in farmland value by Rs.225/hectare ($2.25) (Table 4).

SDII and *WSDI*, the two variables that capture the extremities in climate were also both significant. Higher *SDII* suggests the occurrence of stronger precipitation, and this has an adverse impact on farmland values. On the other hand, higher *WSDI*, which indicates greater number of warmer days, positively affects the farmland values. Heavy rainfall can cause a disruption in crop cycle balance and lead to lower yield which could negatively affect farmland values, as suggested by our findings.

The positive effect of *WSDI* also makes sense since paddy, one of the major crops in Nepal, requires an extended period of the warm growing season. Similarly, maize is another staple crop of Nepal which also requires warm days to grow properly, and thus higher *WSDI* can, in fact, lead to higher farmland values. The results for the average total impacts suggested that intense precipitation lowered farmland values by Rs.42.5/hectare ($0.425); while higher days of warm spell increased land value by Rs.5.70/hectare ($0.057). In terms of the non-climatic variables, PSUs with access to irrigation and market center had land values that were higher by Rs.386/hectare ($3.86) and Rs.361/hectare ($3.61) respectively, compared to the ones that did not have those amenities.

CONCLUSION

Changes in climate and the resulting changes in land use pattern are likely to have a significant impact on sectors like agriculture, forestry, water and food security (Field, 2012). While the severity of climate change impacts on agriculture could be massive, with the right mitigation and adaptation strategies, the negative consequences can be alleviated. This paper used an application of the Ricardian approach to analyze the impact of climate change on farmland values in Nepal. Taking into account the limitation of traditional Ricardian approach in failing to explicitly incorporate the spatial nature of land values, this paper employed a spatial fixed effect model to estimate climate change impacts in the context of Nepal. The results revealed significant evidence of spatial correlation and the effects of climate change were found to be more conservative in spatial models relative to the non-spatial model.

The general findings implied that Nepalese farmlands are sensitive to climate change. The average temperature during the spring and summer season; and average rainfall in the spring, autumn and winter season were found to affect crop yields and thereby, the value of farms. The net effect of annual increases in temperature was negative; while the net impact of higher annual precipitation was a positive outcome on farmland values. In particular, we found that the marginal effect of every degree increase in annual temperature was Rs.180/hectare ($1.80) reduction in farmland values. Likewise, for rainfall, it was found that 1mm increase in average annual rainfall would positively affect farmland value by Rs.225/hectare ($2.25). Additionally, the extreme weather indices suggested PSUs with a greater number of warmer days (WSDI) faced positive effect on farmland values; while PSUs with excessive precipitation (SDII) had lower farmland values.

From a modeling perspective, we found evidence of significant positive spatial correlation, and the aforementioned results are the outcome of a spatial correction model. The implication of our findings from an econometric perspective suggests the need to depart from non-spatial analysis to studies that account for spatial analysis in order to obtain more reliable estimates of climate change impacts on farmland value.

The results from this study also provide an interesting perspective from the policymaking point of view. Agricultural production is one of the major means of livelihood for most people in Nepal and as such, policies should be directed towards helping people combat the impacts of climate change. One solution could be to provide farmers support in the form of loans, access to seeds, and technical advice on crop management and water harvesting so they can better adapt to the changing climatic conditions. The poor farming population are most vulnerable to climate change, particularly because they rely heavily on rain-fed agriculture. As such, policies should be directed towards providing irrigation systems at minimal costs to these populations. Furthermore, policymakers should also provide education and awareness to farmers on the dangers of climate change as well as on the importance of employing irrigation as a way to increase their crop yields and sustain their livelihood.

REFERENCES

1. Acharya SP, Bhatta GR (2013). Impact of climate change on agricultural growth in Nepal. NRB Econ. Rev. 25(2):1-16.

2. Anyamba A, Small JL, Britch SC, Tucker CJ, Pak EW, Reynolds CA, Crutchfield J, Linthicum KJ (2014). Recent weather extremes and impacts on agricultural production and vector-borne disease outbreak patterns. PLoS One 9(3):e92538.
3. Anselin L, Bera AK (1998). Spatial dependence in linear regression models with an introduction to spatial econometrics. Statistics Textbooks Monogr. 155:237-290.
4. Anselin L, Le Gallo J, Jayet H (2008). Spatial panel econometrics. In The econometrics of panel data Springer Berlin Heidelberg. pp. 625-660.
5. Bannayan M, Hoogenboom G, Crout NMJ (2004). Photothermal impact on maize performance: A simulation approach. Ecol. Model. 180(2):277-290.
6. Baylis K, Paulson ND, Piras G (2011). "Spatial approaches to panel data in agricultural economics: a climate change application. J. Agric. Appl. Econ. 43(3):325-338.
7. Bhatt D, Maskey S, Babel MS, Uhlenbrook S, Prasad KC (2014). Climate trends and impacts on crop production in the Koshi River basin of Nepal. Reg. Environ. Change 14(4):1291-1301.
8. Bronaugh D, Hiebert MJ (2015). Pacific Climate Impacts Consortium. climdex.pcic: PCIC Implementation of Climdex Routines [Computer software manual]. Retrived from https://CRAN.R-project.org/package=climdex.pcic (R package version 1.1-6).
9. Central Bureau of Statistics (CBS) (2004). Nepal Living Standards Survey: Statistical Report 2003/04, Volumes I and II. Kathmandu: National Planning Commission/ His Majesty's Government of Nepal.
10. Central Bureau of Statistics (CBS) (2010). Nepal Living Standards Survey: Statistical Report 2010/11, Volumes I and II. Kathmandu: National Planning Commission/ His Majesty's Government of Nepal.
11. Cohen J, Cohen P, West SG, Aiken LS (2013). Applied multiple regression/correlation analysis for the behavioral sciences. Routledge.
12. Chatzopoulos T, Lippert C (2016). Endogenous farm-type selection, endogenous irrigation, and spatial effects in Ricardian models of climate change. Euro. Rev. Agric. Econ. 43(2):217-235.
13. Cline WR (1996). The impact of global warming of agriculture: comment. Am. Econ. Rev. 86(5):1309-1311.
14. Darwin R (1999). The impact of global warming on agriculture: A Ricardian analysis: Comment. Am. Econ. Rev. 89(4):1049-1052.
15. Deressa T, Hassan R, Poonyth D (2005). Measuring the impact of climate change on South African agriculture: the case of sugarcane growing regions. Agrekon 44(4):524-542.
16. De Salvo M, Begalli D, Signorello G (2014). The Ricardian analysis twenty years after the original model: Evolution, unresolved issues and empirical problems. J. Dev. Agric. Econ. 6(3):124-131.
17. Deschenes O, Greenstone M (2007). The economic impacts of climate change: evidence from agricultural output and random fluctuations in weather. The American Economic Review, 97(1), 354-385.
18. Deschenes O, Greenstone M (2011). Using panel data models to estimate the economic impacts of climate change on agriculture. Handbook on Climate Change and Agriculture pp. 112-140.
19. Elhorst JP (2014). Spatial panel data models. In Spatial Econometrics. Springer Berlin Heidelberg pp. 37-93.
20. Field CB (2012). Managing the risks of extreme events and disasters to advance climate change adaptation: special report of the Kunwar and Bohara 159 intergovernmental panel on climate change. Cambridge University Press.
21. Gum W, Singh PM, Emmett B (2009). Even the Himalayas have stopped smiling': climate change, poverty and adaptation in Nepal. 'Even the Himalayas have stopped smiling': climate change, poverty and adaptation in Nepal.
22. Howden SM, Soussana JF, Tubiello FN, Chhetri N, Dunlop M, Meinke H (2007). Adapting agriculture to climate change. Proceed. Natl. Acad. Sci. 104(50):19691-19696.
23. IPCC (2007). Climate Change 2007: Synthesis Report. Contribution of Working Groups I, II and III to the Fourth Assessment Report of the Intergovernmental Panel on Climate

Change [Core Writing Team, Pachauri, R.K and Reisinger, A. (eds.)]. IPCC, Geneva, Switzerland 104 p.
24. IPCC (2014). Climate Change 2014: Synthesis Report. Contribution of Working Groups I, II and III to the Fifth Assessment Report of the Intergovernmental Panel on Climate Change [Core Writing Team, R.K. Pachauri and L.A. Meyer (eds.)]. IPCC, Geneva, Switzerland, 151 pp.
25. Joshi J, Ali M, Berrens RP (2017). Valuing farm access to irrigation in Nepal: A hedonic pricing model. Agric. Water Manage. 181:35-46.
26. Karn PK (2014). The impact of climate change on rice production in Nepal (No. 85).
27. Krishnamurthy PK, Hobbs C, Matthiasen A, Hollema SR, Choularton RJ, Pahari K, Kawabata M (2013). Climate risk and food security in Nepal—analysis of climate impacts on food security and livelihoods. https://cgspace.cgiar.org/handle/10568/34077
28. Kumar KK (2011). Climate sensitivity of Indian agriculture: do spatial effects matter?. Cambridge J. Reg. Econ. Soc. P. rsr004.
29. LeSage JP, Fischer MM (2008). Spatial growth regressions: model specification, estimation and interpretation. Spatial Econ. Anal. 3(3):275-304.
30. LeSage JP, Pace RK (2009). Introduction to Spatial Econometrics (Statistics, textbooks and monographs). CRC Press.
31. Lippert C, Krimly T, Aurbacher J (2009). A Ricardian analysis of the impact of climate change on agriculture in Germany. Clim. Change 97(3):593-610.
32. Lobell DB, Burke MB (2008). Why are agricultural impacts of climate change so uncertain? The importance of temperature relative to precipitation. Environ. Res. Lett. 3(3):034007.
33. Malla G (2009). Climate change and its impact on Nepalese agriculture. J. Agric. Environ. 9:62-71.
34. Massetti E, Mendelsohn R (2011). Estimating Ricardian models with panel data. Clim. Change Econ. 2(04):301-319.
35. Massetti E, Nascimento GRDC, Fortes de Oliveira A, Mendelsohn RO (2013). The Impact of Climate Change on the Brazilian Agriculture: A Ricardian Study at Microregion Level.
36. Mendelsohn R, Nordhaus WD, Shaw D (1994). The impact of global warming on agriculture: a Ricardian analysis. Am. Econ. Rev. pp. 753-771.
37. Meyers RH (2000). Classical and modern regression with applications (Duxbury Classic). Duxbury Press, Pacific Grove.
38. Patton M, McErlean S (2003). Spatial effects within the agricultural land market in Northern Ireland. J. Agric. Econ. 54(1):35-54.
39. Polsky C (2004). Putting space and time in Ricardian climate change impact studies: Agriculture in the US Great Plains, 1969–1992. Ann. Assoc. Am. Geogr. 94(3):549-564.
40. Rosenzweig C, Iglesias A, Yang XB, Epstein PR, Chivian E (2001). Climate change and extreme weather events; implications for food production, plant diseases, and pests. Glob. Change Hum. Health 2(2):90-104.
41. Schlenker W, Hanemann WM, Fisher AC (2005). Will US agriculture really benefit from global warming? Accounting for irrigation in the hedonic approach. Am. Econ. Rev. 95(1):395-406.
42. Seo SN, Mendelsohn R (2008). Climate change impacts on Latin American farmland values: the role of farm type. Rev. Econ. Agron. 6(2):159-176.
43. Tobler WR (1979). Cellular geography. In Philosophy in geography. Springer Netherlands pp. 379-386.

Appendix Table 1. Marginal impacts – spatial lag model

Variables	Direct	Indirect	Total
	Climate Normals		
Spring temperature	-4,059.016*	-2,038.298*	-6,097.314*
Spring temperature sq.	84.954*	42.661*	127.616*
Summer temperature	5,684.092*	2,854.309*	8,338,302**
Summer temperature sq.	-98.004*	-49.283*	-147.288*
Autumn temperature	-1,684.281	-845.776	-2,530.057
Autumn temperature sq.	29.038	14.520	43,559
Winter temperature	1,113.534	559.178	1,672.712
Winter temperature sq.	-47.527	-23.866	-71.393
Spring rainfall	35.573*	17.863	53.437*
Spring rainfall sq.	-0.174	-0.087	-0.261
Summer rainfall	15.467*	7.767	23.234
Summer rainfall sq.	-0.017	-0.008	-0.026
Autumn rainfall	52.987***	26.608**	79.595***
Autumn rainfall sq.	0.361***	0.182***	-0.547***
Winter rainfall	271.849***	136.513**	408.362***
Winter rainfall sq.	-5.320***	-2.671**	-7.991***
	Deviation from Climate Normals		
Spring temperature dev	70.092	35.198	105.290
Summer temperature dev	87.028	43.702	130.731
Autumn temperature dev	-145.110*	-72.872	-217.088*
Winter temperature dev	23.736	11.919	35.656
Spring rain dev	-12.732	-6.393	-19.125
Summer rain dev	6.511	3.269	9.780
Autumn rain dev	-81.524	-40.938	-122.463
Winter rain dev	262.610**	131.873*	394.483**
Climate Extremes			
SDII	-28.350*	-14.236	-42.586*
WSDI	3.795**	1.906**	5.701**
	Controls		
Irrigation facilities	257.101**	129.107*	386.209**
Electricity	125.513	63.028	188.542
Road	-0.497	-0.249	-0.747
Population	0.148***	0.074**	0.223***
Farmer's group	97.084	48.752	145.836
Market center	240.485**	120.763*	361.249**

*** $p<0.01$, ** $p<.5$, * $p<0.1$. This table shows the output of the marginal effects from the SAR model. The first column lists the average direct impact (ADI); the second lists the average indirect impact (AII); while the last column is the average total impact (ATI).

Appendix Table 2. Marginal impacts (Simulated z-value) – spatial lag model

Variables	Direct	Indirect	Total
Spring temperature	-1.906	-1.688	-1.881
Spring temperature sq.	1.872	1.657	1.845
Summer temperature	2.106	1.776	2.052
Summer temperature sq.	-1.791	-1.760	-1.754
Autumn temperature	-0.929	-0.863	-0.916
Autumn temperature sq.	0.603	0.562	0.594
Winter temperature	0.721	0.702	0.721
Winter temperature sq.	-0.902	-0.868	-0.901
Spring rainfall	1.679	1.471	1.638
Spring rainfall sq.	-1.528	-1.340	-1.486
Summer rainfall	1.696	1.498	1.662
Summer rainfall sq.	-1.292	-1.184	-1.275
Autumn rainfall	3.257	2.589	3.182
Autumn rainfall sq.	-5.702	-3.302	-5.157
Winter rainfall	4.108	2.652	3.741
Winter rainfall sq.	-3.741	-2.549	-3.463
Spring temperature dev	1.056	0.993	1.050
Summer temperature dev	1.098	1.015	1.086
Autumn temperature dev	-2.011	-1.707	-1.962
Winter temperature dev	0.318	0.305	0.316
Spring rain dev	-0.641	-0.610	-0.635
Summer rain dev	0.926	0.845	0.910
Autumn rain dev	-1.233	-1.159	-1.228
Winter rain dev	2.179	1.839	2.124
SDII	-1.595	-1.408	-1.564
WSDI	2.393	1.926	2.304
Irrigation facilities	2.075	1.755	2.022
Electricity	0.959	0.902	0.951
Road	0.032	0.026	0.030
Population	3.255	2.442	3.119
Farmer's group	0.676	0.645	0.672
Market center	2.743	2.133	2.631

FOOTNOTES

[1] We excluded Nepal Living Standard Survey 1996/1997 (NLSS-I) from this study due to the lack of common identifiers of NLSS-I with NLSS-II and

[2] The data obtained from NLSS in this study are the self-reported data by a knowledgeable person in a community (PSU).

[3] The PSU identifier for the rural NLSS survey are either individual wards or sub-wards, or group of contiguous wards in the same village development

committee (VDC). Wards are the smallest denomination of administrative units in Nepal and are equivalent to zip codes in the United States. Likewise, VDC is the lower subdivision of a district and is similar to municipalities. Each VDC is further subdivided into several wards.

4 The analysis of the spatial econometric model is more reliable using a balanced panel data, which restricted us to 155 PSUs to create a balanced data. Of the 155 PSUs, 100 PSUs were common in NLSS II and III, while we included another 55 PSUs to increase the sample size. In order to obtain the additional 55 PSUs, we selected only those PSUs that were adjacent to each other in NLSS II and III. So, if there was a particular PSU in NLSS-II and its neighboring PSU was used in NLSS-III, we considered the two neighbors as the same PSU. In this way, we came up with the 55 additional PSUs. It should be noted that we also ran our final analysis with only the 100 original PSUs, and the results did not substantially change from the findings presented in this paper (with 155 PSUs)

5 In order to minimize the distance between PSU and weather stations, we extracted those weather stations that were within 10 km radius from a particular PSU of interest. However, if there were no weather stations within that radius, we extracted the stations that were within 25 km radius.

6 Spring season = March-May; Summer season = June-August; Autumn season = September-November; Winter season = December-February.

7 For example, the standard deviation for summer temperature in the year 2002 was constructed as follows:

8 WSDI and SDII indices were derived using the "climdex.pcic" package available in R.

9 While several Ricardian studies use soil type as another set of the control variable, it has been excluded in this paper due to the nature of the econometric model specification. This paper employed fixed effects model for both the non-spatial and spatial-analyses and as a result, time invariant factors like soil type have been ruled out from the analyses since these estimates are washed away.

10 The Hausman test rejected the null hypothesis of random effects (Chi-square = 49.19; p-value = 0.011). Thus, our model specification takes the fixed effects form.

11 Climate normal: = {spring temperature, summer temperature, autumn temperature, winter temperature, spring precipitation, summer precipitation, autumn precipitation, winter precipitation}

12 Weather deviation and extremes: = {spring temperature deviation, summer temperature deviation, autumn temperature deviation, winter temperature deviation, spring precipitation deviation, summer

precipitation deviation, autumn precipitation deviation, winter precipitation deviation, WSDI, SDII}

13 PSU sociodemographic variables: = {access to irrigation facilities, access to electricity, access to market center, population of the PSU, access to paved road, existence of farmer's group}

14 Interested readers should refer to Elhorst, (2014) for greater detail on spatial panel data model.

15 It is also called marginal value or Ricardian climate sensitivities (Polsky, 2004).

16 We tested for the presence of multicollinearity, and the variance inflation factor of independent variables were less than 10, thus mitigating the concern for collinearity (Meyers, 2000). Additionally, the presence of multicollinearity would lead to unstable regression coefficients and large standard errors (Cohen et al., 2013), neither of which occurred in our analysis.

17 We also tested for the presence of spatial correlation and lag simultaneously using the mixed spatial model, but the spatial effects were not significant in the joint model.

18 The turning point from column (1) and (2) were similar to column (3). We used column (3) for the interpretation since that is the final model used.

19 Appendix Table 2 in the appendix shows the simulated z scores for the Marginal Impacts from the spatial lag model listed in Appendix Table 1.

CHAPTER 8

Spatial market integration and price transmission for papaya markets in Ethiopia

Zewdie Habte

Department of Agricultural Extension, Wolaita Sodo University, Ethiopia.

ABSTRACT

This work aims to study the integration of papaya market, price transmission and price causality patterns with the help of Johansen co-integration test, vector error correction model and Granger causality test using 13 years average monthly prices of papaya. Johansen co-integration tests indicate that four papaya markets significantly co-integrated with each other. Vector error correction (VEC) model test indicates that speed of papaya price adjustment for Arbaminch market was statistically significant at 1% level and the fastest compared to other payaya price adjustment. Its equilibrium price was stable. Whereas, speed of price adjustment for Adama market was insignificant and the slowest as compared to other market prices; its equilibrium was unstable because price change was away from equilibrium price. This implies that there was asymmetric information. The Granger causality test indicates that Arbamnch papaya price had bidirectional relationships with Merkato and Shashemenie markets. Concerned bodies should work on asymmetric information to address slow price adjustment between various papaya markets.

Keywords: Market integration, papaya, vector error correction model, Granger causality.

1. INTRODUCTION

Spatial market integration strengthens successful trade between food-deficit and food-surplus areas. This leads to specialization and economic growth. Moreover, market integration has a great contribution to food security and economic growth. It also improves producers' and consumers' social welfare, particularly in a diverse and highly vulnerable country such as Ethiopia. It accelerates effective price transmission between markets with the help of market reforms (Golettie and Babu, 1994). On the contrary, poor food price integration has unfavorable effects on social welfare of producers and consumers (Goletti et al., 1995). Poor market integration could also reflect the existence of imprecise price information, presence of either state policies or infrastructural and institutional problems that influence producers' market decisions and the efficient flow of goods between markets. The marketable surplus generated by farmers could then result in depressed farm prices and diminishing income (Tahir and Riaz, 1997).

Market integration is expected to ensure a more rapid and effective price adjustment between markets with the help of market reforms (Golettie and Babu, 1994). Thus, government has paid attention to agricultural market integration in developing countries (Abdulai, 2000; Van Campenhout, 2006; Amikuzuno, 2009).

Price is a basic means to tie different stages of a market chain. Price signals could be a good evidence of market segmentation or potential manipulation, distortion of price information which causes inefficient allocation of resources. These situations do not attract farmers to produce marketable surplus because of low price which causes low income (Tahir and Riaz, 1997). Price shocks are passed on from one stage to next stage of market chain and the extent of adjustment to such shocks constitutes important factors reflecting the actions of market participants at different market levels. The speed of price adjustment greatly relies upon the kind of products. Fruit markets have high probability to adjust price rapidly. On the contrary, in some level processed products and non-perishable products have low chance to adjust price. It is deemed that price adjustment between various stages in the market chain is not symmetric (Reziti and Panagopoulos, 2008). This means that positive and negative price shocks are not transmitted in the same way.

Moreover, in developing countries poor infrastructure and transport services results in large marketing margins due to high costs of delivering traded commodities. High transfer costs hinder the transmission of price signals, and they may prevent or discourage goods arbitrage (Sexton *et al.*, 1991). Investigation of market integration is useful to understand the function of market; design and adopt most suitable agricultural price stabilization policies (Seneshaw, 2013).

Thus, it is crucial to conduct this research because there is no empirical evidence about spatial market integration and its price transmission in Ethiopian papaya prices. Information on market integration is thus useful in making agricultural policies, including policies and strategies for price stabilization, price risk management and food security. Moreover, this study helps to come up with the latest, accurate, reliable, and prompt information about market integration of papaya prices across markets and speed of price transmission in Ethiopia. This result is also useful to generate information and fill knowledge gaps in the spatial papaya market integration and its price transmission.

Therefore, this study examines the degree of spatial domestic papaya market integration; its price transmission; and causality patterns using Johansen co-integration test, vector error correction model and Granger causality test.

2. REVIEW OF LITERATURE

Market integration can be classified into three: (1) vertical market integration which includes different stages in marketing and processing channels; (2) spatial integration which relates to spatially distinct markets; and (3) inter-temporal market integration which refers to arbitrage across periods of time (Barret as cited by Uchezuba, 2005). This study focuses on spatial integration of papaya prices among a number of markets in Ethiopia.

The spatial arbitrage condition is that market integration leads itself to a co-integration interpretation which is measured co-integration tests (Fackler and Goodwin, 2002). If two spatially separated price series are co-integrated, then there is a tendency for them to co-move in the long run according to a positive linear relationship. In the short run, the prices may drift apart, as shocks in one market may not be instantaneously transmitted to other markets due to delays in transport or information; however, arbitration opportunities ensure that these divergences from the underlying long run (equilibrium) relationship are transitory and not permanent.

The analysis of spatial market integration attempts to address three main questions about the nature of the price transmission process among spatially separated markets; such as causality patterns, dynamic interactions, or long-run equilibrium. Co-integration analysis is only concerned with the existence of long run equilibrium between markets, and cannot answer questions pertaining to the price adjustment process over time. But, the vector error correction (VEC) addresses question of the price adjustment process over time.

However, various authors postulated and applied various analysis methods to assess market integration. Also, Meyer and Von Cramon-Taubadel (2002) pointed out that there is still no unified approach to evaluate market integration. Market integration can be measured in different ways, including the movement of goods. Previously, price correlation coefficients were used to investigate whether or not markets were linked by price changes (Timmer, 1987; Dadi et al., 1992). However, price correlation coefficients may mislead the results due to the presence of trends (non-stationary data) in the data (Wyeth, 1992). Regression analysis has also been used to analyze integration (Alexander and Wyeth, 1994). This practice was modified using price variables in their first difference form, but this caused the loss of long-run information. Co-integration, on the other hand, allows a way of dealing with time series data that avoids spurious results, thus enhancing the validity of research findings. Johansen's approach to co-integration is now used widely to test the level of integration among markets. Several authors applied threshold autoregressive models to deal with asymmetric price transmission in agricultural marketing. Unit root test and Co-integration test do not have power to check the prevalence of asymmetric adjustment because they assume symmetric and linear adjustment. So, Enders and Siklos (2001) suggest an extension to standard ECM which appears in the literature as threshold autoregressive (TAR) model. However, the TAR models have calculation challenges and impose ex-ante non-theoretical restrictions. Additionally, TAR models are suggested to check the existence of non-linear transaction costs and price bands. Therefore, Johansen's approach to co-integration, VECM and Granger causality are proposed for this study to investigate market integration, price adjustment and causality patterns of papaya markets in Ethiopia respectively.

3. RESEARCH METHODOLOGY

3.1 Source of data

The study aimed to analyze the degree of market integration and price adjustment of papaya markets among three regions in Ethiopia. Average monthly retailer prices with 156 total observations were gathered from CSA data bases from September 2002 to September 2014. Five markets for price of papaya were selected for this study based on the availability of data, surplus and deficit markets.

3.2 Methods of analysis

The time series data are stationarity when conditional mean, variance, and auto-covariance are constant over time. If x and y have unit root, the standard *t*-test is invalid, which leads to equation 1 to come up with a spurious regression.

$$Yt = \alpha + \beta Xt + \varepsilon t \qquad yt = \alpha + \beta x_t + \varepsilon_t \tag{1}$$

Time series is non-stationarity when conditional mean, variance and auto-correlation are not constant over time. If they are not constant over time, then the series is said to be a non-stationary process (i.e. a random walk/has unit root). Differencing a series using differencing operations produces other sets of observations such as the first-differenced values, the second-differenced values and so on.

If a series is stationary without any differencing it is designated as I (0), or integrated of order 0. On the other hand, a series that has stationary first differences is designated I (1), or integrated of order one (1). Augmented Dickey-Fuller test has been suggested by Dickey and Fuller, (1979) while the Phillips-Perron test recommended by Phillips and Perron (1988) has been used to test the stationarity of the variables. The price variables have been tested for unit roots by the Augmented Dickey-Fuller (ADF) test with different specifications, with trend and constant:

$$\Delta Yt = \alpha + \beta Yt - 1 + \delta t + \Delta Yt - 1 + \zeta 2\Delta Yt - 2 + \zeta k\Delta Yt - k + \varepsilon t$$

$$\Delta y = \alpha + \beta y_{t-1} + \sum_{l=1}^{r} \gamma_l \Delta y_{t-1} + \varepsilon_t \tag{2}$$

Where α, β and $\delta\gamma$ are coefficients, r is the number of lagged variables specified and e_t is the random term to be estimated and tested. This test statistic is probably the best-known and most widely used unit root test. It is a one-sided test whose null hypothesis is β=0 versus the alternative β< 0 (and hence large negative values of the test statistic lead to the rejection of the null). Under the null, y_t must be differenced at least once to achieve stationarity; under the alternative, y_t is already stationary and no differencing is required.

If all the variables are stationary, the VAR can be used; OLS can also be used to estimate each equation and standard statistical methods can be employed. If some of the original variables have unit roots and are not co-integrated, then the ones with unit roots should be differenced and the resulting stationary variables should be used in the VAR. If the variables have unit roots and are co-integrated, the vector error correction model should be used.

3.3 Johansen and Juselius cointegration test

Johansen and Juselius (1990) suggest Maximum Eigenvalue test and the Trace test to determine the number of co-integration vectors. The Maximum Eigenvalue statistic tests the null hypothesis of r co-integrating relations against the alternative of r+1 co-integrating relations for r = 0, 1, 2...k-1. This test statistics are computed as:

$$Jmax(r) = -T * ln(1 - \lambda r + 1) \tag{3}$$

Where λ is the estimated Maximum Eigenvalue and T stands for the sample size. The trace test conducts a joint test whereas the maximum Eigenvalue test carryout separate tests for the individual eigenvalues. Trace statistics examines the null hypothesis of r co-integrating relations against the alternative of n co-integrating relations, where n is the number of variables in the system for r = 0, 1, 2...k-1. It is formulated as follows:

$$Jtrace(r) = -T * \sum_{i=r+1}^{k} ln(1 - \lambda_i) \tag{4}$$

The results of trace test are preferred while Trace and Maximum Eigenvalue statistics come up with different results in some case (Alexander, 2001). If a long-term equilibrium relationship exists between time series data, price adjustment is conducted to evaluate the short run properties of the co-integrated series with the help of VECM. VECM is not needed to carry out, if time series data not co-integrate.

3.4 Vector error correction model (VECM)

VECM can be applied to measure price adjustment. Adjustment of prices induced by deviations from the long-term equilibrium (ECT) is assumed to be a continuous and linear function of the magnitude of the deviation from long-term equilibrium. Thus, even very small deviations from the long-term equilibrium will always lead to an adjustment process in each market. If time series data are co-integrated this implies that there exists a long-term equilibrium relationship between them so VECM can be applied to evaluate the short run properties of the co-integrated series. If co-integration is not absent between series, VECM will no be longer required. So, Granger causality tests are directly applied to see causal relationship between variables. Given the following general specification of the VECM model which considered with VAR.

$$y_t = \delta + A_1 y_{t-1} + A_2 y_{t-2} +, \cdots, A_{p-1} y_{t-p+1} + \varepsilon_t \tag{5}$$

Yt = δ+Where Yt is an (n x 1) vector of endogenous variables (prices), δ is an (n x 1) vector of parameters, y and y_{t-p} are lagged values of prices; Ai represents (n x n) matrices of parameters, and εt is an (n x 1) vector of random variables. In this model, the price series for the five papaya markets were endogenous variables and as such no exogenous variable was used. To test the hypothesis of integration and co-integration in Equation 5, we transform it into its vector error correction form.

$$\Delta Y_t = \alpha + \Gamma_1 \Delta Y_{t-1} + \Gamma_2 \Delta Y_{t-2}, \ldots, \Gamma_{k+1} \Delta Y_{t-k+1} + \Pi \Delta Y_{t-k} + \varepsilon$$

$$\Delta Yt = \mu + \Gamma 1 \Delta \tag{6}$$

Where Y_t=[P1t, P2t]', vector of endogenous variables, which are I(1), Δ Yt= Yt- Y_{t-1}, a is a (2×1) vector of parameters, $\Gamma_1, \ldots, \Gamma_{k+1}$ and π are (2×2) matrices of parameters, and εt is a (2×1) vector of white noise errors. Where π is of a reduced rank, that is r≤1, it can be decomposed into π =αβ' and when r=1, α= [α_1, α_2]' is the adjustment vector and β= [β_1, β_2]' is the co-integrating vector.

The existence of co-integration and error correction representation implies the existence of causality in at least one direction (Granger, 1988). However, co-integration itself is not useful to deduce about the direction of causation between series. Thus, Granger causality test is necessary. Granger causality test focuses on the presence of unidirectional causality linkages as an indication of some extent of integration (Gupta and Mueller, 1982). Bi-directional causality implies that both markets depend each other on market price information to set their price while unidirectional causality implies only one market uses other market price information to set its own price. Granger causality specification for co-integrated variables is written as:

$$\Delta P_{i1} = \alpha + \sum_{i=1}^{n} \beta_i \Delta P_{(t-1)i} + \sum_{i=1}^{n} \beta_i \Delta P_{(t-1)j} + \phi ECT_{t-1} + \varepsilon_i \tag{7}$$

Where Δ is the difference operator, P_{i1} and P_{j2} are papaya prices; are white noise error terms, ECT_{t-1} is the error correction term derived from the long-run co-integrating relationship, while r is the optimal lag of the variables which are chosen with the help of Akaike criterion (AIC), Schwarz Bayesian criterion (BIC) and Hannan-Quinn criterion (HQC) (Hendry and Ericsson, 1991).

4. RESULTS AND DISCUSSION

4.1 Stationarity test

For co-integration analysis, it is important to test the unit roots with the help of the Augmented Dickey- Fuller (ADF) at the beginning to check whether modeled variables I (0) at levels and I (1) at first differences were stationary or non- stationary. The tests were applied to each variable over the period of 2001-2013 with and without constant at the variables level and at their first difference.

The result in Table 1 indicates that the null hypothesis of no unit roots for all the time series were rejected at their levels. On the other hand, the all variables were stationary and integrated of same order, that is, I (1) at their first difference for both with and without constant, which means unit roots in the first differences were rejected at 1%. Therefore, the results allow proceeding for co-integration tests for the testing of the long run equilibrium relationship.

Moreover, according to Mesike et al. (2010), any endeavor to determine the dynamic function of the variable in the level of the series based on results of the variables are I (1) and I (0) will be inappropriate and may lead to problems of spurious regression. The econometric results of the model cannot be used for prediction in the long-run in that level of series because it will not be ideal for policy making (Yusuf and Falusi, 1999).

ADF test results enable researcher to conduct Johansen co-integration test which is suitable to see the existence of long-run relationships among variables because they fulfill the precondition for co-integration analysis. In this study, the optimal number of lag for VAR model was determined based on value of Akaike criterion (AIC), Schwarz Bayesian criterion (BIC) and Hannan-Quinn criterion (HQC).

The result in Table 2 shows that candidate of optimal lag of the AIC is lag 4; optimal lag of SBIC and HQIC is lag 1. As we can see in Table 2, there is more than one candidate of optimal lag exist, so the value of R^2 from the VECM analysis was checked with lag 1 and lag 4. Based on the results of VECM analysis, lag 4 is found to be optimal lag for this model because it yields higher R^2.

Table 1. ADF unit root test results for papaya prices.

Variables	Level				Difference			
	Without constant		With constant		Without constant		With constant	
	Test statistic	P-value	Test statistic	P- value	Test statistic	P-value	Test statistic	P-value
Adama	0.9164^{ns}	0.904	-0.486	0.891	-4.778	0.000	-4.933	0.000
Shashemene	0.7639	0.878	0.655	0.855	-6.283	0.000	-6.431	0.000
Awassa	3.196	0.999	1.734	0.999	-5.925	0.000	-6.428	0.000
Arbamnch	1.357	0.956	-0.594	0.869	-7.881	0.000	-8.294	0.000
Merkato	1.980	0.989	0.736	0.993	-5.935	0.000	-6.265	0.000

Source: Computed from data in Central Statistic Agency (CSA) of Ethiopia.

Table 2. Lag-order selection criterion.

Lags	AIC	HQIC	SBIC
1	14.247	14.844*	14.489*
2	14.254	15.348	14.699
3	14.040	15.631	14.686
4	13.900*	15.989	14.749

Source: Compiled from data in Central Statistic Agency (CSA) of Ethiopia.

4.2 Johansen co-integration tests

To state a co-integration model, Johansen's testing procedure was followed. Each co-integrating equation has an intercept and a slope coefficient. The null hypotheses for the trace test are rejected at the 10% level of significance; we reject the null hypotheses that r=0 and r <1, but we failed to reject the null hypothesis that the co-integrating rank of the system is at most two.

Johansen's trace and -max tests rejected first four hypotheses (*r*= 0 to 3) of no co-integrating vector at 1% level of significant; Johansen trace statistic rejected 0-3 hypotheses(r=0, r=1 r=2, r=3) at 1% level of significant. In other words, this trace test result rejected the null hypotheses because these four variables were co-integrated (Table 3). These results suggest that there are four long-run equilibrium relationships between the five price series.

4.3 Vector error correction model

The ADF test results approve that a VEC model is more pertinent than a vector autoregression model to distinguish the multivariate interactions among the three price series (Engle and Granger, 1987). That is, all price series data which have unit roots are more pertinently examined the existence of a number of long run co- integration vectors than a vector autoregression model.

The presence of co-integration between variables suggests a long term relationship among the variables under consideration. The coefficient of price adjustment with negative sign indicates a backward movement towards equilibrium; a positive sign indicates movement away from equilibrium. The coefficient should lie between 0 and 1, 0 suggesting no adjustment one time period later, 1 indicates full adjustment. The coefficients of the error correction term show the speed of convergence to the long run equilibrium as a result of shock of their own prices.

Table 3. Results of Johansen co-integration tests for five papaya market prices

Sample: 2002: 1 - 2014:01 Trend: trend	Number of observation =156 Lag=4				
Rank	Eigenvalue	Trace statistic	p-value	Lmax test	p-value
0	0.35575	172.13	0.0000	67.710	0.0000
1	0.25116	104.42	0.0000	44.542	0.0000
2	*0.20121*	59.876	0.0000	34.597	0.0002
3	0.14904	25.279	0.0009	24.854	0.0005
4	0.0027559	0.42500	0.5145	0.42500	0.5145

Source: Computed from data in Central Statistic Agency (CSA) of Ethiopia.

Table 4. Result from vector error correction model for adjustment vectors of papaya prices.

Places	Cointegrating vectors(β)		Adjustment vectors		Adjusted R^2	Durbin Watson
Arbamnch	1.0000 0.00000	(0.00000) (0.00000)	-0.47***	0.0425***	0.2095	2.1817
Merkato	0.00000 1.0000	(0.00000) (0.00000)	0.22591	-0.033019	-0.00553	2.0898
Adama	-0.010136 1.0898	(0.056557) (0.18389)	0.10344	-0.049172	-0.00324	2.6603
Shashemenie	-0.31587 -4.0867	(0.094245) (0.30643)	0.35 ***	0.222***	0.41484	2.1112
Awassa	-0.27085 0.49390	(0.099869) (0.32471)	0.427***	0.020635	0.1379	2.2619

*** and ** stands for 1 and 5% significance levels, respectively. Source: Computed from data in Central Statistic Agency (CSA) of Ethiopia

In this study, coefficient of dynamic adjustments that is obtained with the help of the VEC model analysis is used to estimate the speeds of price transmission. The results of the speeds of adjustment/adjustment vectors are displayed in Table 4. The speeds of adjustment for Arbamnch retail papaya price were statistically significant at 1% level. The speeds of adjustment for Adama and Merkato the retail price were not statistically significant. The estimate of the error correction coefficients for the selected papaya markets indicates that Shashemenie market was significant at 1% with a wrong sign (positive). This shows that any disequilibrium in the long run retailer price would be corrected in the short run; thus, the short run price movements along the long run equilibrium path may be unstable (Table 4). The coefficient of adjustment vector (α) for Awassa market has a wrong sign (positive) and significant at 1% level showing that the short run price movements along the long run equilibrium path may be unstable.

The speed of adjustment for Arbamnch papaya price has the expected negative sign because of the overreaction of prices in the short run in response to an exogenous shock. The dynamic speed of adjustment for the Arbaminch price was higher (0.47), in absolute value, than other papaya market prices, an indication of asymmetric price transmission with respect to speed. This is an interesting result suggesting that with the safety shock, Arbamnch prices adjust more quickly and are more flexible than farm prices to restoration in the long-run equilibrium. This result is also important for policy makers and agribusinesses and has clear implications for the efficiency and equity of the Ethiopia papaya marketing system. It indicates that the speeds of price adjustment are not the same in different markets. Prices in the Arbamnch market adjust more quickly than prices at other market in response to the safety shock.

Even if we demonstrate market integration through co-integration, there could be disequilibrium in the short run, i.e., price adjustment across markets may not happen instantaneously. It may take some time for spatial price adjustments to occur. The error correction model takes into account the adjustment of short-run and long-run disequilibrium in markets and time to remove disequilibrium in each period. In terms of efficiency, prices are transmitted fully and completely given efficient market conditions. The fact that price dynamics differ might point to noncompetitive market conditions that can lead to market inefficiencies. It is important to note that our analysis cannot directly test for imperfect competition and does not explicitly address imperfect competition. Future research and modeling efforts are required to address this hypothesis directly and appropriately.

4.4 Granger causality tests

Granger causality was also estimated between pairs of papaya market prices. Granger causality means the direction of price formation between two markets and related spatial arbitrage, i.e., physical movement of the commodity to adjust for these prices differences. The findings in Table 5 indicate that there was bidirectional causality between Arbamnch and Merkato papaya prices. That is, the Arbamnch Granger caused price formation in the Merkato papaya market which in turn provided feedback to the Arbamnch base market as well. On other hands, Awassa had unidirectional relationships with both Shashemenie and Adama markets. Arbamnch had also unidirectional relationships with Adama and Awassa, the base markets. This implies that the Arbamnch papaya market relied on Adama and Awassa papaya price information to set its own price. But, Adama and Awassa papaya market prices did not depend on Arbamnch papaya price to fix price.

CONCLUSION AND RECOMMENDATIONS

This paper investigates spatial market integration, price transmission and causality directions in papaya markets with the help of co-integration and VECM and Granger Causality test. Johansen's Trace and -max tests indicate that there was long run co-integration among four papaya prices at 1% level of significant. Papaya price for Arbamnch retailer market removed 47% of disequilibrium, and the remaining was corrected by the external and internal forces. Further research is required to go through the influence of external and internal factors such as market infrastructure, government policy and other factors towards market integration.

In general, speed of price transmission were slow for almost all papaya prices may be for various reasons such as transportation costs, imprecise price information, and lack of good government policies, infrastructural and institutional arrangement. So government should create conducive policy environments that improve good flow of price information; work on infrastructural accessibility and institutional arrangement to reduce transaction costs.

Table 5. Granger causality from error correction model.

Pair wise Granger causal test		F-Statistics	Probability
Merkato	Arbaminch	3.12	0.017
Arbamnch	Merkato	3.487	0.0095
Adama	Arbaminch	2.064	0.0885
Arbaminch	Adama	3.62	0.0077
Shashemenie	Arbaminch	2.56	0.038
Arbaminch	Shashemenie	5.31	0.0005
Awassa	Arbaminch	2.345	0.0557
Arbamnch	Awassa	4.41	0.0021
Adama	Merkato	4.46	0.002
Merkato	Adama	8.97	0.0000
Shashemenie	Merkato	3.51	0.009
Merkato	Shahemenie	9.333	0.0000
Awassa	mekato	4.46	0.002
Merkato	Awassa	6.7	0.000
Shahemenie	Adama	6.033	0.0002
Adama	Shahemenie	1.712	0.149
Awassa	Adama	3.654	0.0072
Adama	Awassa	6.999	0.0000
Awassa	Shahemenie	0.827	0.509
Shahemenie	Awassa	6.82	0.000

Source: Computed from data in Central Statistic Agency (CSA) of Ethiopia.

REFERENCES

1. Abdulai A (2000). Spatial price transmission and asymmetry in the Ghanaian maize market. J. Dev. Econ. 63:327-349.
2. Alexander C (2001). Market Models: A Guide to Financial data Analysis. John Wile Chichester.
3. Alexander C, Wyeth J (1994). Co-integration and market integration: An application to the Indonesian rice market. J. Dev. Stud. 30:303-328.
4. Amikuzuno J (2009). Spatial price transmission and market integration between fresh tomato markets in Ghana: Any benefit from trade liberalization. Department of Agricultural Economics and Extension, University for Development Studies. Tamale, Ghana.
5. Dadi L, Negassa A, Franzel S (1992). Marketing of Maize and Tef in Western Ethiopia, Implications for Policy Following Liberalization: Food Policy 17:201-213.
6. Dickey D, Fuller W (1979). Distribution of the estimators for autoregressive time series with a Unit Root. J. Am. Stat. Assoc. 74:427-431.
7. Enders W, Siklos PL (2001). Co integration and threshold adjustment. Journal of business and Economics Statistics. 20:166-176.
8. Engle RF, Granger CWJ (1987). Cointegration and error correction: representation, estimation and testing, Econometrica 55:251-276.
9. Fackler PL, Goodwin BK (2002). Spatial Price Analysis. In B.L. Gardner and G.C. Rausser, eds. handbook of agricultural economics. Amsterdam: Elsevier Science.
10. Goletti F, Babu S (1994). Market liberalization and integration of maize markets in Malawi. Agric. Econ. 11:311-324.
11. Golettie F, Raisuddin A, Farid N (1995). Structural determinants of market integration: The case of rice market in Bangladesh. Dev. Econ. 33:185-202.

12. Granger CWJ (1988). Some recent developments in the concept of causality. J. Economet. 39:199-211.
13. Gupta S, Mueller RA (1982). Analysing the pricing efficiency: concepts and application. Eur. Rev. Agric. Econ. 9:301-312.
14. Hendry DF, Ericsson NR (1991). An Econometric analysis of U.K. money demand in monetary trends in the United States and the United Kingdom by Milton Friedman and Anna J. Schwartz. Am. Econ. Rev. 81:8-38.
15. Johansen S, Juselius K (1990). Maximum likelihood estimation and inference on cointegration with application to the demand for money. Oxford Bull. Econ. Stat. 52:169-210.
16. Mesike CS, Okoh RN, Inoni OE (2010). Supply response of rubber farmers in Nigeria: An application of vector error correction model. Agric. J. 5:146-150.
17. Meyer J, von Cramon-Taubadel S (2002). Asymmetric price transmission: a survey. In: Proceedings of the Xth EAAE Conference, Zaragoza.
18. Phillips PC, Perron P (1988). Testing for a unit root in time series regression. Biometrika 75:335-346.
19. Reziti I, Panagopoulos Y (2008). Asymmetric price transmission in the Greek Agri-Food Sector: Some Tests. Agribusiness 24:16-30.
20. Seneshaw T (2013). Spatial integration of cereal markets in Ethiopia, Ethiopia Strategy Support Program, Ethiopian Development Research Institute, ESSP Working Paper 56.
21. Sexton R, Kling C, Carman H (1991). Market integration, efficiency of arbitrage and imperfect competition: methodology and application to US celery. Am. J. Agric. Econ. 73:568-580.
22. Tahir Z, Riaz K (1997). Integration of agricultural commodity market in Punjab. Pak. Dev. Rev. 36:241-262.
23. Timme PC (1987). Corn marketing and the balance between domestic production and consumption. Working Paper No. 4, BURLOG-Stanford Corn Project.
24. Uchezuba DI (2005). Measuring market integration for apples on the South African fresh produce market: A threshold error correction model. Master's Thesis. University of the Free State Bloemfoentein, South Africa.
25. Van Campenhout B (2006). Modeling trends in the food market integration: Method and an application to Tanzanian maize markets. Food Policy 32:112-127.
26. Wyeth J (1992).The measurement of market integration and application to food security policies. Discussion paper 314, Brighton: Institute of Development Studies, University of Success.
27. Yusuf SA, Falusi AO (1999). Incidence analysis of the effects of liberalized trade and exchange rate policies on cocoa in Nigeria: An ECM approach. Rural Econ. Dev. 13:3-14.

CHAPTER 9

Impacts of corn price and imported beef price on domestic beef price in South Korea

GwanSeon Kim andTyler Mark

Department of Agricultural Economics, University of Kentucky, Lexington, KY, USA

AUTHORS' CONTRIBUTIONS

GK is the primary architect of this manuscript. He is responsible for collecting the data, modeling of the data, and drafting the manuscript. TM's primary role in the paper is revision of the draft manuscript and guidance on additional modeling questions. Both authors have read and approve the final manuscript.

ABSTRACT

South Korea is one of the largest beef importing countries in the world. Australia, USA, and New Zealand provide the majority of beef they import. The objective of this study is to evaluate the price relationship of Korean cattle and corn, prices of the imported beef and corn, and prices of Korean cattle and imported beef. We examine threshold points using threshold vector autoregressive (TVAR) model by assuming the presence of nonlinearity. We find that there are two thresholds in Australian beef price with first difference data. This study finds that the current Korean beef price is positively affected by increasing in the previous month of Australian beef price in first and second regimes. Changing in previous corn price has no significant impact on current Korean cattle price in the second regime. In the third regime, we find that current Korean cattle price is positively affected by previous corn price whereas no impact from previous Australian beef price. In this study, we could not use data for the USA due to the periods of bovine spongiform encephalopathy (BSE). However, we expect that current Korean beef price is significantly affected by changing in the previous month of US beef price using data after the periods of BSE.

Keywords: Korean beef prices, Structural change, Threshold, BSE

1. BACKGROUND

South Korea is the fifth largest beef import country out of 58 countries, which is equivalent to 7% of the world, followed by USA, Russia, Japan, and Hong Kong (Cook 2016). In 1976, Korea started to import beef, and the domestic market was fully opened due to the tariff and import quotas by Uruguay Round of Multinational Trade Negotiation (UR) in 2001. According to the Korea Institute for Animal Products Quality Evaluation (KAPE), imported beef quantities in 2013 were 56% from Australia, 35% from USA, 9% from New Zealand, and 1% from Canada, respectively (KAPE 2013). However, the USA was the largest exporter of beef to South Korea before 2003. In 2003, a dairy cow in Washington State was discovered to have bovine spongiform encephalopathy (BSE), and Korea banned beef imports from the USA on December 2003 (Giamalva 2013, Park et al. 2008). Based on US Department of Agriculture, Economic Research Service (USDA ERS 2012), beef imports from Australia and New Zealand have increased since December 2003 due to the BSE outbreak in the USA. This ban lasted until July 2008 when the ban was lifted and the USA resumed exporting beef to South Korea.

According to the Ministry of Agriculture, Food, and Rural Affairs in South Korea, the proportion of imported formula feed is 53% in total feed; in addition, a proportion of imported corn is 79% of total feed grains (MAFRA 2015). Corn is one of the most important feed grains used in beef production. Hebert and Anderson (2011) mention the exogenous effect arising as a consequence of a negative relationship between corn and beef prices. In this study, we hypothesize that change in corn prices will significantly and directly affect both imported beef prices from Australia and prices of Korean native cattle. Given that Korea has to import a large percentage of their feed used in Korean native cattle, small changes in the world corn prices could have significant impacts on the cost of production both within and outside of Korea. In addition, Korean beef prices are also affected by changes in imported beef prices. Figure 1 shows how corn, wheat, and soybean prices are changed over time from 1960 to 2015. Based on Fig.1, prices of corn, soybean, and wheat move together, so focusing on changes in corn prices will provide information on how prices of beef will be affected if there is a shock in corn price. Table 1 shows the results from the correlation test between the commodities, and the results show that there is a strong and positive correlation among the commodities.

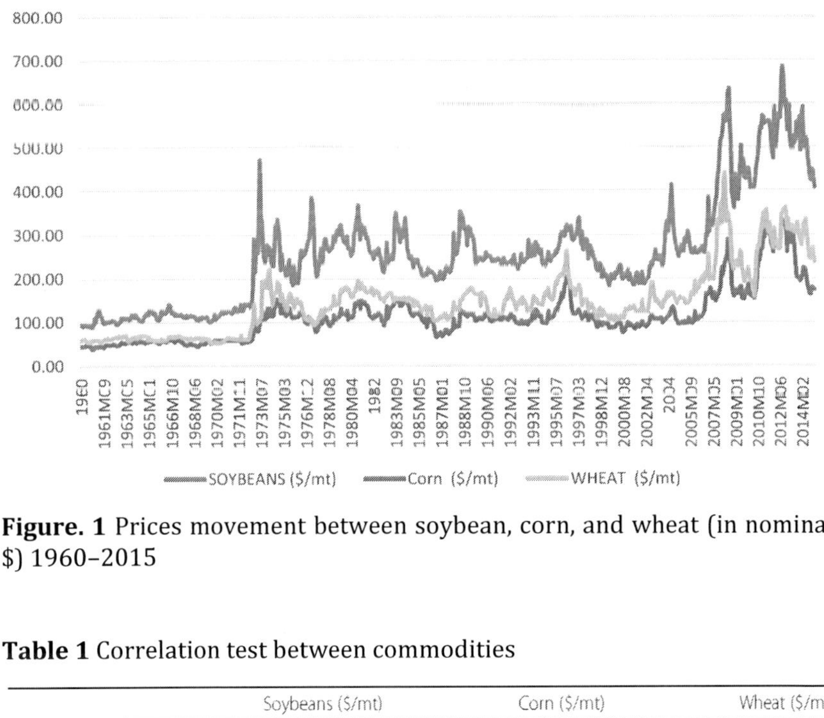

Figure. 1 Prices movement between soybean, corn, and wheat (in nominal $) 1960–2015

Table 1 Correlation test between commodities

	Soybeans ($/mt)	Corn ($/mt)	Wheat ($/mt)
Soybeans ($/mt)	1		
Corn ($/mt)	0.939	1	
Wheat ($/mt)	0.937	0.935	1

In this study, we focus on imported beef prices from the Australia because the imported beef quantities from Australia are relatively higher than other countries: 20.91% higher than USA, 47.12% higher than New Zealand, and

55.22% higher than Canada. In addition, data for the imported beef prices and quantities from USA are not available from 2003 to 2008, due to the ban on US beef (Marsh et al. 2008). By 2013, Korea had increased imports to 35% from the USA, making it the second largest exporter of beef to Korea.

The primary objective of this study is to evaluate the relationship between Korean native cattle prices and imported corn and beef prices. Specifically, we evaluate the presence of structural break points by assuming that there are structural changes due to the BSE outbreak. The outbreak of BSE created a fundamental change in the Korean beef market in 2003. With the announcement of the BSE outbreak, the beef imports from Australia began to increase, changing the relationship between Korean native cattle prices and corn and beef import prices. Therefore, we hypothesize the relationship before and after certain structural breakpoints are statistically different.

Studies that examine BSE impact utilizing structural breaks are limited. Sanjuan and Dawson (2003) investigated producer-retail price transmission for beef, lamb, and pork by considering the BSE crisis in the UK. Jin et al. (2008) examine the impact of North American BSE cases on daily live cattle future prices. However, the literature on overall BSE impact is more developed. Lloyd et al. (2006), Saghaian (2007), and Hassouneh et al. (2010) investigate the impact of BSE on price transmissions in the beef market. Both Lloyd et al. (2006) and Hassouneh et al. (2010) address issues of human health and food scares within the UK and Spain by examining the impact of BSE on the retail and producer levels. They find different impacts between beef producers and retailers. However, the main difference between studies is that Lloyd et al. (2006) did not consider any threshold component. Saghaian (2007) evaluated the implications of the 2003 BSE outbreak on a feedlot, wholesale, and retail beef prices. They find that beef prices overall were negatively affected by the BSE discovery even though the magnitude of price effects are different for each price series. Park et al. (2008) investigate impacts of animal disease outbreaks on Korean meat market. They employ a vector error correction (VEC) model by applying known structural break points where the animal diseases are actually discovered and then conduct out-of-sample forecasting for meat prices after the period of the event. Based on our knowledge, no study investigates how domestic beef price in South Korea is affected by imported beef price and imported corn price by considering structural change due to BSE.

Furthermore, we employ a threshold framework to examine nonlinearity between the variables by hypothesizing that price effects are statistically different at certain threshold points. We assume that the structural break is a consequence of the import ban, based on the literature and increases in Australian beef imports. Nonlinearity is considered from a theoretical standpoint, and empirical studies have also shown its existence in time series data. Goodwin and Holt (1999) and Goodwin and Harper (2000) state the potential issues of nonlinear and threshold type adjustments in error correction models have been recognized in recent time series analysis. In addition, the dynamic price responses may be caused by a nonlinear nature due to the different combinations of adjustments from alternative regimes by threshold specifications. Especially, Goodwin and Harper (2000) point out that many recent time series models examining price transmission consider asymmetric adjustments assume that the price transmission process to be fundamentally linear. Also, Zivot and Wang (2007) mention that nonlinear behavior in economic time series allows not only the presence of different regimes but also the dynamics to be different in different regimes. In addition, Luoma et al. (2004) argue that linear models in a nonlinear situation result in the estimation of incorrect inferences.

Other studies that investigate price relationship of meat utilizing a threshold model are as follows. Abdulai (2002) examines the relationship between producer and retail pork prices in Switzerland using threshold autoregressive (TAR) and momentum-threshold autoregressive (M-TAR) models. Based on TAR and M-TAR, the author finds evidence of asymmetric pricing behavior on the part of pork retailers. Luoma et al. (2004) investigate the asymmetric price transmission between producer and consumer in Finnish beef and pork markets by using monthly price data from January 1981 to May 2003. They employ a Bayesian multiple-regime vector error correction model. Serra et al. (2006) investigate price transmission relationship between EU pork markets using nonlinear methods from 1994 to 2004. They employed a nonparametric technique (LLR) and compared it with a parametric technique (TAR). Based on both parametric and nonparametric techniques, they find that Spain and Germany, which are two leading EU pork producers, have symmetric price adjustment whereas price transmission processes are asymmetric between Germany, Demark, and France.

This study provides the following contributions to the literature. First, it provides economic implications of the domestic price dynamics in response to the 2003 BSE outbreak. Price dynamics are explored for changes in world corn prices and their transmission into domestic Korean beef prices and imported beef prices. Secondly, this study evaluates the presence of threshold nonlinearities. Finally, based on our knowledge, no study investigates and finds the impact of imported Australian beef price and corn price on the Korean beef market with different threshold points under the assumption of the nonlinearity.

1.1 Data

Data for this study comes from multiple sources. Table 2 includes the variable descriptions and summary statistics for the key variables. The primary variables are the Korean native beef price, world corn price, and the imported Australian beef price. Data for Han-Woo prices or Korean native beef price is from the KAPE. Imported beef prices are sourced from Korea Customs Service (KCS). Corn prices are obtained from the World Bank and adjusted using the real exchange rate from the Bank of Korea (BOK). Since there is limited data availability especially for Han-Woo prices, we use monthly time series data from 2000 to 2014. An initial observation is that the price of Korean beef is almost three times higher than the price of the imported Australian beef. There are two possible reasons to explain the price difference between Korean beef and imported Australian beef. First, South Korea does not put the cattle out to pasture due to land constraints compared to cattle in Australia. Second, most feed grains used for cattle are imported. Figure 2 shows how each variable moves over time from 2000 to 2014.

Table 2 Summary statistics

Variable	Description (won/kg)	Obs	Mean	Std. dev.	Min	Max
KORP	Price of Korean native cattle	178	13,291.83	1905.46	8677.00	18,625.00
CORNP	Imported corn price	178	184.98	82.26	83.92	380.46
AUSP	Imported Australian beef price	178	4163.63	1147.44	2034.33	6118.35

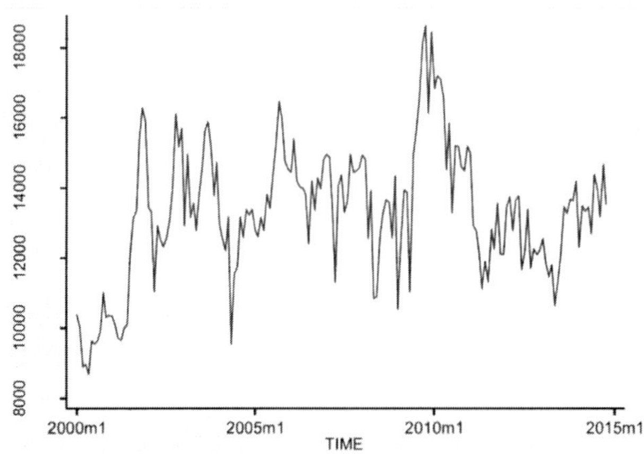

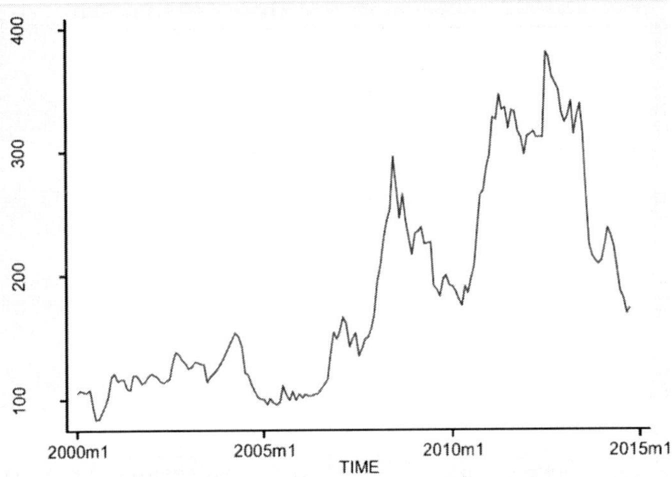

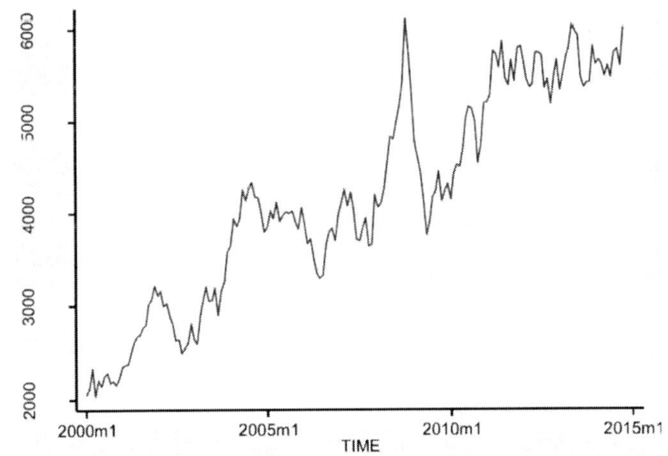

Figure. 2 Monthly data for each variable from 2000 to 2015

2. METHODOLOGIES AND CONCEPTUAL FRAMEWORKS

2.1 Structural break with unit root test

According to Baum (2001), results from tests of Dickey-Fuller (DF), Phillips-Person (PP), and Dickey-Fuller using a generalized least square (DF-GLS) for the unit root may cause bias in that the model excluding structural change leads to misspecification by omitting relevant independent variables. Baum (2001) suggests using a technique based on Perron and Vogelsang (1992) for the unit root test in the presence of one structural break. Perron and Vogelsang (1992) describe that the additive outlier (AO) is used when the change is assumed to affect instantaneously whereas the innovational outlier (IO) model for slowly changing. In this study, we assume the change by BSE affect instantaneously, and the Clemente-Montañés-Reyes unit root test with single mean shift (also called AO) model is used in this study.

For the conceptual framework for the AO model, we follow the original paper by Perron (1990). The null hypothesis that the series y_t, which is a realization of a time series process, is considered by the presence of a unit root, but the standard classification is generalized in order to allow a one time change in the structure of the series at a time T_B where $(1 < T_B < T)$. Then, the null hypothesis can be rewritten as:

$$Y_t = yD(TB)_t + yt_{-1} + wt (t = 1 \ldots, T) \quad (1)$$

where $D(TB)_t = 1$ if $t = T_B + 1$ and 0 otherwise; $y_0 = y(0)$ either a fixed constant or a random variable. In order to use a functional central limit theorem for partial sums $S_t \sum^t_{j=1} wt$, the conditions on the sequence of innovations w_t are specified. With Gaussian innovation, a series w_t is generated by any finite order ARMA (p, q) process. When conditioning upon the initial observation y_0, the mean of the series y_t is given by y_0 up to time T_B and by $y_0 + \gamma$ afterward under the null hypothesis. However, the series do not contain a unit root under the alternative hypothesis, and the model can be defined as follows:

$$Yt = \text{▯▯} yDU_t + e_t \quad (2)$$

where $DU_t = 0$ if $t \leq T_B$ and 1 otherwise. Here, the conditions on e_t are general enough to permit an ARMA $(p + 1, q)$ representation consistent with the process.

2.2 Generalized least square Dickey-Fuller test

The DF-GLS is used to test unit root in time series, and this test was proposed by Elliott et al. (1996) who find that test from the DF-GLS is significantly larger power than the augmented Dickey-Fuller test. Following discussion follow Stock and Watson (2012). The augmented Dickey-Fuller test includes fitting a regression of the form as follows:

$$\text{▯} y_t = \text{▯▯▯} y_{t-1} + \text{▯} t + y1 \text{▯} y_{t-1} + \ldots + Yp \text{▯} Y_{t-p} + \text{▯▯} t \quad (3)$$

The testing null hypothesis is $H_0 : \beta = 0$. The null hypothesis explains that y_t is a random walk or it possibly has drift. The alternative hypothesis is that y_t is either stationary with linear trend or nonzero mean with no trend.

2.3 No cointegration vs threshold cointegration test

We test whether there is a threshold cointegration between the variables or not. It is important to be tested in this paper because threshold vector autoregressive (TVAR) or threshold vector error correction model (TVECM) can be used based on results of the cointegration test. The TVAR is appropriate if there is no threshold cointegration whereas the TVECM is appropriate if there is threshold cointegration. The test can be done by using the Seo test, which was introduced by Seo (2006). The threshold cointegration was initially introduced by Balke and Fomby (1997). According to Seo (2006), a standard linear cointegration is explained by the threshold cointegration due to the fact that it allows adjustment to long-run equilibrium to be nonlinear or not. The conceptual framework for testing the linear no cointegration null in a TVECM follows Seo (2006). The test for the linear no cointegration null hypothesis can be defined as $H_0 : \alpha_1 = \alpha_2 = 0$ from the model below:

$$\emptyset(L)\Delta x_t - \alpha_1 z_{t-1}\{z_{t-1} \leq \gamma_1\} + \alpha_2 z_{t-1}\{z_{t-1} > \gamma_2\} \mid \mu \mid \varepsilon_t \tag{4}$$

Where $t = 1,\dots n$, and $\emptyset\,(L)$ is qth-order polynomial in the lag operator, so it can be written as $\emptyset\,(L) = \mathrm{I} - \emptyset_1 L^1 - \cdots - \emptyset_q L^q$. For a known cointegrating vector β, the term for the error correction is defined as zt $=x\,'_t$⍰ The threshold parameter $\gamma = (\gamma_1, \gamma_2)$ satisfying $\gamma_1 \leq \gamma_2$, which takes values on a compact set Γ. The model above allows for the no-adjustment region in the middle $(\gamma_1 < z_{t-1} \leq \gamma_2)$.

3. EMPIRICAL RESULTS

3.1 Structural break unit root test

Initially, we assume that the short-run and long-run relationships between the variables are substantially different with different structural break points due to the BSE. In addition, the different structural break points come from the nonlinearity. Table 3 shows the result from the Clemente-Montañés-Reyes unit root test with single mean shift (also called AO) model. In Table 3, du1 and rho -1 represent time structural break and unit root, respectively. The optimal structural break points for prices of Korean cattle beef, imported corn, and Australia beef are April 2001, July 2008, and August 2010, respectively. We can interpret the result of the unit root test that we cannot reject the null hypothesis of a unit root in series of three variables despite the structural break. In other words, all three variables are nonstationary even though we take into account the structural break. Figure 3 shows graphical structural break points. Based on results from Table 3, we find that we have approximately 1-year pre-structural break and 4 years of data post the structural breakpoint, which are not enough observations to compare and find the short-run and long-run relationships. Even though this study cannot further investigate short-run and long-run price relationships before and after the structural break points due to lack of observations, this finding provides and supports new consideration and evidence of nonlinearity in times series. This finding also contributes to the existing literature related to price adjustments by BSE crises and the need to use new methodologies to detect nonlinearity problems.

Table 3 Results from the Clemente-Montañés-Reyes unit root test with single mean shift (AO) model

	KORP	CORNP	AUSP
du1	3821.87***	133.68***	1909.20***
	(9.33)	(17.98)	(15.04)
rho -1	−0.28	−0.12	−0.07
	(−3.09)	(−3.05)	(−2.88)
Optimal break point	2001 April	2008 July	2010 August

***1% significant levels
Parenthesis indicates *t* statistics

There are two main sources to cause nonlinearity in time series analysis. According to the Giordani et al. (2007), the presence of nonlinearity is caused by the structural change; however, Goodwin and Holt (1999) mention that the threshold effects occur when larger shocks induce a different response compared to smaller shock, and this dynamic response is due to the nonlinearity. Even though we are unable to observe short-run or long-run relationships by considering different structural changes, we do observe the threshold points by assuming the presence of nonlinearity from the fluctuation of variables as shown in Fig. 2.

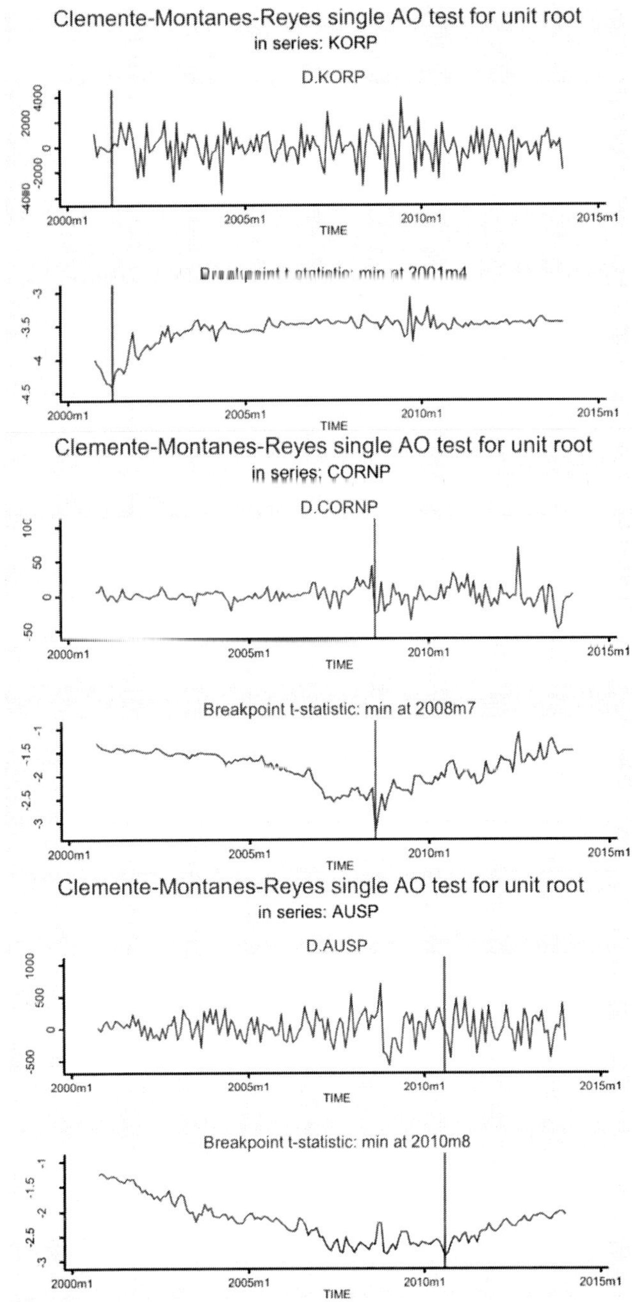

Figure. 3 Results from the Clemente-Montañés-Reyes unit root test with single mean shift (AO) model

3.2 DF-GLS unit root test

First, we conduct unit root test by using DF-GLS proposed by Elliott et al. (1996) who find that DF-GLS is statistically powerful than the commonly known Dickey-Fuller test. Table 4 shows the results from the DF-GLS unit root test. Based on the results, we find that there is no unit root (i.e., variables are stationary) for both constant with the trend and without trend in first differentiated data at 1% significant level. However, we find that prices of Korean cattle and Australian beef have unit root without trend whereas they have no unit root with the trend in original data at 5% significant level. For the price of corn, we find that there is unit root regardless of trend.

Table 4 The results from the DF-GLS Unit Root Test

	Level			First difference		
Variable	Constant w/o trend	Constant with trend	Decision	Constant w/o trend	Constant with trend	Decision
KORP	−1.89	−2.96**	I(1)/I(0)	−17.48***	−18.53***	I(0)
CORNP	−1.01	−1.74	I(1)	−10.75***	−10.87***	I(0)
AUSP	0.32	−3.44**	I(1)/I(0)	−12.36***	−12.54***	I(0)

I(0) indicates no unit root, and I(1) indicates unit root
Estimates are statistically significant at ***1%, **5%

3.3 Number of lag selection

Table 5 shows the result from selecting a number of lags. In the original data, we find that the numbers of lags are 2, 1, and 1 based on Akaike Information Criterion (AIC), Bayesian Information Criterion (BIC), and Hannan-Quinn Information Criterion (HQC), respectively. By using the first difference data, we find that the number of lags is same as 1 for all three different criterions. According to Tsay (1984), Potscher (1989), and Cavaliere et al. (2015), BIC is consistent compared to AIC because AIC overestimates with positive probability. Therefore, we use BIC information.

Table 5 The results from selecting number of lags

Data	AIC	BIC	HQC
Original data	2	1	1
Differentiated data	1	1	1

3.4 Threshold cointegration test

We use Seo (2006) test whether there is no cointegration or threshold cointegration between the variables. The null of no cointegration is used in Seo test. Table 6 shows the result from the test of no cointegration vs threshold cointegration. Based on the result from Seo test, we find that there is threshold cointegration between prices of Korean beef and Australian beef and between the price of Australian beef and corn with 1 lag based on the BIC. However, there is no cointegration between the price of Korean beef and corn. Therefore, we conclude that there is no cointegration among the three variables. In the Seo test, we are unable to test whether no cointegration or threshold cointegration for all the variables at the same time. For this reason, we conduct threshold cointegration test for each variable separately. Based on the result of Seo test, we conclude TVAR with first differentiated data is preferred to TVCEM since there is no cointegration between the variables.

Table 6 Test of no cointegration vs threshold cointegration

	Test statistic	*P* value
KORP vs AUSP	12.34	0
AUSP vs CORNP	19.80	1
KORP vs CORNP	27.83	0

3.5 Testing threshold linearity

Two sources primarily create nonlinearities in time series analysis. First, the structural change can create nonlinearity, according to the Giordani et al. (2007). Secondly, when larger shocks induce a different response compared to smaller shocks, it creates a dynamic response due to the presence of nonlinearity (Goodwin and Holt 1999). Goodwin and Harper

(2000) found that recent time series models examining price transmission consider asymmetric adjustments linear. However, Luoma et al. (2004) argue that linear models in a nonlinear situation result in incorrect inferences. In this case, we expect there to be nonlinearities, so utilizing a linear model would not be appropriate.

To find the number of thresholds, we test threshold nonlinearity by using the Lo and Zivot test that was proposed by Hansen (1999) and developed by Lo and Zivot (2001). Utilization of this test allows for the testing of nonlinearities between price regimes, in this case, pre- and post-BSE. Table 7 shows test statistics, which are results from the threshold nonlinearity test with 1 lag based on BIC from Table 5. In Table 7, the null hypothesis of 1 vs 2, for example, is that linear VAR model with no threshold is true model compared to the model with one threshold since 1, 2, and 3 indicate no threshold, 1 threshold, and 2 thresholds, respectively. Therefore, if we reject the null hypothesis of 1 vs 2, we conclude that the linear VAR model with one threshold is a true model. Based on the results in Table 7, we find that only differentiated Australian beef price has 2 thresholds at 1% significant level whereas other variables have no thresholds. Therefore, we focus only on TVAR with 2 thresholds in first differentiated Australian beef price even though we want to examine the effect on corn price.

Table 7 Results of test statistics from the threshold nonlinearity test

Threshold hypothesis	DAUSP	DCORNP	DKORP
1 vs 2	35.30***	16.69	19.41
1 vs 3	62.30***	28.86	42.41
2 vs 3	27.01***	12.17	23.00

D indicates first difference

Significant at ***1%

3.6 Threshold vector autoregressive model

Since we find that there are 2 thresholds in first differentiated Australian beef price. The TVAR model can be defined as the equation below:

$$y_t = \begin{cases} \delta_1 + \rho_{1,1} y_{t-1} + \cdots + \rho_{1,p} y_{1-p1} + u_t \\ \delta_2 + \rho_{2,1} y_{t-1} + \cdots + \rho_{2,p} y_{1-p1} + u_t \\ \delta_3 + \rho_{3,1} y_{t-1} + \cdots + \rho_{3,p} y_{1-p1} + u_t \end{cases} \quad (5)$$

where y is price vector of Korean beef, corn, and Australian beef, δ_1, δ_2, δ_3 refer to the intercepts in each regime, $\rho_{j,1}, \ldots, \rho_{j,m-1}$ are the number of lags in regime, θ_L, θ_H are the thresholds, d is the delay of transition variable, and x_{t-d} is the transition variable from one of three variables. To find 2 threshold points, a grid search is used in this study. Based on Stigler (2010), the grid search is to find threshold points (θ) that minimize the sum of squares and maximize the log-likelihood. The objective function is defined as follows:

$$\hat{\theta} = argmin\ SSR(\theta) \tag{6}$$

The values of the variables are sorted, and a minimal number of observations in each regime are ensured through a grid search by excluding a certain percentage of the first and last values. Then, the SSR is estimated for each selected value, and the one that minimizes the SSR is taken as the estimator.

Table 8 shows the results from the TVAR model with 2 thresholds, which are −275.15 and 198.29, in Australian beef price, in addition, 11.4% of observations in the first regime, 69.3% in the second regime, and 19.3% in the third regime, respectively. Figure 4 visually shows the results from threshold points and grid search. From here, the name of each variable represents first difference for simplicity.

Table 8 Results from the TVAR model in Australian beef price

	Regime	Intercept		DAUSP (-1)		DCORNP (-1)		DKORP (-1)
DAUSP (11.40%)	1	−157.462		−0.840		3.172		0.030
	2	10.883		−0.053		−0.001		1.489
	3	33.261		−0.163		−1.797		−0.019
DCORNP (69.30%)	1	−14.893		−0.036		0.090		−0.001
	2	0.183		−0.001		0.183	***	−0.003
	3	−5.032		0.019		0.152		−0.001
DKORP (19.30%)	1	1992.633	***	5.037	***	−39.993	***	−0.599
	2	130.931		1.489	***	1.426		−0.297
	3	−335.071		0.067		22.135	***	−0.397

Threshold value: −278.15 and 198.29. Percentage of observations in each regime: 11.4, 69.3, and 19.3%. Note: (-1) indicates first lag
Significant level: ***1% and *10%

In the first regime, which indicates decreasing in Australian beef price (less than −278.15 threshold), we find that current Korean beef price is positively and significantly affected by a previous month of Australian beef price at 5% significant level, and it can be explained by substitutability between Australian beef and Korean beef. For corn, if corn price increased in the previous month, there is an incentive to increase current Korean beef price, but the incentive is canceled out (i.e., buffered) due to weak Australian beef price.

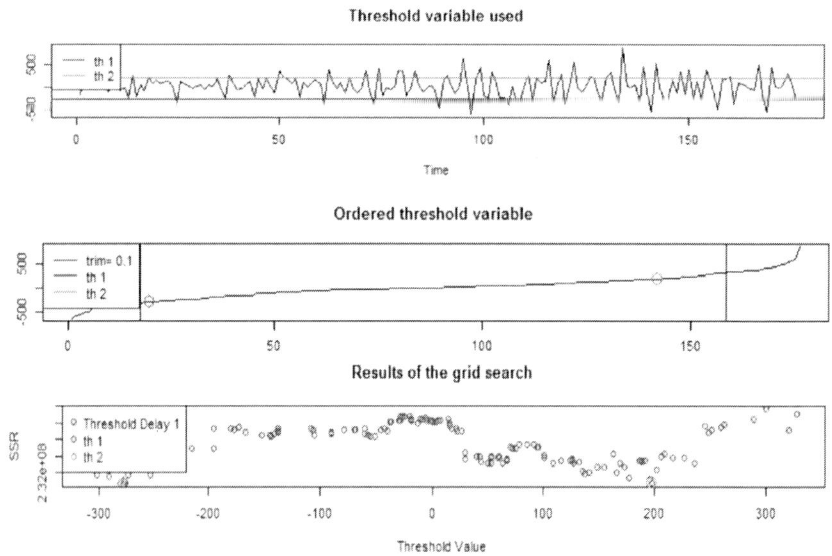

Figure. 4 Results from threshold points and grid search

In the second regime, which indicates less volatility in Australian beef price compared to other two extreme cases (i.e., first and third regime), we find that change in previous corn price has no significant impact on Korean beef price compared to extreme cases; in addition, change in previous Australian beef price has significant impact on Korean beef price at 5% level.

In the third regime, which indicates increasing in Australian beef price (greater than 198.23 threshold), we find that current Korean beef price is affected positively by previous corn price at 5% and negatively by previous Korean beef price at 1% significant level respectively whereas no impact from previous Australian beef price. Results from the third regime can be explained that there is an incentive to increase Korean beef price as corn price increases, and the incentive cannot be canceled out in this regime due to increasing price in substitute good (Australian beef).

4. DISCUSSION

Korea is one of the largest beef import countries in the world with the majority coming from Australia, USA, and New Zealand. The primary objective of this study is to evaluate the relationships between the prices of the Korean native cattle, world corn price, and imported Australian beef price. In 2003, the outbreak of BSE in the USA, the second largest exporter of beef to South Korea, resulted in a ban on US beef imports until July 2008, thus created an expected structural break in Korean native beef prices. Secondly, given the dynamic changes that take place in the prices used in the model, the potential for nonlinearities in regimes thresholds are evaluated.

The structural breakpoints are estimated by using Clemente-Montañés-Reyes unit root test with single mean shift and AO model. A structural break is found for Korean native beef price in April 2001, world corn price in July 2008, and Australian imported beef price in August 2010. Even considering the structural breaks, the data is still nonstationary. Additionally, given the data limitations on observations pre or post the structural breaks, the estimation of a long-run relationship is not possible. However, the existence of these structural breaks does provide evidence that there could be nonlinearity in the data (Giordani et al. 2007). Given this evidence, a threshold framework is employed to investigate the presence of nonlinearity from fluctuations in the original data.

This study employs a threshold framework to examine nonlinearity between the variables by hypothesizing that price effects are statistically different at certain threshold points. Based on threshold cointegration test and the Lo and Zivot test, we find that there is threshold cointegration between prices of Korean beef and Australian beef and between the price of Australian beef and corn. In addition, differentiated Australian beef price has two thresholds at 1% significant level. The existence of a threshold in price variable supports the presence of nonlinearity and infers the different effect of the parameter (i.e., price effect) in three different regimes based on two threshold points.

We find that there are two thresholds in Australian beef price using the first difference. Also, we find there is no cointegration among three variables, so we use TVAR model. We find from the TVAR that the current Korean native beef price is positively affected by increasing in the previous month of Australian beef price in each regime even though Australian beef is a substitute for Korean native beef. This can be explained that Korean beef price is relatively higher than Australian beef price, so shocks in Australian beef price are buffered by the price difference. Changing in previous corn price has no significant impact on current Korean native beef price in the second regime. If corn price increased in the previous month, there is an incentive to increase current Korean beef price, but the incentive is canceled out (i.e., buffered) by low Australian beef price in the first regime. However, the incentive is not buffered in the third regime where increasing in previous Australian beef price is high.

CONCLUSIONS

In summary, this study finds evidence that a BSE outbreak in the USA did impact the price relationships between Korean native beef price, world corn price, and Australian import beef price. This finding is consistent with previous studies for other countries that find BSE outbreaks fundamentally change price relationships. Furthermore, this study found that three different price regimes exist for Australian beef prices. Korean native beef prices are found to be highly sensitive to declining Australian beef prices, as shown in regime one. Therefore, Korean native beef producers and policy makers need to consider Australian beef import price as an indicator of Korean native beef production, given their high cost of production, land constraints, and the price premium for high-quality Han-Woo beef.

REFERENCES

1. Abdulai A (2002) Using threshold cointegration to estimate asymmetric price transmission in the Swiss pork market. Appl Econ 34:679–687
2. Balke NS, Fomby TB (1997) 'Threshold cointegration". Int Econ Rev 38(3):627–645
3. Baum CF (2001) Stata: the language of choice for time series analysis. Stata J 1:1–16
4. Cavaliere G et al (2015) Lag length selection for unit root tests in the presence of nonstationary volatility. Econ Rev 34:512–536
5. Cook R (2016). "World beef imports: ranking of countries." [online] http://beef2live.com/story-world-beef-imports-ranking-countries-0-106900. (Accessed 01 Dec 2016)
6. Elliott G, Rothenberg TJ, Stock JH (1996) Efficient tests for an autoregressive unit root. Econometrica 64:813–836
7. Giamalva J (2013) "Korea's demand for US Beef." Journal of International Commerce and Economics. http://www.usitc.gov/journals
8. Giordani P, Kohn R, van Dijk D (2007) A unified approach to nonlinearity, structural change, and outliers. J Econ 137:112–133
9. Goodwin BK, Harper DC (2000) Price transmission, threshold behavior, and asymmetric adjustment in the US pork sector. J Agric Appl Econ 32:543–553
10. Goodwin BK, Holt MT (1999). "Price transmission and asymmetric adjustment in the US beef sector." American Journal of Agricultural Economics 81:630–637
11. Hansen B (1999) Testing for linearity. J Econ Surv 13:551–576
12. Hassouneh I, Serra T, Gil JM (2010) Price transmission in the Spanish bovine sector: the BSE effect. Agric Econ 41:33–42
13. Hebert CJ, Anderson BC (2011) "The exogenous effect of corn prices on beef prices in the US using R&D as an instrumental variable." Economics Honors Thesis, Colgate University

14. Jin YH, Power GJ, Elbakidze L (2008) The impact of North American BSE events on live cattle futures prices. Am J Agric Econ 90:1279–1286
15. Korea Institute for Animal Products Quality Evaluation (2013) [online] http://www.ekape.or.kr/view/user/main/main.asp. (Accessed 10 Apr 2015)
16. Lloyd TA et al (2006) Food scares, market power and price transmission: the UK BSE crisis. Eur Rev Agric Econ 33:119–147
17. Lo MC, Zivot E (2001) Threshold cointegration and nonlinear adjustment to the law of one price. Macroecon Dyn 5:533–576
18. Luoma A, Luoto J, Taipale M (2004) Threshold cointegration and asymmetric price transmission in Finnish beef and pork markets: Pellervo Economic Research Institute. https://www.researchgate.net/profile/Arto_Luoma/publication/242238244_Threshold_cointegration_and_asymmetric_price_transmission_in_Finish_beef_and_pork_markets/links/5735bc5a08ae298602e08bfb.pdf
19. Marsh JM, Brester GW, Smith VH (2008) Effects of North American BSE events on US cattle prices. Appl Econ Perspect Policy 30:136–150
20. Ministry of Agriculture, Food and Rural Affairs (2015) [online] http://english.mafra.go.kr/main.jsp. (Accessed 05 Apr 2015)
21. Park M, Jin YH, Bessler DA (2008) The impacts of animal disease crises on the Korean meat market. Agric Econ 39:183–195
22. Perron P (1990) Testing for a unit root in a time series with a changing mean. J Bus Econ Stat 8:153–162
23. Perron P, Vogelsang TJ (1992) Nonstationarity and level shifts with an application to purchasing power parity. J Bus Econ Stat 10:301–320
24. Potscher B (1989) "Model selection under nonstationarity: autoregressive models and stochastic linear regression models." The Annals of Statistics. 17:1257–1274
25. Saghaian SH (2007) Beef safety shocks and dynamics of vertical price adjustment: the case of BSE discovery in the US beef sector. Agribusiness 23:333–348
26. Sanjuan AI, Dawson PJ (2003) Price transmission, BSE and structural breaks in the UK meat sector. Eur Rev Agric Econ 30:155–172
27. Seo M (2006) Bootstrap testing for the null of no cointegration in a threshold vector error correction model. J Econ 134:129–150
28. Serra T, Gil JM, Goodwin BK (2006) Local polynomial fitting and spatial price relationships: price transmission in EU pork markets. Eur Rev Agric Econ 33:415–436
29. Stock JH, Watson MW (2012) Introduction to Econometrics: Global Edition: Pearson Education
30. Tsay RS (1984) "Order selection in nonstationary autoregressive models." The Annals of Statistics. 12:1425–1433
31. U.S Department of Agricultural, Economic Research Service. (2012) [online] https://www.ers.usda.gov/topics/animal-products/cattle-beef/trade.aspx. (Asscessed 15 Apr 2015)
32. Zivot E, Wang J (2007) Modeling financial time series with S-Plus®: Springer Science & Business Media

CHAPTER 10

Determinants of participation in fertilizer subsidy programme among rice farmers in Ogun State, Nigeria

OBI-EGBEDI Ogheneruemu, BANKOLE Olaide Abdul-hameed*

Department of Agricultural Economics, University of Ibadan, Oyo State, Nigeria.

ABSTRACT

Nigerian farmers, including rice farmers, still record very low levels of fertilizer use thereby limiting productivity. Subsidies have been known to encourage fertilizer use among farmers. This paper examined the factors influencing rice farmer participation in the government's fertilizer subsidy programme. Data was collected through the aid of a well-structured questionnaire from 263 rice farmers. Descriptive and Logistic regression analyses were used to analyze the data. Statistical mean differences were found in age, household size, years of farming experience, farm size, output and total annual income between participants and non-participants. Also, participation was significantly and positively influenced by marital status, household headship, membership of farmer association/groups, motorcycle ownership, mobile phone ownership, access to credit and total farm size. The paper concludes that efforts should be geared towards encouraging membership of farmer groups, availability and timely distribution of subsidized fertilizer and the establishment of more redemption centres.

Keywords: Fertilizer, growth enhancement scheme (GES), participation, rice, subsidy.

1. INTRODUCTION

Increased fertilizer use played a significant role in the success of the green revolution in Latin America and Asia. It helped raise agricultural productivity and farm incomes, thus laying the foundation for broader economic growth. As much as 50% of yield growth in these regions could be attributed to increased fertilizer use (Toenniessenn et al., 2008). Despite the growing evidence that fertilizers can substantially increase yields in Sub-Saharan Africa (SSA) as well as slow down soil degradation, farmers in SSA still lag far behind other developing countries in fertilizer use. The average fertilizer use in Sub-Saharan Africa (SSA) is estimated at 16 kg/ha; much lower than other parts of the world with 90 kg/ha in Middle East and North Africa, 126.6 kg/ha in North America, 127.9 kg/ha in Latin America and Caribbean, 158.5 kg/ha in South Asia and 344.3 kg/ha in East Asia and Pacific. In Nigeria, the fertilizer use was estimated at 4.5 kg/ha in 2002 and 10.9kg/ha in 2014, below the average for SSA (World Bank, 2014).

Furthermore, the results of a Food and Agricultural Organization (FAO) study spanning 1983-2000 along with some other studies (FAO and ITPS 2015, Sheldrick and Lingard, 2004; Lesschen et al., 2003; Stoorvogel and Smaling, 1990) which assessed soil nutrients (Nitrogen, Phosphorus and Potassium-NPK) balances by land use systems revealed a general depletion in Africa characterized by annual negative nutrient balances. For Nigeria, the nutrient balances were -34 kg/ha in 1983 and -37 kg/ha in 2000 for N; -4 kg/ha in 1983 and -4 kg/ha in 2000 for P; and -24 kg/ha in 1983 and -31 kg/ha in 2000 for K. These figures are indicative of unrelenting nutrient mining over time (Bationo et al., 2012). The gap in fertilizer use in SSA and Nigeria relative to the rest of the world is given as one reason for the failure of the region to achieve its green revolution objectives. This failure raises the question of what types of policies and programme are needed for the region to realize the potential benefits from fertilizer usage (Kelly, 2006).

In 2006, African leaders in the context of the Comprehensive Africa Agriculture Development Programme (CAADP) through the Abuja Declaration resolved to improve the use of fertilizer as a means to achieving the region's green revolution objectives. As a follow up, the Federal Government of Nigeria (FGN) decided to disengage from direct procurement of fertilizer in favor of promoting private sector participation. This was done via the Growth Enhancement Support (GES) Programme; a fertilizer subsidy programme under the Agricultural Transformation Agenda (ATA) which set ambitious goals of increasing fertilizer use from the year 2010 level of approximately 13 to 50 kg/ha (FMARD, 2011). The GES was different from previous subsidy schemes in that it targeted beneficiaries through vouchers and the handing over of subsidized fertilizer distribution from the government to private dealers. This contrasts with previous subsidy schemes in which the government directly participated in the procurement and distribution of subsidized fertilizer through the agricultural development project (ADP) and other agencies (IFPRI, 2012).

In 2011, the Nigerian government made an effort to find a long-lasting solution to the problem of food insecurity by raising agricultural productivity and boosting food production. In order to achieve this objective, the Agricultural Transformation Agenda was launched in the same year. This was anchored on the philosophy of treating agriculture as a business rather than a development programme. The goal was to add 20 million metric tonnes (MT) of food to domestic food supply and create 3.5 million jobs by year 2015.

The GES Programme was designed as a component of the Agricultural Transformation Agenda of the Federal Government (ATA). The Federal Government of Nigeria introduced the GES which was designed to deliver government subsidized farm inputs directly to farmers via mobile phones. The GES scheme was powered by e-Wallet, an electronic distribution channel which provided an efficient and transparent system for the purchase and distribution of agricultural inputs based on a voucher system. The scheme guarantees registered farmers e-Wallet vouchers with which they could redeem fertilizers, seeds and other agricultural inputs from agro-dealers at half the cost, the other half being borne by the federal government and state government in equal proportions (FMARD, 2011). Individual farmers were registered in a national database. Each farmer was entitled to a 50% subsidy on the price of two 20 kg bags of fertilizer. This intervention became necessary as a result of the crisis that riddled the agricultural sector in the past, given its critical role for food security and economic diversification.

On inception, the aims of the GES was to migrate smallholder farmers from subsistence farming to commercialized systems over a 4 to 10 year period in order to facilitate trade and competitiveness. According to Takeshima and Liverpool-Tasi (2013), the potential in the fertilizer subsidy reform under the ATA include improved targeting through voucher and crowding-in of the commercial fertilizer sector. By June 2014, agricultural productivity and food production had increased by 17 million MT and was expected to reach 21 million MT by the end of the year and exceed the 20 million MT target set for 2015. However, challenges remain in farmer access to redemption facilities, entitlement risk (mobile phone), fertilizer quality regulation and the speed at which the private sector respond.

Generally, fertilizer demand still depends on broader agricultural policies, factor endowments and farming systems.

The Federal Government under the current administration has decided to build on the achievements of the ATA by launching a new strategy known as the Agricultural Promotion Policy (APA). The plan is to solve the problems associated with the previous attempt at ensuring an efficient fertilizer distribution system. Therefore, the current policy objective is to increase productivity by ensuring timely access to high quality and price competitive inputs (FMARD, 2016). Thus, encouraging more farmer participation in the program is key to the policy success.

This study seeks to contribute to existing literature on the factors responsible for participation in the fertilizer subsidy programme. In order to achieve this, answers were provided to the following questions:

1) What differences exist in rice farmers' characteristics by their level of participation in the fertilizer subsidy programme?
2) What factors influences the participation of farmers in the fertilizer subsidy programme in the study area?

2. MATERIALS AND METHODS

2.1 Descriptive statistics

Descriptive statistics include the use of frequencies, percentages, means and standard deviation to analyze the socio-economic characteristics of respondents. It was also used to describe the reasons for the non-receipt of subsidized fertilizer.

2.2 Empirical estimation

The decision whether or not to participate in the fertilizer subsidy programme can be explained as a discrete binary variable, 1 for participants and 0 for non-participants. The simplest possible binary regression model is the linear probability model (LPM) in which the binary response variable is regressed on the relevant explanatory variables by using the standard Ordinary Least Square (OLS) methodology. However, it suffers from several estimation problems; one of which is that it can produce predicted probabilities outside the (0; 1) bounds (Gujarati, 2004). Other appropriate models that can be used are logit and probit. Logit and probit models usually yield similar results. Hence; the choice is not too critical, even though the logit distribution has more density mass in the bounds. Estimating participation is to define an adequate measurable indicator that will distinguish between participants and non-participants.

A binary variable indicates whether or not the farmer participates in the programme. When one is interested only in comparing outcomes for those participating (T = 1) with those not participating (T = 0), this estimate can be constructed from a probit or logit model. In this study, a participant is defined as a rice farmer that has received subsidized fertilizer in the last rice production season. The sample of participants and nonparticipants was pooled, and then participation T was estimated on all the observed covariates X in the data that are likely to determine participation. Traditional instruments used in the literature include the distance between the farm and the fertilizer selling points, or social capital proxied by how long the farmer has lived in the community (Seck, 2015). The vector of explanatory variables includes farm characteristics that may influence the probability of getting subsidized fertilizer such as farm size, access to credit, mobile phone ownership, and ownership of a means of transportation and affiliation to farmers' union.

In this analysis, participation (Z) is defined as the dependent variable which takes the value of 1, if a rice farmer participates in the fertilizer subsidy programme and 0, otherwise, that is, Z = 1 , if a rice farmer participates in the fertilizer subsidy programme and Z = 0, otherwise. The logistic model postulates the probability that participation is a function of an index where:

(Z_i) is an inverse of the standard logistic cumulative function of P_i that is, $P_i(y) = f(Z_i)$; (Z_i) is also an inverse of the standard logistic cumulative function of P_i:

$$P_i\ (y = 1) = f\ (Z_i)$$

The probability of participation is given by:

$$P_i\ (y = 1) = \left(\frac{1}{1+e}\right) - Z_i \qquad (1)$$

e represents the base of natural logarithms (2.718).

The probability of no partiiipation is given by:

$Q_i = (y = 0) = 1 - P_i\ (y = 1)$

Since, $1 - P_i\ (y = 1) = 1 - \frac{1}{1+e^{-Z_i}}$; $1 - P_i\ (y = 1) = \frac{1+e^{-Z_i}}{1+e^{-Z_i}} -$ $1\ and\ \ 1 - P_i\ (y = 1) = \frac{e^{-Z_i}}{1+e^{-Z_i}}$

But,

$$\frac{1}{P_i\ (y=1)} = 1 + e^{-Z_i} \qquad (2)$$

Thus: $\frac{P_i\,(y=1)}{1-P_i(y=1)} = \frac{1}{e^{-Z_i}}$ and

$$\frac{P_i(y=1)}{1-P_i(y=1)} = e^{Z_i} \qquad (3)$$

We take as comparison category, farmers who did not participate in the fertilizer subsidy programme. This means that the changes in relative risk will represent the improvement of a non-participating rice farmer given the impact of a specific variable.

The explanatory and dependent variables that were used in the econometric model (logit) are defined as follows:

$$\ln\left(\frac{P_i}{1-P_i}\right) = Z_i = \alpha + \beta_1 X_1 + \beta_2 X_2 + \ldots + \beta_m X_m + \varepsilon \qquad (4)$$

Where, Z_i = Participation (1 = participants, 0 = non-participants); X_1 = Age in years; X_2 = Marital status (1 = single, 2 = married, 3 = divorced, 4 = widowed); X_3 = Household headship (0=female, 1=male); X_4 = Farming experience in years; X_5 = Years of education; X_6 = Ownership of a means of mobility (motorcycle) (1=Yes, 0=No); X_7 = Ownership of a mobile phone (1 = Yes, 0 = No); X_8 = Access to credit (1 = Yes, No = 0); X_9 = Membership of farmers' association/group (1 = Yes, 0 = No); X_{10} = Total farm size (in hectares);X_{11} = Ownership of land (1 = Personal, 0 = otherwise) and = Error term.

2.3 Data collection

This study was carried out in Ogun State in the South-Western geo-political zone of Nigeria. The state has 21 Local Government Areas (LGAs) and a projected population of 4,424,066 (NPC, 2011). The state is located in the moderately hot, humid tropical climate zone of Southwestern Nigeria and it favours the production of food crops such as maize, cassava, yam, cocoyam, soybean and rice. The major occupation of the people is farming (OGADEP, 2015). There are four Agricultural Development Project (ADP) zones in the state as categorized by the Ogun State Agricultural Development Project (OGADEP) namely Ilaro, Ijebu-Ode, Abeokuta and Ikenne zones. Thus, a peculiar nature of OGADEP is that zones are further divided into blocks and cells.

The data for the study was collected in 2015 through the use of structured questionnaires by employing a multi-stage sampling technique. Three agricultural zones were purposely selected from a total of four due to the availability of rice farmers who participated in the fertilizer subsidy programme. They are Abeokuta, Ikenne and Ilaro zones. The second stage involved the random selection of three local government areas from the selected zones, these included Ewekoro (Abeokuta Zone), Obafemi Owode (Ikenne Zone) and Yewa North (Ilaro Zone). Next, cells were randomly selected in each of the zones. Lastly, a total of 270 questionnaires were distributed to the farmers; 263 were used for analysis consisting of 113 and 150 participating and non-participating farmers respectively.

3. RESULTS AND DISCUSSION

3.1 Socioeconomic characteristics of participating and non-participating farmers

The description of farmer characteristics is presented in Table 1 and it reveals that both groups (participants and non-participants) have similar characteristics with only slight differences recorded. Rice farming was a male dominated activity in the study area.

Table 1. Socio-economic characteristics of rice farmers (n=263).

Variable	Participants	Non-participants	Mean difference
Sex			
Male (%)	82.30	82.67	
Female (%)	17.70	17.33	
Age (mean)	47.92	44.80	3.12***
Household size (mean)	5.58	5.11	0.46*
Years of education (mean)	5.92	6.36	-0.44
Years of farming experience (mean)	24.98	22.33	2.66**
Rice farm size in *ha* (mean)	1.726	1.448	0.28***
Output in kg (mean)	2022.57	1526.60	495.97***
Total annual income (mean)	542,272.56	367,800.00	174472.57***

Source: Field Survey, 2015. *, **, *** implies that coefficients are statistically significant at 10, 5 and 1%, respectively.

Generally, there were more households headed by males than females participating in the programme. Most of the farmers were middle-aged, economically active and productive with a mean age of forty six years. The implication of this ist hat they are still within the productive class.

According to Okoruwa and Ogundele (2004), being in the productive class would have a positive effect on rice production in the country. There was a significant difference in the mean ages of participants and non-participants with the average age of the participants higher than that of the non-participants.

The average household size for both groups is between five and six persons per farming household. This closely follows Okoedo-Okojie and Onemolease (2009) finding that larger household size of participants could imply that they have enough free labour for farm activities. A significant difference occurs between the mean household sizes of both groups of farmers at the 10% level.

A majority of the farmers spent an average of six years in school. There exists no significant difference in the number of years spent by farmers in school. This is consistent with the results of Azhar (1991) who reported that elementary education (4 - 6 years of schooling) does not have much effect on agricultural productivity in traditional farm settings. Other authors who lend support to this notion include Bravo-Ureta and Evenson (1994), Ajibefun and Aderinola (2003) and (Okoruwa et al., 2006).

With respect to the farm characteristics of the farmers, the average years of farming experience for participants was found to be significantly higher than that of the non-participants. There was also a significant difference in the farm size allocated to rice production between both groups of farmers with the participants having the larger sizes. This could also encourage the participating farmers to access more input for usage on their farms. Table 1 also shows that the mean output were about 2,023 and 1,527 ha for participants and non-participants while the total annual income for both groups were about ₦542,272.56 and 367800, respectively.

3.2 Factors affecting participation in the fertilizer subsidy programme

This section reports the results from the binary logistic model used to evaluate the determinants of participation of rice farmers in the fertilizer subsidy programme. The result of the regression analysis is presented in Table 2. The diagnostics reveal the model has a log likelihood of 158.59 and a chi-square statistics of 42.18; which is significant at 1%. This shows that the model is a good fit for the data. Seven of the eleven variables were statistically significant. All of the significant variables have positive signs. The variables are marital status (married), household headship (male), ownership of motorcycle, ownership of mobile phone, access to credit, membership of farmers' association/groups, and total farm size; positively associated with the probability of participation in the subsidy programme.

Table 2. Logit regression result of factors influencing participation in the fertilizer subsidy programme.

Variable	Coefficients (Std. Error)	t-value	Marginal effect
Age	0.0231 (0.0208)	1.11	0.0055
Marital status	2.2237** (1.0380)	2.14	0.5333
Household headship	3.3094*** (1.9778)	1.67	0.4251
Years of farm experience	-0.0023 (0.0171)	-0.13	-0.0005
Years of education	-0.0232 (0.0360)	-0.65	-0.0056
Ownership of motorcycle	0.5581* (0.2910)	1.92	0.1327
Ownership of mobile phone	1.4307*** (0.6865)	2.08	0.2745
Access to credit	0.7732** (0.3277)	2.36	0.1891
Membership of farmers' association	0.5063* (0.2920)	1.73	0.1191
Total farm size	0.1000*(0.0538)	1.86	0.0240
Ownership of land	-0.1394 (0.2974)	-0.47	0.0336
Constant	-11.3470*** (4.1236)	-2.75	

Source: Generated by Authors using Stata. *, **, *** implies that coefficients are statistically significant at 10, 5 and 1%, respectively. Number of observation = 263; LR chi^2 (13) = 42.18; Prob> chi^2 = 0.0000; Log likelihood = -158.59451; Pseudo R^2 = 0.1174.

The coefficient of marital status (married) is significant at 5%. Thus a 1% increase in the number of married farmers may likely increase the likelihood of famers' participation by 0.53%.

With respect to household headship, the coefficient is significant and positively influences the probability of participation. Households headed by females were less likely to have received a coupon in the sample than those headed by males (consistent with the results of Chibwana et al., 2010). The marginal effect result implies that a 1% increase in the number of male headed households is likely to increase the probability of participation by about 0.43%.

The coefficient for the ownership of a means of transportation (motorcycle) was positive and statistically significant at 10%. Redemption centres are usually some kilometers away from the farmers residence, therefore, a motorcycle increases the probability of participating in the programme. This result is consistent with the study of Takeshima and Liverpool-Tasie (2013) which reported that farmers who lived closer to town are more likely to receive subsidized fertilizer. In this case, ownership of motorcycle could get a farmer to town in a timely manner. The result of the marginal effect reveals that a 1% increase in the ownership of motorcycle increases the probability of participation increases by 0.13%.

With respect to the ownership of a mobile phone, there exists a positive and significant relationship between the variable and participation. One of the main components of the GES was that farmers must own mobile phones through which they can be alerted to retrieve their voucher. Therefore, this result is consistent with the objective of the programme as the marginal effect has shown that a 1% increase in the ownership of mobile phone was likely to increase the probability of participation increases by about 0.27%.

Access to credit also has a positive and significant relationship with participation. It is expected that a farmer might be encouraged to take advantage of the subsidy to relieve the burden of the credit facility. The result of the marginal effect shows that there is a likelihood of about 0.19% to participate in the subsidy programme with every 1% increase in access to credit facility.

Furthermore, the coefficient of the membership of a farmer association has a positive and significant effect on participation. This result is consistent with the studies of Ricker Gilbert and Jayne (2008) and Liverpool-Tasie (2012) which reported that social networks increases the probability of participation. Also, the result of the marginal effect reveals a 0.20% likelihood of a socially connected farmer to participate in the fertilizer subsidy programme.

Lastly, farm size has a positive and significant coefficient. It is expected that the bigger the farm, the more inputs that are needed to sustain production. Therefore, it provides an incentive for the farmer to take advantage of cost reduction in form of a subsidy. The result of the marginal effect shows that a 1% increase in farm size induces a 0.02% likelihood that a farmer participates in the subsidy programme.

3.3 Reasons for non-receipt of subsidized fertilizer

Figure 1 show the distribution of reasons why farmers did not participate in the fertilizer subsidy programme. About 57% of the farmers could not receive subsidized fertilizer either due to lack of cash/long distance, 16% because of missing names, 10% because of long distance and about 7% because of delay in supply. Also, 8% of the farmers did not receive an alert to redeem their vouchers while about 3% did not register.

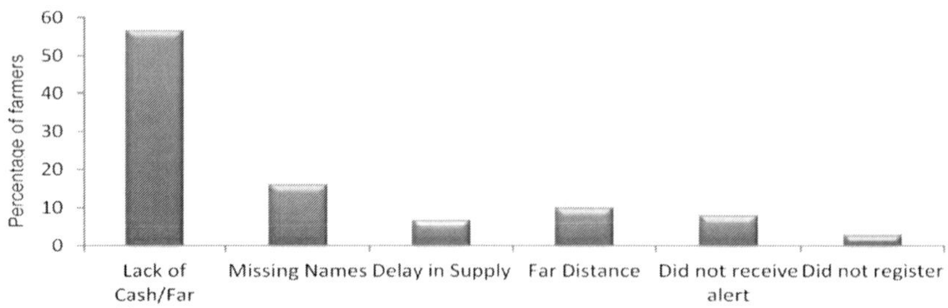

Figure 1. Reasons for non-receipt of subsidized fertilizer. Source: Field Survey, 2015.

4. CONCLUSION

This study investigated the factors/determinants responsible for rice farmer participation in the fertilizer subsidy programme using Ogun State of Nigeria as a case study. There exist statistical mean differences in age, household size, farming experience, farm size, output and total annual income between participants and non-participants. Also, the factors which significantly influence participation include marital status (married), household headship (male), ownership of a means of mobility (motorcycle), mobile phone ownership, access to credit, membership of farmers' association and total farm size. The study hereby recommends that Stakeholders (government and the private sector) should ensure the establishment of more redemption centres or make available means of mobility for farmers. Also, membership of a farmer association and other social groups should be encouraged to avoid information asymmetry. In addition, availability and timely delivery of fertilizer should be ensured to avoid farmer apathy towards the programme.

REFERENCES

1. Ajibefun IA, Aderinola EA (2003). Determinants of Technical Efficiency in Traditional Agricultural Production: Application of Stochastic Frontier Modeling to Food Crop Farmers in South-Western Nigeria. AJEP. 10(2):31-56.
2. Azhar R (1991). Education and Technical Efficiency during the green revolution in Pakistan. Econ. Dev. Cult. Change 39:651-665.
3. Bationo A, Waswa B, Kihara J, Adolwa I, Vanlauwe B, Saidou K (2012). Lesson learned from long-term soil fertility management experiments in Africa. Springer Life Sci. Agric.
4. Bravo-Ureta BE, Evenson RE (1994). Efficiency in agricultural production: the case of peasant farmers in eastern Paraguay. Agric. Econs. 10: 27-37.
5. Chibwana C, Fisher M, Jumbe C, Masters W, Shively G (2010). Measuring the Impacts of Malawi's Farm Input Subsidy Programme. Paper presented at the 2010 annual meeting of the African Association of Agricultural Economists in Cape Town, South Africa.
6. FAO, ITPS (2015). Food and Agricultural Organization of the United Nations and Intergovernmental Technical Panel on Soils, Rome, Italy. Status of the World's Soil Resources (SWSR) – Main Report.
7. FMARD (2011). Federal Ministry of Agriculture and Rural Development. Agricultural Transformation Agenda: Blueprint on Agriculture and Rural Development; A Presentation to the National Economic Management Team by the Honourable Minister of Agriculture and Rural Development, Abuja, Nigeria.
8. FMARD (2016). Federal Ministry of Agriculture and Rural Development. The Agricultural Promotion Policy. Building on the Successes of the ATA, Closing Key Gaps. Policy Strat. Document P 5.
9. Gujarati DN (2004). Qualitative Response Models in Basic Econometrics, African Edition 15:541-591.
10. IFPRI (2012). International Food Policy Research Institute-Africa Lead Report on Assessment of Nigeria Agriculture Transformation Agenda (ATA) and Capacity Building Needs. Africa Leadership Training and Capacity Building Programme.

11. Kelly VA (2006). Factors Affecting Demand for Fertilizer in Sub-Saharan Africa. Agriculture and Rural Development Discussion Paper 23. Washington, D.C.: The World Bank.
12. Lesschen JP, Asiamah RD, Gicheru P, Kanté S, Stoorvogel JJ, Smaling EMA (2003). Scaling Soil Nutrient balances. Rome, FAO.
13. Liverpool-Tasie LSO (2012). Targeted Subsidies and Private Market Participation. An Assessment of fertilizer demand in Nigeria. IFPRI Discussion Paper 01194.
14. NPC (2011). National Population Commission. Projected Population (2008 – 2011).
15. OGADEP (2015). Ogun State Agricultural Development Programme. Annual Report.
16. Okoedo-Okojie DU, Onemolease EA (2009). Factors affecting the adoption of yam storage technologies in the Northern Ecological zone of Edo State, Nigeria. J. Hum. Ecol. 27(2):155-160.
17. Okoruwa VO, Ogundele OO (2004). Technical efficiency differentials in rice production technologies in Nigeria. World J. Agric.Sci. 3(5):53-58.
18. Okoruwa VO, Ogundele OO, Oyewusi BO (2006). Efficiency and productivity of farmers in Nigeria: A study of rice farmers in North Central Nigeria. Poster paper prepared for presentation at the International Association of Agricultural Economists Conference, Gold Coast, Australia P 12.
19. Ricker-Gilbert J, Jayne TS (2008). The Impact of Fertilizer Subsidies on National Fertilizer Use: An Example from Malawi, Paper presented at the American Agricultural Economics Association Annual Meeting, Orlando, FL.
20. Seck A (2015). Fertilizer Subsidy and Agricultural Productivity in Senegal. Department of Economics Cheikh Anta Diop University Dakar, Senegal.
21. Sheldrick WF, Lingard J (2004). The use of nutrient audits to determine nutrient balances in Africa. Food Policy Vol. 29(2):61-98.
22. Stoorvogel JJ, Smaling EMA (1990). Assessment of Soil Nutrient Depletion in Sub-Saharan Africa: 1983-2000. Main Report, 2nd Edition. Winand Staring Centre, Wageningen. Netherlands 1:28.
23. Takeshima H, Liverpool-Tasie LSO (2013). Fertilizer subsidy, political influence and local food prices in sub-Saharan Africa: Evidence from Nigeria. Selected paper prepared for presentation at the Agricultural and Applied Economics Association's 2013 AAEA & CAES Joint Annual Meeting, Washington, DC, August 4-6, 2013.
24. Toenniessenn G, Adesina A, DeVries J (2008). Building an Alliance for a Green Revolution in Africa. Annals of the New York. Acad. Sci. 1136:233-242.
25. World Bank (2014). Fertilizer Consumption (kilograms per hectare of arable land). Food and Agricultural Organization, electronic files and web site.

CHAPTER 11

Values-Based Supply Chains as a Strategy for Supporting Small and Mid-Scale Producers in the United States

Gail Feenstra[1,*] and Shermain Hardesty[2]

[1]*Deputy Director at the Sustainable Agriculture Research and Education Program, University of California, Davis, CA 95616, USA*

[2]*Department of Agricultural and Resource Economics and UC Small Farm Progam, University of California, Davis, CA 95616, USA*

ABSTRACT

Consumers are increasingly interested in the "values" associated with the food they eat and are often willing to pay more for food for which there is a "story" that links farm to fork. The "values" associated with these foods may be that they are locally produced, by small or mid-scale farms, or use production practices that enhance the environment. Wholesale channels that provide marketing options for small and mid-scale producers and support these values are referred to as "values-based supply chains" (VBSCs). Goals of VBSCs are to: (1) provide greater economic stability for producers and others along the supply chain; and (2) provide high quality, regional food to consumers. After a brief overview of VBSCs, we then describe three cases—a specialty food manufacturer, a natural food cooperative and a regional “food hub”—representing different entry points along the food supply chain. We analyze them regarding the common benefits of VBSCs: transparency, fair prices to farmers and ease of purchasing from small and mid-scale producers. We conclude with common themes that emerge for VBSCs of the future and what it will take to strengthen them within regional food systems.

Keywords: specialty foods; values-based supply chains; small scale farms; mid-scale farms; agriculture of the middle; values-based marketing; local foods

1. INTRODUCTION

Consumer interest in local and regional food in the United States (US) has increased substantially over the last four decades. For the most part, this has been measured by the growing demand in direct-to-consumer marketing, which increased from $551 million in 1997 to $1.3 billion in 2012 [1].

Several factors are motivating consumers to purchase locally grown foods. Their top reasons for buying locally grown foods in grocery stores were: freshness (83 percent), supporting the local economy (68 percent), and taste (53 percent) [2]. They are also interested in the environmental impacts of growing food and supporting family farms [3]. The nutritional value of local food was also cited as a common reason. Additionally, local foods are part of a food system that supports social relationships that may or may not be geographically nearby [4]. In short, consumers want “food with a face on it”.

Institutional food service is a growing market channel for local foods. Schools, colleges, universities, corporate cafeterias, hospitals, and government food service sites are all expanding their local foods programs. Feenstra et al. [5] found that colleges and universities in California wanted produce that was locally grown and sustainably produced, but that was also reasonably priced. The US Department of Agriculture (USDA) now has an entire program devoted to supporting Farm to School programs nationwide and tracks progress through a Farm to School census every two years. The most recent census for school year 2013–2014 found that schools purchased $790 million in local food [6]. Institutional buyers are responding to consumer demand.

Supply chains for purchasing local or regional food for retail or institutional buyers are often different than direct-to-consumer channels.

The scale of farmers and ranchers using intermediated channels also differs. Some producers are utilizing food supply chains that are characterized by *partnerships* throughout the supply chain from producers to buyers, and who share a commitment to environmental, social and economic values. The "values" associated with the foods may be that they are locally produced, grown by small or mid-scale farms, or production practices enhance the environment and/or worker welfare. Farmers, in particular, receive a premium for these values. Such food supply chains *are commonly called "values-based supply chains" (VBSCs)* [5,7,8]. Unfortunately, there are no data available regarding the numbers of farms and ranches using such supply chains, nor are there data regarding the sales they generate.

The farmers and ranchers supplying these values-based supply chains (VBSCs) are more likely to be part of a sector called "Agriculture of the Middle"—too large for the short chain direct markets, but too small for the large commodity markets. They generally fall into the $50,000–$500,000 gross sales category. The number of farms in this "middle" size category has been in decline for several decades. They rely largely on their farm income to survive and contribute substantially to the economic viability of rural communities and land stewardship nationwide. Concern has been mounting that if these farms continue to fail, the structure of agriculture will become even more bifurcated (with mostly very small or very large farms on the landscape), and the environmental, social and economic values that mid-scale farmers embody will disappear [9].

VBSCs are increasingly seen as opportunities to create strategic business models for farmers and ranchers of the middle. These farmers and ranchers are most likely to fit this model if they produce significant volumes of high quality, differentiated food products; can operate on a regional scale; and are willing to distribute profits among strategic partners [10,11,12,13,14]. Key benefits of VBSCs for participating farmers include: (1) they are more *transparent* than conventional supply chains; values are communicated throughout the chain, providing buyers and consumers with information they need to pay more for these foods; (2) they provide *higher prices* to participating farmers due to the chain's strategic partnerships and the fact that buyers are willing to compensate farmers for particular values; and (3) buyers in these supply chains are more willing to negotiate with farmers and often absorb some of the transaction costs and *work with farmers or ranchers to source products* on an on-going basis [14]. Overall, the goals of VBSCs are to: (1) provide greater economic stability for producers and others along the supply chain; and (2) provide high quality, regional food to consumers.

Other researchers have described supply chains similar to VBSCs as "alternative food networks" [15] or "nested markets" [16]. There are strong similarities between VBSCs and alternative food networks. VBSCs were developed to enhance the viability of mid-scale food producers and rural communities in the US. There is also considerable research regarding the development of alternative food networks to revitalize small farms and rural economies in Europe. Sonnino and Marsden [15] present three case studies of alternative food networks in England that were similarly developed to reconfigure/re-localize food producers who could not compete on a price basis in commodity markets. Sonnino and Marsden [17] (p. 181) state that alternative food networks are "variously and loosely defined in terms of 'quality', 'transparency', and 'locality'.

Stevenson and Pirog [8] describe how VBSCs rely on shared information among supply chain partners (transparency) to improve productivity, enable rapid response to market changes and effectively engage customers. Sonnino and Marsden [17] (p. 195) refer to transparency within two types of spatial relationships in alternative food systems. First, under "processing and retailing", they describe alternative food networks as "... traceable, and transparent; spatially referenced and designed qualities" (p. 195). Second, under "associational frameworks", they refer to alternative food networks as being "relational, trust-based ... competitive but sometimes collaborative". In the case studies examined by Roep and Wiskerke [18], the governance process involves transparency, strategic partnerships and shared decisionmaking—elements that also characterize VBSCs.

For farmers, earning a premium above commodity prices for the environmental and/or social benefits that they generate through their production practices is an objective of VBSCs. These benefits also contribute to the higher quality of foods in the alternative food networks analyzed by Goodman [19], Ilbery et al. [20], and Sonnino and Marsden [15,17]. Sonnino and Marsden [17] (p. 183) state that, unlike the globalized conventional food system, a key characteristic of alternative food networks is "... their capacity to re-socialize or re-spatialize food" by either its production area or even a specific farm, which provides a quality dimension. van der Ploeg et al. [16] (p. 139) define nested markets as "... alternative agri-food chains which are embedded or (nested) in normative frameworks (and associated forms of governance) which are rooted in the social movements, institutional frameworks and/or policy programmes out of which they emerge ... they are markets with a particular focus (sometimes underpinned by a specific brand, or a specific quality definition, or by relations of solidarity, or specific policy objectives, etc.)". Nested markets, alternative food networks and VBSCs, then, all seem to have several common elements that engage food system partners in environmental, social and economic sustainability in regional food systems.

Since the concept of VBSCs has emerged, various case studies have examined how value chain businesses have fared using this model. Lev and Stevenson [13] highlight the advantages of acting collectively and joint learning within and across the value chains in four cases—(1) Country Natural Beef, a beef rancher cooperative in the NW US; (2) Cooperative Regions of Organic Producer Pools (CROPP)/Organic Valley, a multi-regional organic farmer cooperative; (3) Shepherd's Grain, a business marketing sustainably grown wheat in the NW United States; and (4) Red Tomato, a nonprofit, domestic fair trade business marketing regional produce in the northeastern US. Feenstra et al. [21] describe key findings of five VBSCs in California, namely that: (1) VBSCs are not neat and linear, but complex and more like networks; (2) customers all want to know the story of the farm, its scale, where it is located and how food was produced; (3) aggregators or regional food hubs are emerging to distribute products for many smaller producers; and (4) telling an authentic story is often more important than "local". Two of these cases and four others were part of an analysis of how access to capital, regulatory/business policy and business acumen influence the development of food VBSCs in the Western US [7]. Although VBSCs can support farmers and ranchers of the middle in various contexts, more attention is needed regarding the extent to which values are communicated along the supply chain and how to develop appropriately scaled infrastructure for this mid-scale sector.

One emerging business structure that can be part of VBSCs is a "food hub". The USDA defines a "food hub" as "a business or organization that actively manages the aggregation, distribution, and marketing of source-identified food products primarily from local and regional producers to strengthen their ability to satisfy wholesale, retail, and institutional demand" [22]. Low et al. [1] (p. 13) noted that "regional food hubs have emerged as collaborative enterprises for moving local foods into larger mainstream markets, providing scale-appropriate markets for mid-sized farmers ...". Food hubs can and do support "Agriculture of the Middle" farmers and ranchers in accessing local markets.

As the concept of VBSCs matures, it is imperative to continue to investigate further to understand the ongoing challenges faced by different supply chain partners across the value chain. This paper will use a case study approach to examine three intended benefits (and potential challenges) of VBSCs: (1) transparency in communicating values throughout the supply chain; (2) the ease with which VBSCs can coordinate purchases from small and mid-scale producers; and (3) the extent to which VBSCs provide fair prices to farmers versus remaining competitive in the marketplace.

2. METHODS

For this study, we chose three different VBSCs in Northern California, focusing on key partners at different entry points within each chain (see Figure 1). Each chain is a regional supply chain, consistent with the spatial limitations included in the definition of values-based supply chains. We intentionally chose supply chains with different types of products in order to explore potential differences between meat and produce supply chains.

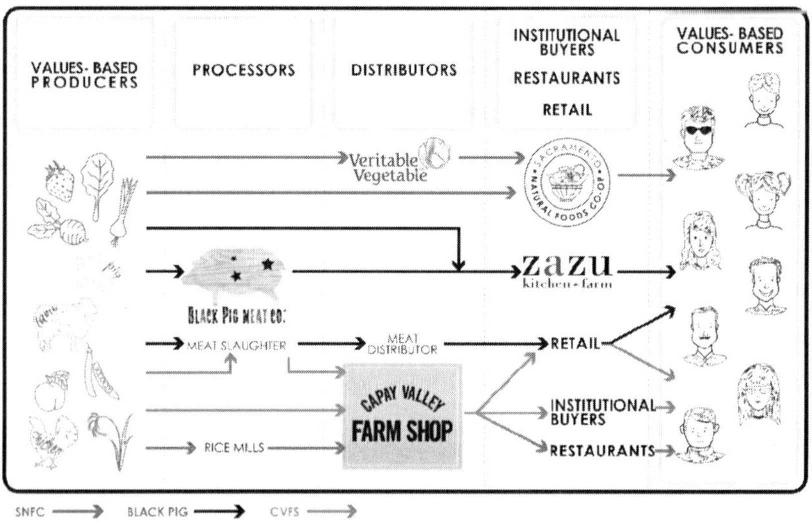

Figure 1. Partners in a values-based supply chain.

The first—Black Pig Meat Company—can be characterized as a processed meat/specialty food business. They process their own salumi (Italian cold cuts predominantly made from pork, of which salami is one variety) and work with partners throughout the supply chain to purchase, process, distribute and sell specialty bacon. The second case—Sacramento Natural Foods Co-op—is a cooperative grocery store that works closely with partners throughout the supply chain—from farmers to distributors—to sell only organic food. We also include information about the primary organic produce distributor used by the cooperative grocer as part of this case since their values are so complementary. This VBSC is very mature; it was established in 1978. Our third case—Capay Valley Farm Shop—is a small, regional "food hub"—a business that aggregates food products from small and mid-scale farmers in the area and distributes them to the broader region of Northern California. These three cases allowed us to view VBSCs from different perspectives along the chain and to note differences in supply chains for different product types (e.g., meat versus produce).

To elicit in-depth information from these supply chains, we opted to use case studies and followed Yin's approach [23]. We created an interview guide to collect data about each of these businesses and their supply chains, including: basic information about the size and nature of the business; values the business espouses; pricing considerations regarding procuring and marketing products from small and mid-scale farms; logistics of sourcing specialty/local/organic food from their suppliers; information/values about their suppliers and how that information is communicated throughout the chain; challenges and benefits of working with small and mid-scale suppliers in particular; the motivations for doing so; decisions or challenges about fair pricing; and final thoughts about the future of their business as part of a VBSC. We conducted several interviews with key partners in each case study during winter/spring of 2016, either in person or by phone to collect the primary data for this study. Follow-up data were collected by phone or email. Since we have known the leaders in each of these case studies for years, we also used supplementary data from previous conversations. Some interviewees provided written materials in which their business had been showcased. We also used material from all of their websites.

We now describe the key elements of each of the three case studies.

3. FINDINGS

3.1. Black Pig Meat Company/Zazu Kitchen and Farm—(Chefs Duskie Estes and John Stewart, Sebastopol, California)

The Black Pig Meat Company (Black Pig) was created in 2005 by Chefs Duskie Estes and her husband, John Stewart, of Zazu Kitchen and Farm [24] in Sebastopol, California, USA. Duskie and John decided to focus on sustainable meat and process their own salumi for the restaurant. They source sustainably grown hogs that they use to make their signature bacon and other pork products. In 2015, Black Pig's revenues totaled approximately $300,000; sales were split fairly evenly, between distributors and direct to consumers at the restaurant. The products are sold at retailers across the country from California to New York.

Today, about 30% of the restaurant's produce comes from a farm a few minutes from the restaurant. It is also the source of some of Zazu's chicken, duck, turkey, pork, sheep, rabbits and goats.

3.1.1. Making Sustainable Bacon and Salumi

Since 2005, Duskie and John have been making bacon, and selling it through retail and wholesale outlets and through their restaurant. They have used two co-packers (the Sonoma Meat Company which is only six miles from their restaurant, and Carlton Farms in Oregon (about 600 miles to the north) that are committed to quality processing and using Duskie and John's curing recipe. Both co-packers operate USDA inspected meat processing facilities. Duskie maintains that Black Pig's bacon has superior flavor, while bacon usually found at stores is "wet cured" and "smoked" with liquid smoke in less than a day, their bacon is dry-cured with brown sugar for up to 21 days and then finished with applewood smoking for another 12 h.

Every Tuesday, the Zazu restaurant closes so Duskie and John can turn the kitchen into a state-inspected "custom exempt" meat processing facility. This means that the meat (salumi and bacon made from California-grown hogs) they process can only be served or sold by the pound "to go" at their restaurant. To facilitate this process, John became a California Department of Food and Agriculture processing inspector. Regional inspectors visit their operation two to three times per month, often unannounced. Duskie and John serve the salumi they make on their menu at the restaurant the other six days of the week.

The values Duskie and John promote are selling quality processed meat (particularly bacon) from animals that are sustainably raised on family farms. The heritage breed pigs they source are raised without antibiotics or hormones, have vegetarian feed, roam free in pastures, and the sows are not put into farrowing crates to give birth to their piglets.

3.1.2. Transparency/Communication about Values

Duskie and John educate their patrons about the food they eat through what they serve at their restaurant and the information they provide. Many of the menu items identify the farm, ranch, bakery, dairy or cheesemaker from which the items came. However, Zazu does not identify the hog farm suppliers for their processed pork products on their menu because there are multiple farms involved. The Zazu website also lists many of their farm suppliers.

Duskie also educates her patrons about how Zazu uses the whole animal. To do that, she offers "pig tastings" at the restaurant, which showcase various parts of the pig, from tail to snout, cooked in different ways. The "pig tastings" are clearly another way to educate consumers about the importance of using the whole animal in order to reduce waste and support smaller producers in a regional food system.

Zazu staff are educated about how the business supports sustainability values throughout the supply chain. They then share this information with customers at the restaurant. In addition, Duskie trains her kitchen staff about how to cut up and prepare all of the pig. She insists that all of her kitchen staff witness a pig being slaughtered so they understand and respect the meat they are preparing and serving.

3.1.3. Sourcing from Small and Mid-Scale Farms

Duskie buys her hogs from three small, regional farms and one large farm in Iowa. These farms are all careful about what they feed their pigs. The pigs are then processed at two meat processing companies. After the bacon is processed at Sonoma Meat Company, Duskie and John take some of it to Golden Gate Meat Company to distribute to local grocers and restaurants.

Duskie says she frequently loses her small hog ranchers who have pigs on pasture; they find it challenging to pasture-raise their hogs because few consumers are willing to pay the price for the end product. Duskie also keeps losing co-packers who are willing to use her lengthy dry cure method. She is constantly looking for suppliers (hog farms and meat processors) who will meet her specifications (of sustainable production practices and pricing). If she could surmount the barriers, she would buy more from small- and mid-scale farms.

Another challenge of working with small farms and processors is the regulatory environment. Duskie says that the USDA regulations currently give preference to large farms, especially with regard to food safety. The cost of meeting all of the regulations is difficult for smaller-scale farmers and processors alike.

3.1.4. Pricing

Putting the time and care into bacon and salumi processing is costly. In fact, Duskie says that their meat is four to five times more expensive than commodity products. For a 12-ounce package of bacon, she charges $13 to $14 locally, and $18 in East Bay stores to cover the costs of the processes necessary to change the way animals are raised in the US and to ensure the viability of small pork producers. Her first priority is choosing producers who give their hogs access to pasture. She tries not to "beat down farmers on price" but she does want them to meet the price she can get from another similar producer.

The average dinner meal at Zazu is about $39 currently, including drinks and tax. It should be closer to $60 if Duskie were to pay everyone (farmer, pork producer, etc.) the way she would like. However, she does nor think customers would be willing to pay that much.

To help make their whole business viable, Duskie and John also produce pork-based value-added products which they sell at the restaurant, such as "swine sweets" (peanut butter cups with fried pork belly or pork fat), "rodeo jax bacon caramel corn", and "bacon piggy pops". These novelty products also use other parts of the whole pig and promote the Black Pig Meat brand.

3.2. Capay Valley Farm Shop (Thomas Nelson, President/General Manager, Esparto, California)

Capay Valley Farm Shop (CVFS) is a food hub that operates both a community supported agriculture (CSA)-style subscription program for consumers and a wholesale aggregation/distribution business targeted at corporate cafeterias, restaurants, online grocers, and a few specialty food markets in San Francisco and the South Bay Area (primarily the Silicon Valley). Its revenues in 2015 totaled just over $700,000, with 80 percent generated from the wholesale business. Thomas Nelson serves as the firm's president and General Manager. He has existing relationships with many producers in the small Capay Valley community in western Yolo County where he is a long-time resident.

CVFS is marketing products from 54 family farms that are all located in western Yolo County, except for one farm that grows rice in a neighboring county. Since rice is not perishable, this farm delivers to CVFS about once a month. When CVFS was founded in 2007, it was a retail operation; it has since transitioned into other market channels. It began selling wholesale in 2011. It is currently phasing out its CSA-style program because of intense competition among online local food delivery services. CVFS expects to grow its wholesale business, and will add a limited offering of grower services (such as deliveries), which Thomas expects will generate about 2% of its overall revenues in 2016. About fifty percent of CVFS's wholesale revenues are generated from sales to restaurants and corporate cafeterias, forty percent from sales to online grocers, and about 10% from sales to specialty retailers. Neither CVFS's customers nor its farms have formal contracts with CVFS.

Most of the original 27 farms that were CVFS suppliers in 2007 are still marketing through CVFS. The food hub handles a wide variety of fresh fruits and vegetables, pasture-raised meats, eggs, and value-added products—including dried herbs, flour, rice, legumes, jams, honey, nuts, olive oil and handmade lavender beauty care products. CVFS has a co-branding program with Capay Valley Vision, a local noprofit organization created to preserve the valley's heritage, natural resources and farmland, and enhance its community health and economic well-being [25]. Capay Valley Vision operates a promotion program for local farms called Capay Valley Grown [26].

3.2.1. The Logistics Behind CVFS

Some CVFS members were already familiar with the logistics of operating a food hub. During the 1980s, they founded a cooperative food hub called YoCal in their community in order to have produce delivery services into the San Francisco Bay Area [27]. Although YoCal was not as well-operated as CVFS, it enabled many of the members to expand their farms and sales considerably. Currently, CVFS is making deliveries three days a week. Typically, the food hub's farms email or phone CVFS at least once a week to identify what products they have available. CVFS sends out an availability list to its customers, and receives their orders (usually by email) the following day. CVFS then relays the order information to each farm. On the following day, a CVFS truck picks up most of the products being ordered and brings them to the warehouse in Esparto to be sorted and consolidated with other items being delivered to each customer. Meat cuts are usually processed at the small USDA-inspected meat plant that is in the same building with CVFS.

3.2.2. Transparency/Communication about Values

Thomas knows the production practices of all the CVFS farms; it is vital that each farm provides high quality product through CVFS. He understands the importance of sharing customer feedback—both good and bad—with a specific farm. He also brings CVFS producers together annually to discuss CVFS's overall successes and challenges. Last year, an additional meeting was held to examine the complications created by the drought in California. When assessing a farm as a new CVFS supplier, he inquires about its products and production practices, which are the attributes that CVFS's customers are most interested in. Customers want to know if the product is organic, and if not, what is different about it (such as pasture-raised for meats and eggs). Thomas considers farm identity to hold more resonance in the long term, since organic production can be industrial-scale.

Thomas believes that cultivating strong relationships with customers is also very important to CVFS' success. He has not yet developed a regular farm visit program for CVFS customers, but he recognizes that farm visits could be a critical tool to building strong relationships. He usually visits each new wholesale customer, and shares information about the small

Capay Valley farming community and the CVFS farms' values and production practices. He noted that if the customer does not share CVFS's values and is very price-sensitive, the relationship will not work out.

When CVFS runs short of a product, it contacts the CVFS farms to determine if they have more of the product. If not, the customer(s) will be advised about the shortage. CVFS does not bring in products from other wholesalers because their product is not likely to be from the Capay Valley.

CVFS reinforces its values and brand identity through its everyday practices. Its trucks have the CVFS logo visible prominently on their doors. When the food hub sends its availability list to its wholesale customers, the farm is identified for each product, along with descriptors such as organic or pasture-raised. The products are shipped in standardized boxes identifying the farm and the product. CVFS's invoices include the farm identity of each product sold.

Certain CVFS customers, such as the on-line produce delivery supplier Good Eggs, always identify the farm on their products as part of their corporate strategy. Other customers only identify the Capay Valley farms occasionally. Cafeterias at the Silicon Valley companies do not appear to identify the farms producing the produce and meats they serve.

For its CSA program, CVFS has been emailing a weekly newsletter to its CSA customers that identifies the farm for each product in the CSA box. The newsletter usually features a profile of one of its network of 54 farms. The profiles are then posted on CVFS' website, along with a link to the farm's website (if it has one).

3.2.3. Sourcing from Small and Mid-Scale Farms

There is considerable size diversity among the farms that supply CVFS. Most of them are small (with less than $250,000 in gross annual revenues), some are mid-sized, and a few are large (with gross annual revenues of at least $1 million). Since many of the CVFS farms grow some of the same crops, CVFS typically includes only the farm with the lowest price on its availability list. However, it will list two farms selling a product (such as heirloom tomatoes) with different prices during peak availability periods. Two farms are also included on CVFS' availability list when some customers have strong preferences for a specific farm.

Thomas advises the CVFS producers to market through multiple channels and to not rely exclusively on CVFS for their sales. He also informs newer CVFS producers about customers' expectations and requirements, particularly regarding packing standards (such as sorting crops by size and appropriate box sizes), proper postharvest handling of their crops, and food safety practices, including the soon-to-be-implemented Food

Safety Modernization Act. Similarly, Thomas will often walk newer producers through an assessment of their production costs. He recognizes that CVFS's viability depends on the farms being financially sustainable.

Thomas has begun conducting some crop planning with specific farms, in order to ensure a better match between specific customers' needs and the farms' production. He noted that CVFS is "... actively open to working with all local farms that are a good fit with [its] customers' needs. In terms of products, it is always great to diversify the offering". He identified dry beans, more fresh fruit such as kiwis, and meats as welcome additions to CVFS's product list.

3.2.4. Pricing

Each farm sets its own prices. Thomas informs a producer if he thinks the price is too high, which the producer adjusts. Conversely, Thomas sometimes tells a producer "you can get more" and the producer usually raises the price; he wants to avoid price-cutting among the farms.

CVFS's prices to its customers usually reflect a 30 percent margin to cover its costs. If a particular product has a high price (such as meats), Thomas will cap the margin contribution per sale to $15 or a similar value.

Thomas noted that the wholesale produce business is very competitive, even for organic crops. There are many large-scale organic farms in California; most CVFS farms lack the economies of size to compete with them. Additionally, CVFS is a small business and it cannot compete with large distributors, such as Sysco which operates with a 12 percent margin. Therefore, CVFS must differentiate itself through its values, and promote its farms' brand identities. This effort requires constant education by the food hub. Chefs and other buyers need to be educated about the CVFS farms and the more costly practices they engage in to produce organic crops, pasture-raised or free-range poultry. Thomas believes that the recently passed legislation to increase California's minimum wage by 50% to $15 per hour will be challenging to the farms, CVFS, and other small rural businesses, particularly when the same crops can be sourced at lower price from other states or Mexico.

3.3. Sacramento Natural Foods Co-op Produce Supply Chain (Kerri Williams (KW)-Produce Manager and Paul Cultrera- General Manager, Sacramento, California and Bu Nygrens (BN)-Co-Owner, Veritable Vegetable, San Francisco, California)

Sacramento Natural Foods Co-op (SNFC) is a member-owned grocery cooperative store founded in 1973, near downtown Sacramento, California. With sales of $33 million in 2015, it is the largest single-store grocery cooperative in California. Sometime this fall, SNFC expects to move into a larger store (42,000 square feet) that is being built. It carries a full line of grocery products, including meat, produce, deli items, shelf-stable foods, health and beauty products and cleaning products.

For over forty years, SNFC has been owned by thousands of members of the community. Consumers become members of SNFC because they share its values which are: (1) cooperation; (2) sustainable practices; (3) support for healthy choices; (4) belief in cooperative economics; and (5) open, honest and trustworthy business practices [28].

With the largest 100 percent organic produce section in the United States, SNFC buys about 30% of its produce direct from local farmers. However, most of its produce has been supplied since 1978 by a San Francisco-based organic produce distributor, Veritable Vegetable (VV). VV supplies organic produce from over 600 farms primarily to grocery stores, and some restaurants, corporate campuses, juice cafes, and schools throughout California, and parts of the southwestern US and Hawaii. VV's revenues have grown from $45,000 in 1975 to over $50 million in 2015.

As a national leader in the organic produce industry, VV has contributed to the development of organic certification standards and composting practices, and has advocated for sustainable food and agriculture legislation. VV's core values are integrity, community, excellence, innovation and sustainability [29].

All but one of SNFC's core values listed above (belief in cooperative economics) are reflected in its produce sourcing and promotion practices. It is likely that the longevity of SNFC's relationship with VV is largely attributable to their shared core values and the similarity in the business practices—particularly those related to transparency in communicating their values, sourcing from small- and mid-scale producers, and paying fair prices to producers.

3.3.1. Transparency/Communication about Values

SNFC actively supports its core values to differentiate itself from other stores in the Sacramento region. Its produce department's commitment statement describes three of SNFC's values—cooperation, support for healthy choices, and sustainable practices: "Sacramento Natural Foods Co-op is part of a larger food community that includes farmers, wholesalers, manufacturers, shoppers and the Earth. We purchase only the highest quality certified organically grown produce and emphasize direct buying relationships with local small family farms".

VV's values of integrity and community are related to transparency. Its value of integrity is exemplified through its founding principles of collaboration, cooperation and interconnectedness and belief in cultivating and nourishing long-term relationships with vendors. As a founder of the US organic industry, VV worked with many struggling growers to distribute their produce, promote sustainable farming practices and policies, and extend knowledge about organic food and agricultural issues. Its commitment to sustainability also includes operating a "green fleet" of hybrid tractors and trailers to handle and transport produce to its customers. BN, one of VV's owners, admits that although VV does not know specifically about the core values of the farms it sources from, "we tend to buy from people we know and trust". There are dozens of farms that she can identify as being suppliers to VV for twenty years or more.

VV does not have written contracts with any of the farms that are its suppliers; instead, VV tries to be consistent and reliable in its purchasing practices. As a testament of its value "open, honest and trustworthy business practices", SNFC similarly does not have written contracts with any of the farms or distributors from whom it buys produce. If there are problems with the delivery, KW, SNFC's Produce Manager, will contact the grower or distributor immediately. Similarly, if VV receives poor quality product, it rejects the delivery and promptly informs the grower.

Local growers (within 200 miles of the store) that SNFC buys from deliver their produce to the store. Additionally, KW makes a special effort to order produce that is local from VV. She estimates that at least 80 percent of the farms providing produce to SNFC through VV are local. SNFC identifies the farm and its location for every produce item on display. SNFC's Produce Manager noted that she and her staff have a personal relationship with each of the local growers. Its website's produce page has links to the websites of the local growers who sell direct to SNFC.

VV's availability list has been source-identified since 1979. VV sends a customized product availability sheet daily to SNFC. A typical list has about 750 items; each item is identified by variety, size/grade, pack type, price, certifier, and grower label. When ordering online, customers can utilize filters to sort by farm name, produce item, distance from VV's warehouse (such as 150, 250 or 400 miles), and Fair Trade Items. Customers also have the option to order by farm name.

3.3.2. Sourcing from Small- and Mid-Scale Producers

SNFC buys about one-third of its produce direct from small- and mid-scale farms. Its Produce Director intentionally spreads her purchases among smaller growers, and she gets a couple of new growers each year.

Similar to SNFC, VV strives to spread its business around to support the long-term financial viability of the growers who it works with. It offers multiple grower labels for multiple produce categories. Currently, VV is actively trying to match its mid-scale farm suppliers with independent grocers, because they have smaller volume requirements. VV rarely buys produce from small-scale growers because their product volumes are not usually enough to meet most customers' order quantities. In addition, carrying a large variety of labels creates logistical issues for VV.

Food safety is becoming a major concern for VV even though it is certified by USDA for Good Agricultural Practices (GAPs) and Good Handling Practices. BN is concerned that the implementation of the national Food Safety Modernization Act (FSMA) will cause VV to lose some of its customers and/or some of its farm suppliers, particularly the smaller ones. She remarked that VV is "... increasingly hampered by certain customers who only want product from GAP certified farms, and if there are farms we work with that do not have GAPs then we can't sell to those customers. This [problem] will be increasing, and is already an added expense to us for managing our vendors, and their products".

SNFC's Produce Manager admits that it can be challenging sometimes for SNFC to get smaller farm operators to understand there are differences in the requirements of produce sold at the co-op compared to farmers markets. For example, SNFC is not able to display product that is overly large or misshapen. However, she is always encouraging newer farms to grow unusual crops, such as finger limes, salsify and other unique herbs.

3.3.3. Paying Fair Prices to Producers

SNFC's Produce Department has a gross margin rate target of 33.75 percent. KW noted that the Co-op is committed to paying our growers a fair wage ..." She said that "Cheapest is not usually the best quality". It will pay premium prices for certain well-known farm labels, including some which are delivered direct. Nevertheless, SNFC is able to claim that its organic produce is almost always priced lower than the same item at other natural food markets in the area.

VV averaged a gross margin rate of 32 percent in 2015. It does not regularly give smaller-scale farms a premium price. BN remarked "We readily share market price information with growers ... But there are situations when VV will pay a premium price for a specific label and then take a lower mark-up for a product" [30]. Additionally, some VV customers will pay premium prices for certain labels.

4. DISCUSSION

As we have discussed elsewhere [21], these VBSCs cannot be characterized by neat, linear relationships. Rather, they are more like networks that interact with each other and with conventional supply chains in complex ways. Figure 1 shows the key partners in each of the three cases representing different points along the supply chain, and each case involves several marketing channels for small and mid-scale producers to move their products through VBSCs. What distinguishes the particular supply chains in our cases is that they strive to uphold and communicate a set of social and environmental values that prioritize quality, cooperation, inclusiveness, equity, sustainability and health. As previously noted, the "values" in VBSCs may be that the products are produced by small- or mid-scale farms, are locally/regionally produced, and/or the farms and/or ranches use production practices that enhance the environment and/or worker welfare. Profitability is also important; rather than standing alone, it is connected to the social and environmental values.

Rather than relying on market transactions, these VBSCs represent long-term relationships among firms with shared values committed to the long-term sustainability of all of the partners in the supply chain. The buyers and farmers socialize at local events, as well as reconnecting at an annual farm conference. They can be considered to be socially overlaid networks [31].

Each of the three VBSC cases faces unique, yet similar challenges. Similar to Mikkola's analysis [31] of three different vegetable supply chains, we found distinct differences but also overall similarities in challenges on the chain level. In this section, we highlight the key challenges faced by supply chain partners and some innovative ways they are managing to overcome some of these challenges. We make cross-cutting comparisons of the strengths and challenges displayed by each of the three cases related to three key VBSC characteristics: *transparency/communication* about values, *sourcing from small and mid-scale producers* and *fair pricing practices*. Our discussion is summarized in Table 1.

Table 1. Strengths and challenges of values-based supply chains[1].

	Strengths	Challenges
Transparency/ Communication about Values	Farms identified for consumers on menus (BP) and store displays (SNFC). Farms profiled on website (BP, CVFS, SNFC). Retailer holds 'Meet the Farmer' event for customers (SNFC).	Difficult to identify ranches for ingredients that are processed by multiple facilities (BP). Workplace cafes do not identify farms supplying ingredients (CVFS).
	Retail co-op members need less education about farmers' values (SNFC).	Consumers need to be educated why values-based products cost more (BP, VV, SNFC)
	Farms identified for staff on invoices, and with location included on availability list and shipping cartons (CVFS, VV). Farms profiled in newsletters for staff (VV). Holds monthly staff luncheon with a grower (SNFC).	Some restaurants and retailers advertise that they are sourcing from growers who they rarely buy from (BP, SNFC).
	Buyers share customer feedback (positive and negative) with growers (CVFS, VV).	
	Share values with culinary, agricultural and/or cooperative community (BP, CVFS, VV, SNFC).	
Sourcing	**Strengths**	**Challenges**
	Vertically integrated into processing (BP).	Inconsistent livestock supply (BP).
	Purchase entire animal/unpopular cuts from ranchers (BP).	Lack of nearby food safety compliant processors (BP).
	Spread out purchases among multiple growers (CVFS, SNFC).	Unable to wholesale self-processed products (BP).
	Coach growers about understanding customer requirements (CVFS, SNFC) and guide growers on assessing their production costs, crop planning and diversifying markets (CVFS).	Large customers' volume requirements limit grower options (VV).
		Compliance and recordkeeping for insurance certification and/or with new food safety regulations may be too difficult or costly for small growers (CVFS, SNFC, VV).
Fair Pricing	**Strengths**	**Challenges**
	Understand need to pay growers a fair price (BP, CVFS, VV, SNFC).	Paying small growers enough to cover their costs may be unaffordable (BP).
	Lower financial return acceptable due to cooperative structure (SNFC).	Customers must be willing to pay higher price for products with values (BP, VV, SNFC).

[1] BP, Black Pig Meat Company. SNFC, Sacramento Natural Foods Co-op. CVFS, Capay Valley Farm Shop. VV, Veritable Vegetable.

4.1. Transparency/Communication about Values

The partners in all three cases have made a supreme effort to be transparent about who they buy from (farmers/ranchers), particularly when the relationships are direct. Information about the producers' values must flow through each link along the supply chain in order for producers to be paid fairly.

In our cases, suppliers identify the farms with their locations on their availability lists and their shipping cartons, and they include the farms' names on their invoices. They profile farms in newsletters for their customers. However, the distributor sometimes has difficulties obtaining information from a farm about its values. In addition, information and feedback are shared up and down the supply chain to improve the efficiency or quality of the business relationships. Managers share feedback from customers and buyers, and internally with staff. The retailer holds a monthly luncheon with a grower for its staff to become more familiar with the farms. In each case, the type of communication is similar to "socially overlaid networks" described by Mikkola [31] in which actors knew each other from local contexts and had a common history as opposed to the "strategic networks" described by Jarillo [32]. In addition to sharing information and values *within* these supply chains, the partners also share their values by being involved in broader agricultural and culinary communities.

Written forms of communication are most common for sharing information with consumers. Farms are identified on restaurant menus and the retailer's produce displays (along with locations). Farms are profiled on websites. The restaurant and retailer also employ innovative forms of communication involving one-on-one human interaction, such as "pig tastings" to educate consumers about the importance of using the whole animal. The retailer hosts "Meet the Farmer" events for its shoppers and sponsors on-farm events organized by local culinary organizations.

Communication is a very high priority for all of our cases because being part of a VBSC means that values must be communicated throughout the supply chain. However, the process of continually educating consumers about the higher prices they are paying for the values embedded in the foods that they are being served or buying can be costly and time consuming. In some cases, a website cannot convey information about values as effectively as visiting a farm and talking with the farmer about his or her production practices. Nevertheless, this consumer education process is inherently easier when the retailer is a cooperative with consumer members. As discussed by Hingley et al. [33], consumer food cooperatives are ideally suited to participate in VBSCs because their members have already committed to supporting environmental sustainability and responsible and ethical behavior, as well as being experienced in "... combining business and social/community networks"(p. 352).

However, there can be logistical challenges conveying information along multiple links in a supply chain. For example, it is difficult for the restaurant to identify the source on an ingredient when multiple producers and processors are involved. Workplace cafes in the Silicon Valley do not identify the CVFS farms supplying the ingredients in the artisanal dishes they prepare.

Receiving a premium for products (fresh produce, meat, or other product) requires a high degree of integrity to maintain trust between supply chain partners. However, in her thesis regarding the benefits of farm-identified produce in Bay Area grocery stores, Lerman found [34] that perceptions about the extent of these benefits differed between farmers and buyers, and specific contexts mattered. The challenges included: building inter-organizational trust beyond interpersonal relationships; dealing with the volatility of the marketplace; and maintaining authenticity in values claims. Our three VBSCs have experienced instances where buyers have taken advantage of well-known farm names, using them in advertising to obtain more business, when, in fact, they may only have purchased from the farm once or twice. Such malfeasance is likely to destroy socially overlaid networks, but is likely to be tolerated in a market or network relationship for reasons of efficacy. All of the VBSC partners know how important integrity and honesty are in creating authentic relationships with customers or suppliers.

4.2. Sourcing from Small and Mid-Scale Farms

These VBSC cases promote sourcing products from small and mid-scale producers to help them stay financially viable. Food retail cooperatives in the US have a long-established commitment to sustainability, including supporting small- and mid-scale farms [35,36,37]. The SNFC cooperative has a duplex produce supply structure; while VV is its primary supplier, SNFC also sources direct from local farms and regularly adds new farm suppliers. Both the cooperative and the distributor spread their purchases among several smaller growers instead of buying from just one or two, so that everyone has some sales. The majority of producers involved in CVFS are small and mid-scale farms. This food hub is even guiding the producers in assessing their production costs. The restaurant (Black Pig/Zazu) buys the whole animal instead of just the most popular parts because it is difficult for small- and mid-scale farmers to sell certain cuts. Additionally, both the cooperative and the food hub have taken on didactic roles, coaching newer farmers in understanding buyer requirements, such as what type of product is acceptable, packing standards, postharvest handling and food safety. These interactions have long histories and are similar to the socially overlaid networks described by Mikkola [31].

One of the primary challenges of sourcing from small and mid-scale farms is volume requirements. Given the large size of its business, the distributor often utilizes a network relationship structure in order to operate in an effective and efficieint manner. Since the administrative costs to source from small-scale farms are higher, it does not do business with them. However, it is focusing on matching its mid-scale farm suppliers with its independent retailer customers.

Regulatory constraints, including food safety, have been a challenge for producers for a long time. The USDA's meat processing regulations have reduced the number of facilities in California that will process livestock for smaller-scale ranchers; this constrains the supply of sustainably-raised hogs. The restaurant has addressed this challenge by vertically integrating upstream; it closes its restaurant once a week to become a meat processing plant, and it also produces snack foods from the byproducts. However, the regulations prohibit wholesaling the processed meat products; they can only be served or sold at the restaurant.

The biggest challenge on the horizon related to sourcing from small- and mid-scale producers appears to be compliance with food safety requirements. The first difficulty is that it has become a standard business practice for produce distributors, restaurants and grocery retailers to carry product liability insurance because they could face bankruptcy if they are involved in a food safety outbreak. The insurance companies are requiring their policyholders to only use food suppliers that are audited and certified by a specific food safety certification company. Secondly, the compliance costs associated with the soon-to-be-implemented regulations for the federal Food Safety Modernization Act (FSMA) are likely to threaten participation of some small and mid-scale farms in these supply chains. It may be more profitable for them to stop selling wholesale to reduce their annual sales to less than $500,000, which would then exempt them from most of the FSMA's requirements.

4.3. Fair Pricing for Farmers

In market relations, farmers are considered to be interchangeable suppliers; if one farmer goes out of business, a buyer can find another farmer to supply the same commodity. In a VBSC, the farmer is valued for his or her production practices, and is considered to be an equal partner in the supply chain. The VBSC partners are committed to the long-term sustainability of all of the partners.

All three cases are willing to pay farmers a fair price, or as much as they can afford. They cannot always pay a premium to the smallest producers (VV), but they expect to pay their values-based producers what similar producers are making in these values-based markets. They are willing to reduce their own margins sometimes to help out small growers. They are able to offer fair prices because they have distinguished themselves as VBSCs in the marketplace and their customers are also willing to pay more (marginally).

Through their transparency, the VBSC partners have helped the farmers establish their "brand identities" with consumers—their farms' names—which differentiate them by representing the farms' values. Both VV and

SNFC will pay premium prices for certain well-known farm brands, including some farms that deliver direct to SNFC.

Unlike most food retailers, consumer-owned cooperatives in the US are part of socially-embedded networks. They have a long-established commitment to sustainability, including supporting small- and mid-scale farms by paying them fair prices [36]. Unlike shareholders of grocery companies with publicly traded stock, members of consumer cooperatives do not have expectations of high returns and have more flexibility with pricing. They are satisfied as long as their co-op fulfills its social and environmental missions, while adding a small increase to its financial reserves each year.

Lerman [34] found that VBSCs can be challenged by tradeoffs between consumers' expectations about prices and what prices farmers need. She noted "Product differentiation and consumers' willingness to pay is a significant factor in determining the price for products sold not only through VBSCs but in any market. A growing number of consumers recognize the value in supporting farmers growing products embedded with particular values and will pay a price premium for those values"(p. 86). However, she concluded that much more consumer education is needed about the values. US consumers are still price-sensitive when purchasing food because the globalized and highly industrialized food system has been providing them with "cheap" food.

Information about the producers' values must flow through each link along the supply chain in order for the producer to be paid fairly. The key challenge on pricing is negotiating a price that works for everyone. Sometimes farmers get less than they would like; other times, buyers get less. As long as all parties feel that their best interests are considered, sales will continue. The bottom line is that partners throughout these supply chains trust the others to be as fair as possible. This level of authenticity requires constant communication and is intricately linked to the transparency discussed earlier.

5. CONCLUSIONS

Although the experiences of these three case studies are not necessarily generalizable to other VBSCs, the case descriptions and the cross-case comparisons in the preceding discussion give us deeper insight into how VBSCs are working and the challenges they face. They also provide insights about the potential viability of VBSCs as a relatively new organizational structure for developing regional food systems.

In the last decade, much has been accomplished to expand and strengthen these VBSCs, but many VBSCs are still tenuous financially. As SNFC's relationship with VV approaches forty years, it is clear that its durability is largely attributable to how closely their values are aligned. We expect more such success stories as newer VBSCs mature.

Food hubs and values-based produce distributors—such as VV—can and do support "Agriculture of the Middle" farmers and ranchers in accessing local markets. The number of food hubs in the U.S. has increased markedly the last five years. CVFS and other food hubs are working with smaller-scale growers to improve the growers' production efficiency. By aggregating product, the food hubs are also reducing distribution costs. Such economies of size are critical when consumers are prone to comparing prices. However, food hubs face many challenges, including capitalization, liability issues, compliance with food safety regulations and managing human resources [38]. However, the potential role for food hubs within VBSCs and their impacts on producers and others may be promising. More study is needed to understand these challenges and how they can be addressed.

The partners in all three cases would like to find ways to make changes to the larger food system regarding food safety requirements and regulations, investment in mid-scale infrastructure, and higher level of consumer awareness—so they can be more competitive in the marketplace. Research is needed to measure the costs—particularly for small- and mid-scale farms—to comply with their customers' food safety certification requirements and the new FSMA regulations. The number of smaller-scale farms in the US that decide to "downsize" in order to become exempt from most FSMA regulations should also be monitored. Some funding is currently available to conduct feasibility studies for building mid-scale infrastructure, such as processing facilities. However, policymakers need to be shown how such investments can have significant economic development impacts by localizing their region's food supply. Such support could then entice the creation of state and/federal government programs to finance mid-scale infrastructure through the formation of community development districts, as well as providing tax credits and other incentives for local private investment.

Perhaps the most important issue is whether and how VBSCs can enhance their communication about their values and build demand for their values-based products. More retailers are needed to participate in VBSCs. They will join in if their customers demand more value-based products and are willing to pay for them. Therefore, transparency both within the supply chain and within the larger communities of which they are a part (agriculture, distribution, culinary) needs to be expanded. Although the chef community is now fairly conversant with the "farm to fork" theme, most consumers are still unaware of where their food comes from or the values it carries. They need to understand the true cost of their food. Starting with the youth, considerable effort is needed to educate consumers about the connections between the health of the environment, the economy, the community, and their own health. Building strategic alliances outside these immediate supply chains to programs such as "Farm to School", "Farm to Hospital", corporate or institutional wellness programs, ready-to-cook meal kit programs, urban gardens and farms, and community health events, could go far in expanding the understanding and commitment of the public to a more sustainable food system. Such programs should include rigorous evaluation components, in order to identify the most long-term cost-effective education structures.

ACKNOWLEDGMENTS

This project was supported in part by Agriculture and Food Research Initiative Competitive Grant No. 2015-68006-22906 from the USDA National Institute of Food and Agriculture. We also wish to thank our colleague on this project and the lead Principal Investigator, Larry Lev, as well as all the supply chain partners who collaborated with us on the project and interviews. We also wish to thank Claire DeCamp for designing Figure 1, describing the many elements and actors of several values-based supply chains.

REFERENCES

1. Low, S.A.; Adalja, A.; Beaulieu, E.; Key, N.; Martinez, S.; Melton, A.; Perez, A.; Ralston, K.; Stewart, H.; Suttles, S.; et al. Trends in U.S. Local and Regional Food Systems; U.S. Department of Agriculture, Economic Research Service: Washington, DC, USA, 2015.
2. Food Marketing Institute. U.S. Grocery Shopper Trends; Food Marketing Institute: Arlington, VA, USA, 2011. 3. Brown, C. Consumers' preferences for locally produced food: A study in southeast Missouri. Am. J. Altern. Agric. **2003**, 18, 213–224.
3. Selfa, T.; Qazi, J. Place, taste, or face-to-face? Understanding producer–consumer networks in "local" food systems in Washington State. Agric. Hum. Values **2005**, 22, 451–464.
4. Feenstra, G.; Allen, P.; Hardesty, S.; Ohmart, J.; Perez, J. Using a supply chain analysis to assess the sustainability of farm-to-institution programs. J. Agric. Food Syst. Community Dev. **2011**, 1, 69–85.
5. 2015 Farm to School Census. Food and Nutrition Service, United States Department of Agriculture. Available online: https://farmtoschoolcensus.fns.usda.gov/farm-school-works-stimulate-local-economies (accessed on 1 May 2016).
6. Hardesty, S; Feenstra, G.; Visher, D.; Lerman, T.; Thilmany-McFadden, D.; Bauman, A.; Gillpatrick, T.; Nurse Rainbolt, G. Values-based supply chains: Supporting regional food and farms. Econ. Dev. Q. **2014**.

7. Stevenson, G.W.; Pirog, R. Values-based supply chains: Strategies for agrifood enterprises of the middle. In Food and the Mid-Level Farm: Renewing an Agriculture of the Middle; Lyson, T.A., Stevenson, G.W., Welsh, R., Eds.; MIT Press: Cambridge, MA, USA, 2008; pp. 119–143.
8. Kirschenmann, F.; Stevenson, S.; Buttel, F.; Lyson, T.; Duffy, M. Why worry about the agriculture of the middle. In Food and the Mid-Level Farm: Renewing an Agriculture of the Middle; Lyson, T.A., Stevenson, G.W., Welsh, R., Eds.; MIT Press: Cambridge, MA, USA, 2008; pp. 3–22.
9. Agriculture of the Middle: Characterizing Ag of the Middle and Values-Based Food Supply Chains. Available online: http://www.agofthemiddle.org/archives/2012/01/characterizing. html#more (accessed on 3 May 2016).
10. Bloom, J.D.; Hinrichs, C.C. Moving local food through conventional food system infrastructure: Value chain framework comparisons and insights. Renew. Agric. Food Syst. **2011**, 26, 13–23.
11. Brown, C. Values-Based Food Supply Chains, Powerpoint Presentation in Value Chains Teaching Materials for a Course in Agricultural Marketing. Available online: http://www.agofthemiddle.org/archives/2011/07/value_chain_tea.html (accessed on 3 May 2016).
12. Lev, L.; Stevenson, G.W. Acting collectively to develop mid-scae food value chains. J. Agric. Food Syst.Community Dev. **2011**, 1, 119–128.
13. Stevenson, G.W.; Clancy, K.; King, R.; Lev, L.; Ostrom, M.; Smith, S. Midscale food value chains: An introduction. J. Agric. Food Syst. Community Dev. **2011**, 1, 27–34.
14. Sonnino, R.; Marsden, T. Beyond the divide: Rethinking relationships between alternative and conventional food networks in Europe. J. Econ. Geogr. **2006**, 6, 181–199.
15. Van der Ploeg, J.D.; Jingzhong, Y.; Schneider, S. Rural Development through the construction of new, nested, markets: Comparative perspectives from China, Brazil and the European Union. J. Peasant Stud. **2012**, 39,133–173.
16. Sonnino, R.; Marsden, T. Alternative food networks in the south west of England: Towards a new agrarian eco-economy? Res. Rural Sociol. Dev. **2006**, 12, 299–322.
17. Roep, D.; Wiskerke, J. On governance, embedding and marketing: Reflection on the construction of alternative sustainable food networks. J. Agric. Environ. Ethics **2012**, 25, 205–221.
18. Goodman, D. Rural Europe redux? Reflections on alternative agro-food networks and paradigm change. Sociol. Rural. **2004**, 44, 3–16.
19. Ilbery, B.; Maye, D. Food supply chains and sustainability: evidence form specialist food producers in the Scottish/English borders. Land Use Policy **2005**, 22, 331–344.
20. Feenstra, G.; Visher, D.; Hardesty, S. Developing Values Based Distribution Networks to Enhance the Prosperity of Small and Medium Sized Producers: Full Study. Available online: http://asi.ucdavis.edu/programs/sarep/publications/foodandsociety/developingvalues baseddistributionnetworksfullstudy.pdf (accessed on 23 May 2016).
21. Barham, J.; Tropp, D.; Enterline, K.; Farbman, J.; Fisk, J.; Kiraly, S. Regional Food Hub Resource Guide; U.S. Department of Agriculture, Agricultural Marketing Service: Washington, DC, USA, 2012.
22. Yin, R.K. Case Study Research: Design and Methods, 4th ed.; Sage Inc.: Thousand Oaks, CA, USA, 2009.
23. Black Pig Meat Company. Available online: http://www.blackpigmeatco.com/ (accessed on 3 May 2016).
24. Capay Valley Vision. Available online: http://www.capayValleyvision.net (accessed on 3 May 2016).
25. Capay Valley Grown. Available online: http://capayvalleygrown.net (accessed on 3 May 2016).
26. Leff, P.; Hardesty, S. YoCal Produce Cooperative—The Growers' Story and the Cooperative Principles. Available online: http://sfp.ucdavis.edu/files/143596.pdf (accessed on 28 July 2016).

27. Sacramento Natural Foods Coop. Available online: https://sacfood.coop (accessed on 3 May 2016).
28. Veritable Vegetable. Available online: http://veritablevegetable.com (accessed on 3 May 2016).
29. Nygrens, B. (Veritable Vegetable, San Francisco, CA, USA). Personal communication, 6 April 2016.
30. Mikkola, M. Coordinative structures and development of food supply chains. Br. Food J. **2008**, 110, 189–205.
31. Jarillo, J.C. On strategic networks. Strateg. Manag. J. **1988**, 9, 31–41.
32. Hingley, M.; Mikkola, M.; Canavari, M.; Asioli, D. Local and sustainable food supply: The role of European retail consumer co-operatives. Int. J. Food Syst. Dynam. **2011**, 2, 340–356.
33. Lerman, T. Who Benefits from Marketing Farmers? A Case Study of Farm-Identified Produce Distribution in Bay Area Grocery Stores. Masters' Thesis, Community Development Graduate Group, University of California, Davis, CA, USA, 2013.
34. Jarosz, L. Understanding agri-food networks as social relations. Agric. Hum. Values **2000**, 17, 279–283.
35. Katchova, A.; Woods, T. Local foods and food cooperatives: ethics, economics and competition issues. In The Ethics and Econonics of Agrifood Competition; James, H., Ed.; Springer: New York, NY, USA, 2013; pp. 227–242.
36. McLaughlin, E.; Hardesty, S.; Gomez, M. Can local food markets expand? In Growing Local: Case Studies on Local Food Supply Chains; King, R., Hand, M., Gomez, M., Eds.; University of Nebraska Press: Lincoln, NE, USA, 2014; pp. 313–329.
37. Matson, J.; Sullins, M.; Cook, C. The Role of Food Hubs in Local Food Marketing; Service Report 73; U.S. Department of Agriculture, Rural Development: St. Louis, MD, USA, 2013.

CHAPTER 12

The Common Agricultural Policy 2014–2020: scenarios for the European agricultural and rural systems

Concetta Nazzaro* and Giuseppe Marotta

Department of Law, Economics, Management and Quantitative Methods, University of Sannio, Benevento, Italy

ABSTRACT

The reform of Common Agricultural Policy for 2014–2020 aims at promoting greater competitiveness, efficient use of public goods, food security, preservation of the environment and specific action against climate change, social and territorial equilibrium, and a more inclusive rural development. The economic crisis of the last years, the globalization, the increasing pressure on natural resources, the increasing diversity of agriculture and rural areas in the new enlarged Europe, and citizens' expectations regarding the environment, safety and food quality, health and welfare, preservation of the countryside, biodiversity and climate change, and the unequal distribution of resources are some of the major issues that this policy has faced, in a production context directed by principles of fairness and efficiency.

The policy instruments to address these challenges and achieve specific objectives appear greatly transformed and made up of two complementary pillars (direct payments and market measures, the first, and rural development, the second).

The perspectives that the new Common Agricultural Policy provides in order to reach a competitive repositioning of the European agricultural and rural systems have been the theme of the 51st SIDEA Conference aimed at promoting a scientific debate on implementation of this policy in Italy.

The papers selected for this special issue focus on the new perspectives for European farms, agri-food supply chains, and rural territories of the new Common Agricultural Policy.

Keywords: Sustainability supply chain, Farm diversification, Common Agricultural Policy

1. BACKGROUND

Agriculture is an essential component of European economy and society. For this reason, the new Common Agricultural Policy (CAP) aims at promoting greater competitiveness, efficient use of public goods, food security, respect of the environment and action against climate change, and social and territorial equilibrium. These goals are included within the main objective of a sustainable, smart, and inclusive growth of the rural Europe in the "Europe 2020 Strategy."

To this end, the CAP 2014–2020 is based on a first pillar, promoting "greener" and more equally shared support, and on a second one, more focused on competitiveness, innovation, climate change, and environment. Reserving support only to "active" farmers and remunerating the collective services (public goods) which they provide to society enhance the effectiveness and efficiency of policy instruments and help to legitimize the CAP and, in general, to strengthen the social role of agriculture (Marotta and Nazzaro 2013).

Essentially, the CAP reform for the period 2014–2020 aims at making the European agricultural sector more dynamic, competitive, and effective in pursuing the objective of the EC Communication "The CAP towards 2020: meeting the food, natural resources and territorial challenges of the future" (European Commission, 2010). The reform therefore highlights many unresolved issues, some of which are strategic and need a broader scientific debate discussing the main themes, to which the policy refers, and its implications for European agriculture. It seeks to address in an unprecedented global scenario economic, environmental, and social challenges.

The economic crisis, the market globalization, the increasing pressure on natural resources, the increasing diversity of agriculture and rural areas due to the enlargements of the EU, and citizens' expectations regarding the environment, safety and food quality, health and welfare, preservation of the countryside, ecosystem services, biodiversity and climate change, and the unequal distribution of resources are some of the major issues that this policy has faced, in a production context directed by principles of fairness and efficiency.

The necessary instruments to address these challenges and achieve specific objectives appear greatly transformed and made up of two complementary pillars; the first is focused on direct payments and market measures and the second on the multi-annual measures of rural development.

About the direct payments, the European Commission outlines the importance of a redistribution, redesign, and better targeting of the support, based on objective and equitable criteria, easy to understand by the taxpayer. These criteria should be both economic (noting the "income support" element of direct payments) and environmental (reflecting the public goods provided by farmers), with a better support targeted towards active farmers. A more equitable distribution of funds should be organized in an economically and politically feasible way with a transition to avoid major disruption.

The rural development policy (second pillar of the CAP) aims at promoting competitiveness, sustainable management of natural resources, and balanced development of rural areas by means of more specific measures, giving the Member States, through co-financing, the flexibility to meet their priorities. This policy is strategic to face the European environmental challenges; therefore, the scientific debate on these issues is wide and articulated.

In the context of the present-day crisis, the innovations introduced in the rural development policy in terms of integrated intervention strategies and creation of local partnerships for a bottom-up governance of rural areas, realized through the Leader approach, become the strategic mission of new programming instruments. The multilevel governance processes of the integrated local development, with the active involvement of local stakeholders to mobilize material and immaterial resources, as part of a wider sharing of participatory development projects (Peterson 2013), are at the center of Community-Led Local Development for 2014–2020, as a model for collective and integrated action for programmed intervention on a territorial basis (European Commission 2014).

In summary, the new CAP should contribute to the development of a socially responsible and more balanced European agriculture, at the territorial and environmental levels.

Given such a framework, the 51st SIDEA Conference (Italian Society of Agricultural Economics), held in September 2014 at the University of

Sannio, in Benevento (Italy), focused on the theme about "The Common Agricultural Policy 2014–2020: scenarios for the European agricultural and rural systems." This Conference aimed at promoting a scientific debate on the perspectives that the new CAP provides in order to reach a competitive repositioning of the European agricultural and rural systems, in particular, through the deep study of the following themes, organized in three plenary sessions: *CAP 2014–2020 and new challenges for agriculture and rural areas*; *CAP 2014–2020 and market*; and *CAP 2014–2020 and rural development*.

From one side, it seems appropriate to wonder whether the general policy decisions and the measures planned are really able to achieve the objectives set and the challenges of globalization. From the other side, the papers presented at the Conference develop analysis and reflections about predictable effects of the new Common Agricultural Policy configuration.

Five papers presented at the 51st SIDEA Conference have been selected for the *Agricultural and Food Economics* special issue focusing on the new European agricultural model and on the sustainability of the new rural policy. These were subjected to a peer review process before being accepted for publication.

2. REPORT

The growing consumer sensitivity about environmental and health attributes of food products has led Galli, Bartolini, Brunori, Colombo, Gava, Grando, and Marescotti (Agricultural and Food Economics 2015, 3:21) to analyze, in the paper "Sustainability assessment of food supply chains: an application to local and global bread in Italy," the environmental impact which comes from traditional food supply chains. The study analyzes the chain of wheat-to-bread, focusing on the relationship between the geographical scope of the food chain and its sustainability, obtained by comparing the 19 attributes that express the different dimensions of sustainability. After a review of the literature on the sustainability of the supply chain processes (of wheat-to-bread chains), the paper proposes a qualitative assessment based on two Italian study cases, identifying critical aspects and providing a comparison between local and global supply chains.

In the paper "Empirical survey on business models of kindergarten farms," Torquati, Tancini, Paffarini, and Illuminati (Agricultural and Food Economics 2015, 3:25) analyze two cases of Italian kindergarten which took part in a special project promoted by the Marche Region, to improve the quality of life in rural areas. The authors highlight the importance of the synergy between agriculture and the education sector and local and regional institutions, in order to carry out social functions for the creation of collective well-being. The Business Model Canvas has allowed to analyze the economical results of educational farms, the farm's relevance as a place of education, and the role of decision-makers, demonstrating how this diversification of activities in rural areas allows farms to increase income, promote generational change and, in particular, women's employment, and reduce the costs of public social services.

The price asymmetries along the chain of pasta are subject of attention in the paper "Antitrust intervention and price transmission in pasta supply chain" by Luca Cacchiarelli and Alessandro Sorrentino (Agricultural and Food Economics 2016, 4:2). The paper analyzes the pasta market, a strategic product in the Italian agro-food industry. During the last several years, many events have affected the Italian pasta supply chains, the CAP Reform, and price instability. The study analyzes a case of anti-competitive practices against makers of pasta, who have been identified and sanctioned by the Italian Antitrust Authority for the period between October 2006 and March 2008. The authors' objective was to determine whether the antitrust action has produced some substantial effects on the market of Italian pasta, restoring a state of considerable competition among businesses. For the study, the Kinnucan and Forker model has been employed which is used in the literature to analyze the impact of a policy intervention on farm-to-retail price transmission. Results show that analysis of the transmission price between semolina and pasta has highlighted that antitrust intervention would have produced some substantial effects in the Italian pasta market by restoring a state of appreciable competition among companies.

Global population growth and the consequent increase in consumption have led to an increase in the demand of food and energy both in developed and developing countries. It has produced an intensification of land investments that, in most cases, has not respected the rules and did not contribute to the development of local communities. The target of Carroccio, Crescimanno, Galati, and Tulone in the paper "The land grabbing in the international scenario: the role of the EU in land grabbing" (Agricultural and Food Economics 2016, 4:12) is to identify, through a cluster analysis, the drivers of investor countries and assess if EU Member States take a common behavior in the context of the so-called land grabbing. The analysis, carried out from 2000 to 2013, led to the identification, based on specific socio-economic and environmental variables, of four homogeneous groups of countries. The results show that, especially for the EU Member States for which energy efficiency and reducing CO_2 emissions are priority objectives (as in the "Europe 2020 Strategy"), the adopted strategies are driven by the need to achieve energy and food security and to reduce pollutant emissions into the atmosphere.

Social agriculture is able to generate socio-economic benefits that are attracting, in Europe, the attention of several stakeholders. This is demonstrated by Bassi, Nassivera, and Pianiin in the study "Social farming: a proposal to explore the effects of structural and relational variables on social farm results" (Agricultural and Food Economics 2016, 4:13) which is part of a project carried out by the health authority of Friuli Venezia Giulia in order to analyze, through a structural equation model, the socio-economic impact of social farms in the area and, in particular, how this depends on the company's internal resources (corporate assets) and the external environment (relations). The analysis shows that the performance of social farms is mainly influenced by relational variables, while structural ones do not show direct influence, pointing out that to improve social impacts generated by farms, structural investments should be supported by a strong relational network at the local level.

CONCLUSIONS

Taken together, this collection of articles underlines the importance of the social role that the European agricultural model performs nowadays, through the production of public goods and services in favor of a new community welfare model. The studies presented explore the effects on sustainability and welfare of local communities which come from practices of diversification and multi-functionality, also in terms of land grabbing, highlighting efficient and effective results of the different policies.

ACKNOWLEDGEMENTS

The authors would like to thank the *Agricultural and Food Economics* journal for having given visibility to the five papers presented at the 51st SIDEA Conference, contributing to promote the international diffusion of Italian agricultural economists' debate about strategic themes for future development of rural policy.

The papers selected for this special issue focus on the new perspectives for European farms, agri-food supply chains, and rural territories of the new Common Agricultural Policy.

REFERENCES

1. European Commission (2010) The CAP towards 2020: meeting the food, natural resources and territorial challenges of the future., Bruxelles.COM(2010) 672 final
2. European Commission (2014) Guidance on Community-Led Local Development. Bruxelles
3. Marotta G, Nazzaro C (2013) Approcci integrati e governance locale nelle strategie di sviluppo rurale: un framework per la futura programmazione. Riv Econ Agrar 3:89–120
4. Peterson C (2013) Fundamental principles of managing multi-stakeholder engagement. International Food and Agribusiness Management Review. vol. 16 A, special issue. pp. 11-22

CHAPTER 13

The social pillar of sustainability: A quantitative approach at the farm level

Anna Gaviglio, Mattia Bertocchi, Maria Elena Marescotti, Eugenio Demartini and Alberto Pirani

ABSTRACT

The present research proposes a model for the assessment of the social pillar of sustainability at the farm scale. Contrary to what is available for the environmental and economic pillars, there is a considerable lack of exhaustive approaches able to evaluate the social dimension of sustainability in rural areas. Thus, the idea was to create a mean by which a quantitative evaluation of the social characteristics of farms could be made. The study involved farms of the South Milan Agricultural Park, located in northern Italy. Thirty sampled farms were selected in order to represent the different livestock systems, land areas, economic dimensions and levels of multifunctionality of the area. The framework is based on a set of 15 indicators able to evaluate five main social "components": (i) quality of the products and the region, (ii) short supply chain and related activities, (iii) work, (iv) ethical and human development and (v) society, culture and ecology. The work was structured using the following steps: identification of the relevant variables for the social sustainability of farms, determination of the framework of indicators, assignment of their range scores, data collection, calculation of the score for each farm, data analysis and visualization. The method allows different types of analysis in relation to the objective of the research. Three main approaches were individuated: (1) the comparison among farms is the "farms' ranking" and the "aggregate ranking"; (2) the evaluation of single themes of sustainability is the "single indicator evaluation" approach and (3) the temporal comparison of the farm's result is the "score evolution" approach. The method showed a high sensitivity to the multifunctionality and the type of farm production, especially organic vs conventional, while other characteristics, such as the type of livestock and the land area, seem to differentiate the sample less or to characterize it in only a few social components. The work has underlined the importance and the advancement in the study of the social dimension that, however, needs further in-depth analysis through comparison with the other two pillars and among various social states in different rural areas.

Keywords: Social sustainability, Agricultural sustainability, Indicators Farm, performance, Multi-attribute analysis

1. BACKGROUND

Over the past 30 years, interest in the evaluation of agricultural activities' sustainability has grown considerably, many studies have been proposed and several assessment methods developed. The literature offers approaches at different spatial scales, ranging from field and farm to regional, national, and even international scale (Jacobs 1995, Hansen 1996, Smith and McDonald 1998). Researchers that opted for a farm/local scale in their studies (Van der Werf and Petit 2002, Häni et al. 2003, Pacini et al. 2004, Rasul and Thapa 2004, Van Cauwenbergh et al. 2007, Meul et al. 2008, Vilain 2008, Gómez-Limón and Riesgo 2009, Reig-Martinez et al. 2011, Paracchini et al. 2015, Thiollet-Scholtus and Bockstaller 2015) took advantage of the possibility of an in-depth investigation of farm dynamics, while research studies that used a regional/territorial scale (Paracchini et al. 2011, Mazzocchi et al. 2013, Demartini et al.2015) could limit the cost of analysis, ensuring transparency of data and repeatability of measurements (Demartini et al. 2015).

Although the environmental, social and economic pillars of sustainability are linked to each other, a single integrated approach seems to be difficult (Wells 2001, Zimmerer and Basset 2003). This reflects problems of data requirement and incommensurability between different facets or dimensions of sustainability that become stronger as the analysis moves to the system beyond the farm boundaries (Rigby et al. 2001). As a result, research studies tend to include these three pillars in their studies, treating them separately or with a different relevance (Singh et al. 2009). Furthermore, the different pillars have attracted varying levels of attention.

The environmental assessment is more studied, because of the growing social sensitivity of the community to ecological issues, and therefore, many approaches have been discussed and developed and researchers have multiple analytical opportunities. On the other hand, the evaluation of economic and, especially, social sustainability suffers from a lack of accepted and well-grounded frameworks (Chatzinikolaou and Manos 2012).

Although contemporary society recognizes agriculture as having an important responsibility in safeguarding the region, its culture and traditions (Gaviglio et al. 2014a), the measurement of social sustainability seems to be less studied. Moreover, most of the approaches that seek to involve a complete evaluation of the sustainability use qualitative assessments of the social pillar, based on observations and opinions or indicators that require difficult to find data (see Häni et al. 2003, Van Cauwenbergh et al. 2007, Meul et al. 2008, Vilain 2008).

As a matter of fact, finding a match between the social sustainability objectives and their corresponding indicators is a challenging task. According to Omann and Spangenberg (2002), this assessment concerns some relevant problems. Firstly, the perception of social issues is heterogeneous in different territorial contexts and this causes a lack of conceptual clarity. Secondly, when scientists suggest a great number of social indicators they still hesitate to formulate normative targets. Thus, the complexity of the concept might not be manageable in the current institutional settings. All this considered, the assessment of social sustainability seems to be particularly dependent on the local context and the socio-political goals proposed by policy-makers (Littig and Griessier 2005). Thus, the choice of indicators used in the analysis probably represents the most critical step of the research. Particularly, the scientific literature underlines the importance of the balance between a validated approach, reliability and significance of the indicators and the objectives to be achieved (Girardin et al. 1999), under the constraints of data availability.

The aim of this paper is to present a method for the evaluation of social sustainability of farms that uses quantitative indicators. The framework consists of 15 indicators able to assess the social performances of the farms that could have relevant reflection on their role for the rural area. The method offers useful easy-to-read results for farmers, decision-makers and researchers. Furthermore, a wide selection of indicators is presented which seems to represent a good (if not exhaustive) review that could be useful for further studies on this theme.

The remainder of the text is organized into three sections. The "Methods" section presents the features of the method: the approach to the social evaluation, the framework, its indicators and the case study. The "Results" section presents the results and their discussion. Finally, a concluding paragraph offers a summary of the research and some reflections on its potentialities and limitations.

2. METHODS

2.1 Framework

The present method aims at the assessment of social sustainability at farm level using quantitative indicators. It is worth noting that the proposed approach excludes the evaluation of some social dimension that could be described by qualitative analysis; nonetheless, the definition of a system of quantitative measures allows researchers and farmers for direct comparisons between themes and farms that would not be possible otherwise.

Table 1 and Fig. 1, respectively, summarize the indicators used in the method and the component derived, and provide a brief idea of the framework adopted (in-depth details are available in Appendix).

Table 1 Social sustainability indicators and components used in the analysis

Code	Indicator	Max score	Code	Component	Max score
S_1	Quality of the products	20	CS_1	Quality of the products and the region	50
S_2	Rural buildings	12			
S_3	Landscape and territory	18			
S_4	Short food supply chain	30	CS_2	Short food supply chain and related activities	50
S_5	Related activities	20			
S_6	Work	25	CS_3	Work	50
S_7	Sustainability of the employment	15			
S_8	Training	10			
S_9	Livestock management	25	CS_4	Ethical and human development	50

S_10	Associations and social implications	15			
S_11	Cooperation	10			
S_12	Waste management	15	CS_5	Society, culture and ecology	50
S_13	Accessibility to the farm spaces	10			
S_14	Sustainable use of materials	15			
S_15	Education	10			

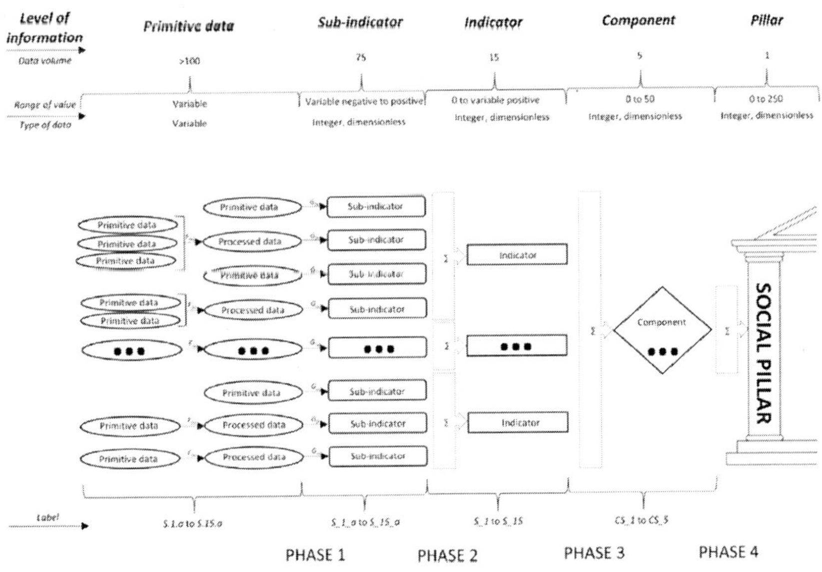

Figure. 1 The framework proposed for social sustainability assessment

The choice of indicators is based on the literature review, even if the lack of a shared framework for the social assessment at the farm scale often forces the researchers to an arbitrary choice of indicators (Van Cauwenbergh et al. 2007). The selection process was carried out through the collection of the indicators detected from currently available methods. Among these, the choice was based on a combination of the best characteristics of simplicity, data requirements and significance for the case study.

The following components of agricultural activities were observed, considering their social dimension (Table 1): (*CS_1*) the products and regional management; (*CS_2*) the short food supply chain and related activities; (*CS_3*) employment and working conditions; (*CS_4*) the ethical and human development about animal welfare and the cooperation and association; and (*CS_5*) cultural and ecological themes. Note that the evaluation of these components involves the main social themes of the agricultural activities; nevertheless, some topics reported in the literature are not treated in order to avoid the use of qualitative indicators of data. In particular, we refer to food hygiene and safety and quality of life.[1]

In the proposed method, the measurements of farm characteristics are converted into dimensionless values that represent an easy-to-read score of the raw data according to the desirability of the measured performance. For example, when measuring the indicator "S_1 Quality of the products" higher scores rationally correspond to higher measures. Furthermore, each indicator can range from a minimum or a maximum score (single scores are reported in Appendix); while the minimum score is always zero, the maximum scores vary depending on the social relevance attributed to the indicator, and therefore, more relevant indicators have higher maximum scores. This weighting procedure derived from a subjective evaluation typical of these types of studies that assigned the scores in accordance to the relevance attributed by the literature (when available) and the characteristics of the case study and its objectives (von Wirén_Lehr 2001). This process, which is convenient for adaptation to the local context, involves the typical risks connected to subjective norms. In order to reduce this possible source of errors, some studies used the principle of equality among the indicators (Meul et al. 2008); nonetheless, other studies argue that indicators cannot be considered equally relevant with reference to sustainability assessment (Vilain 2008, Zahm et al. 2008). In this sense, researchers should be aware of the trade-off between the two options and carefully adopt the one that they consider the best in the research context.

The method is characterized by an aggregative structure (Fig. 1) aimed at reducing data from farm characteristics to a unique value of the social pillar. As shown in Fig. 1, the process is divided into four basic phases:

- Phase 1: collection and analysis ($F_{(x)}$ and $G_{(x)}$) of the farms' characteristics, classified from *S.1.a* to *S.15.a*, in order to obtain a raw data set;
- Phase 2: elaboration of 75 sub-indicators, classified from *S_1_a* to *S_15_a*, leads to integer and dimensionless values that range from negative to positive values, according to their maximum scores;
- Phase 3: calculation of the 15 indicators, classified from *S_1* to *S_15*, reported in Table 1, obtained through the sum of two or more sub-

indicators. A minimum [0] and a maximum [variable] score is applied depending on the case (see Appendix);

- Phase 4: the sum of two or more indicators provides the value of five components of the social sustainability pillar, classified from *CS_1* to *CS_5*. A minimum [0] and a maximum [50] score is applied. In turn, the sum of the components leads to the overall value of the social dimension of sustainability, which can range from 0 to 250.

2.2 The components

In the following sections, we describe the indicators listed in Table 1, classified by component.

2.2.1 Component 1 (CS_1): quality of the products and the region

The social inclusion of rural areas highly depends on the connection between them and the citizenship. One of the most important means is the consumers' perception of the farm's products. Often people assign good environmental standards to high-quality products that contribute to a higher social acceptance of the agriculture and its production systems.

There are wide ranges of categories of consumers, defined as ethical consumers or citizen-consumers, who associate a very high value to the attribute of the quality of the product and the region where they are produced. Among these products, the *CS_1* component identified two main categories: quality-certified commodities and food products (*S_1_a* and *S_1_b*) and organic products (*S_1_c* and *S_1_d*).

In the first case, the literature review on consumers' perceptions shows some social relation between protected denomination of origin (PDO) products and the attributes of support in order to sustain regional manufacturers (Van Ittersum et al. 2007, Verbeke et al. 2012) and to contribute to the survival of the social identity of the region (Vilain 2008). Thus, nowadays, these recent purchasing motivations are assigned a comparable importance with the typical attributes such as high standards (Van Ittersum et al. 2007), the tradition (Verlegh and Steenkamp 1999, Dimara and Skuras 2003), the taste (Platania and Privitera 2006, Vanhonacker et al. 2010) and food safety (Dimara and Skuras 2003).

Regarding organic food consumption, many researches stated the importance of the socio-economic traits of consumers (Hamm and Gronefeld 2004, Falguera et al. 2012), in particular tradition (Chinnici et al. 2002) and animal welfare (Magnusson et al. 2003, Makatouni 2002). In this case too, these attributes seem to have a primary relevance, besides those historically associated with organic consumption: first of all, environmentally friendly behaviour and also, as stated by some recent studies, the importance of the intrinsic attributes of the products (Gaviglio and Pirani 2015a), such as the healthiness (Pieniak et al. 2010), the high quality (Chinnici et al. 2002) and the taste (Fotopoulos and Krystallis 2002, Zanoli and Naspetti 2002, Kihlberg and Risvik 2007).

While the establishment of organic and labelled products is easy because of certification, there is a wide range of attempts to define local products (Hand and Martinez 2010). Because of the difficulties in defining the standard of quality of products without certification, this component only considers labelled products. The evaluation of local and typical products is treated by the *CS 2* component, without taking into account the attribute of their quality.

Finally, conscious social behaviour also involves issues not closely linked to the products, such as the functional and aesthetic roles of rural buildings *(S_2)* and the farm landscape *(S_3)*. These are important features that characterize the architecture (Meul et al. 2008) and they represent positive or negative externalities in the social acceptance of rural areas (Van Cauwenbergh et al. 2007).

2.2.2 Component 2 (CS_2): short food supply chain and related activities

Among the most important motivations in buying local products, the literature found some social attributes such as tradition (Bessiére 1998), supporting local economies and trust in producers (Seyfang 2006, Lockie 2009). These are indicated by consumers as relevant means able to connect the citizenship with the countryside.

Selling products through short chain systems involves different types of opportunities, such as direct sales (*S_4_a*, *S_4_c* and *S_4_d*), online sales (*S_4_b*), ethical purchasing groups (*S_4_e*), farmers' markets (*S_4_f*), restaurants and shops (*S_4_g*) and canteens (*S_4_h*). The direct sales formula is mainly dedicated to local products and creates a close relationship between producers and consumers that cannot be explained just within an economic rationality (Gaviglio et al. 2015b).

The implications on other activities besides the agricultural production, such as the maintenance and management of public spaces and resources (*S_5_a*), related activities like agritourism, restaurants, bed and breakfast (*S_5_b*) and educational farms (*S_5_c* and *S_5_d*) are other important means of linking the town and the countryside (Vilain 2008). Citizens often use these systems to get to know the rural world and to learn about production processes and agro-food systems (Santini and Paloma 2013).

2.2.3 Component 3 (CS_3): work

Employment in the agriculture sector has fallen considerably in the last decades; therefore, the maintenance of a sustainable level of employment (*S_7*) is relevant for the social and economic development (*S_6*) of the agricultural area (Häni et al. 2003, Meul et al. 2008, Vilain 2008, Gómez-Limón and Fernandez 2010, Reig-Martinez et al. 2011, Bonneau et al. 2014). In this context, training (*S_8*) is a key aspect for the growth of the agricultural sector (Vilain 2008) by which farms play a leading role in development and innovation because of the requirement for high-profile skilled jobs involved in related activities, research and breeding.

2.2.4 Component 4 (CS_4): ethics and human development

The human and ethical development of agriculture involves multiple issues. Among these, animal welfare is today a primary requirement of society (Fortun-Lamothe et al. 2009, Broom 2010). Livestock management involves animal health and the farm's ability to implement innovations in the agricultural sector. This is a very complex issue, and the use of a single approach able to evaluate different types of livestock is difficult. Therefore, the indicator (*S_9*) is based on the diversification of the most common species of animals (cattle, pigs, poultry, sheep/goats) bred in the area. In this way, we only evaluated the management of the most important livestock of the farm (through the calculation of the LSU, "livestock units").

Cooperation and association are relevant instruments of innovation of agricultural systems (Vilain 2008) and they are important indicators of human development in rural areas. Among these factors, the social dynamism and vitality of an area heavily depend on membership in associations (*S_10_a* and *S_10_b*), consortia (*S_10_c*) and cooperation with other farms in the surrounding area in direct sales (*S_11_a* and *S_11_b*), agritourism activities (*S_11_c*), the production structures (*S_11_d*) and the workforce (*S_11_e*).

2.2.5 Component 5 (CS_5): society, culture and ecology

The inclusion of rural areas involves the recognition of the ecological, cultural and social effort of farms in their production process. In this sense, the component takes into account four main aspects. A proper waste management through recycling processes (*S_12_a*) and the use of recycled materials (*S_12_b*, *S_12_c* and *S_12_d*) has a great environmental importance but it also involves the social acceptance of the agriculture systems. The farm's open spaces (*S_13_a* and *S_13_b*) have important recreational functions useful for the population of a rural area.

Moreover, a proper landscape management can be a relevant source of income for the multifunctional farm's activities (Vilain 2008). The agricultural production systems involve the use of materials, such as feed (*S_14_a*), fertilizers (*S_14_b*), water (*S_14_c*) and seeds (*S_14_e*) that characterizes the sustainability of the farm management (Vilain 2008). A high dependence from the outset, even the buying of animals (*S_14_c*), often causes a reduction of autonomy in making production, marketing and management decisions. It also alters the resilience of the system and the ability to adapt to economic, environmental and social changes.

The educational level (S_15_a) of the farm personnel is important for the cultural and social growth of agricultural areas (Van Cauwenbergh et al. 2007). As found by Elfkih et al. 2012, there is a probable positive effect of educational level on the achievement of overall sustainability. The education of the workforce and the entrepreneur is also able to encourage openness to new knowledge and innovation in agriculture.

2.3 Case study

The survey was carried out on the region of the South Milan Agricultural Park. The park is an area located on the Po plain (northern Italy), in one of the most intensively agricultural regions in Europe (INEA 2014). This is an agricultural park that involves the peri-urban area around the city of Milan. It covers more than 40,000 ha of lowland where farms are characterized by intensive production systems, a wide range of land areas, livestock and economic dimensions. The main crops are maize, rice and grassland (Migliorini and Scaltriti 2012), while the main livestock are cattle, poultry and pigs. The high population density confers the typical attributes of peri-urban areas, such as fragmentation and high economic value of the land (Gaviglio et al. 2014b).

We carried out the sample stratification considering farm size, the main production and geographical location. The selection was also linked to the willingness of farmers to respond to interviews and to provide some administrative data (Briquel et al. 2001; Viglizzo et al. 2006).

Thirty farms with different breeding systems were selected: cattle (for milk or meat production), poultry, pigs, sheep and goats. Table 2 reports some of the main features of the sampled farms.

Table 2 Characteristics of the farms in the sample

Farm characteristic	**Quantity *N***	**Percentage %**
Type of breeding		
Bovine (meat prod.)	7	23.33
Bovine (milk prod.)	15	50.00
Poultry	4	13.33
Pigs	3	10.00
Sheep/goat	1	3.33
Land size—utilized agricultural area		
<50 ha	14	16.66
50–100 ha	9	30.00
>100 ha	7	23.33
Multifunctionality		
Non-multifunctional	9	30.00
Multifunctional	21	70.00
Type of production		
Conventional	25	83.33
Organic	5	16.66
Economic size—standard output (SO)		
SO < 100	13	43.33
100 < SO < 300	9	30.00
SO > 300	8	26.66
Total	30	100.00

Data were collected using (i) interviews through questionnaire to farm personnel that provided an easy method to detect information, (ii) the Sistema Informativo Agricoltura Regione Lombardia (SIARL) database and (iii) estimations when not available from the other sources.

3. RESULTS

The output of the method provides the values of each sub-indicator, indicator and component that contribute to the overall score of the social pillar. In relation to the objective of the research, the method still allows the use of different types of analysis.

3.1 Farm comparison by ranking of indicators

When the case study is focused on social sustainability aspects of individual farms, the method allows an in-depth analysis of farms through the evaluation of the basic indicators. Figure 2 shows a possible application of the results of two farms. Farm 1 has a conventional production system, the livestock are cattle for milk production and the land area is large. Farm 2 has the same characteristics but it also practises multifunctional activities, while farm 1 can be defined as non-multifunctional. These two farms are discussed as an example of how the method performs and how data can be interpreted to identify key actions to be adopted to improve farm performance. In general, farmers who have direct contact with consumers achieve higher social results that can also lead to higher environmental performance (Gafsi and Favreau 2010). In our sample, farm 2 obtained higher scores in the *CS_2*, *CS_3* and *CS_5* components while farm 1 showed better scores in the *CS_4* component. The *short food supply chain related activities* component result is a direct consequence of the different characteristics on multifunctionality. The third component underlined the higher contribution of multifunctionality to employment in the rural context. On the contrary, the other components seem to be less dependent on the multifunctionality level. Their scores are probably more influenced by other farm characteristics such as land area, type of production or the management choices of each farm. However, both farms could improve their sustainability level in the *quality of the products and the region* (*CS_1*) and *work* (*CS_3*). Farm 1 seems to show no interest in diversification of income and activities. Both farms showed high performance in ethical and human development (*CS_4*) and social, cultural and ecological sustainability (*CS_5*).

3.2 Farm comparison by aggregate component ranking

Radar diagrams are able to represent the overall performance of farms in the five components. This method is able to provide different approaches of farm ranking in relation to the sample features. The following figures show the farms' results, classified by livestock type (Fig. 3), land area (Fig. 4) and type of production (conventional vs organic) (Fig. 5).

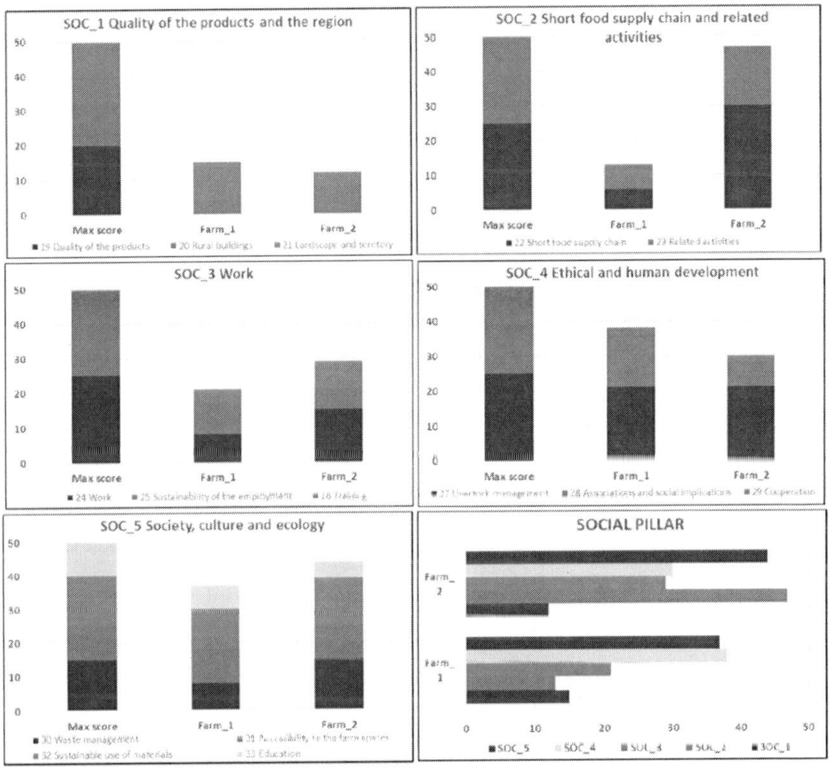

Figure. 2 Decomposition of the social pillar into five components and the related indicators for two cattle farms (milk production)

In Fig. 3, the sample is divided into four categories of livestock type. Cattle farms (Fig. 3) showed generally lower values, in particular for meat production. These are often large farms that are not interested and do not have the necessary conditions to diversify production and to develop systems of short chain and related activities. In the *CS_2* component (*short food supply chain and related activities*), they reached an overall score of 19 and 20, respectively, for milk and meat production, while poultry achieved 29.5 out of 50. This is probably because their product types are often sold to only a few large clients. On the contrary, pigs and especially poultry farms achieved the highest average values in almost all components, because of the diversification of production and income and the greater ability to offer sales services and other social functions. Similar results were noticed in the *CS_3* component (*work*). This is probably due to the higher level of multifunctionality of these farms, rather than a greater demand for workers with livestock, that leads to a higher contribution to employment. On the other hand, the *CS_4* (*ethical and human development*) and the *CS_5* (*society, culture and ecology*) components are characterized by higher scores in the cattle, particularly for milk production, and the pig samples. This result is rather interesting and suggests the need for further evaluation. The fact that the high level of knowledge requested for these types of systems could also influence education and culture or that, traditionally, these farmers are more involved in innovation processes are only two of many potential interpretations.

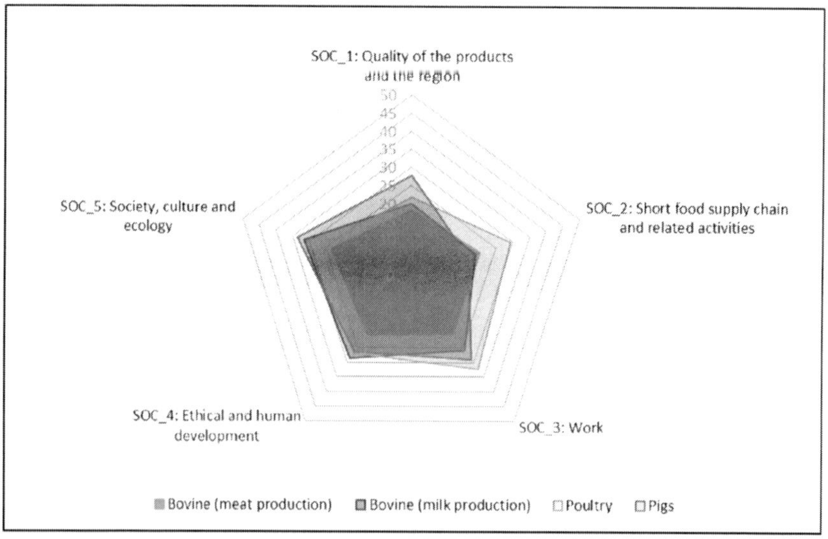

Figure. 3 Radar diagram of bovine (meat production) vs bovine (milk production) vs poultry vs pig farms' average performance

Figure 4 shows the results of the farms, aggregated by land area classes. Only a few differences were noted. In particular, the *CS_2* component seems to allow some relevant considerations; smaller farms tend to achieve higher results (22.9 out of 50) because of their predisposition to finding different types of income. The results of the *CS_3* component (*work*) seems to not be influenced by the land area of the farms. In the literature, these results have contradictory feedbacks. Our data is in accord with that of Häni et al. 2003, which did not find any substantial difference in the corresponding social component (*work condition*) of the RISE method. On the other hand, it is in disaccord with Gavrilescu et al. 2012 who, applying the IDEA method, attributed lower scores to family farms compared to legal entities mostly because of their rigidity in creating new jobs, low professional training and a lower minimum wage. Reig-Martinez et al. 2011 found an overall high level of sustainability, even social, of larger farms that could be explained through the opportunity to develop a more diversified range of crops and the generation of sufficient income that has permitted the continuity of agricultural activity. It can be supposed that these great differences are probably due to the high dissimilarity of the samples involved in the studies.

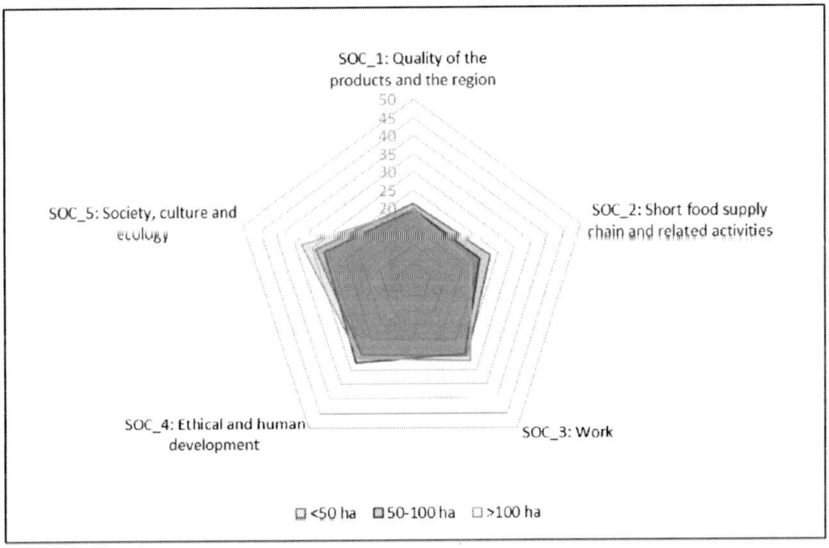

Figure. 4 Radar diagram of small (up to 50 ha) vs medium (50 to 100 ha) vs large (more than 100 ha) farms' average performance

Finally, in Fig. 5, the performances of conventional and organic farms are compared. The differences between the two types of production were considerable: organic farms achieved higher scores in every component. The result of the *CS_1* component was expected, in consideration of the relevance of organic certification. The excellent results in the *short chain and the related activities* (*CS_2*) and *work* (*CS_3*) components are a direct consequence of their higher regard for multifunctionality. These aspects also involve the *CS_4* and *CS_5* components, as confirmation of the higher level of culture, education and training required for the practice of organic production.

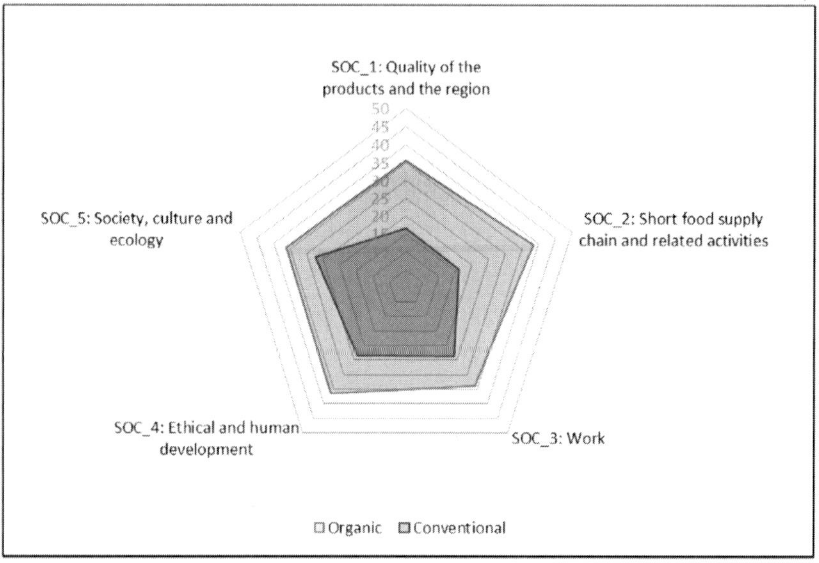

Figure. 5 Radar diagram of conventional vs organic farms' average performance

4. DISCUSSION AND CONCLUSIONS

A new method for the evaluation of the social sustainability of farms has been proposed. It provides easy-to-read results and information at different scales of interpretation. The method seems adequate to perform an evaluation of the social sustainability of agricultural activities allowing for different approaches, such as:

1) *Farm comparison by ranking of indicators* and *farm comparison by aggregate component ranking*: the comparison among farms, or groups of farms, that seems to be useful for policy-makers to identify agriculturally sustainable practices and the farm features that influence social sustainability, such as the type of animal production, land area and multifunctional activities, individual and detailed aspects of sustainability, as well as a high variability of aspects aggregated together to a more comprehensive vision of the evaluation;

2) *Single indicator evaluation*: the evaluation of a single or few aspects of social sustainability is able to provide synthetic information to farmers who want to evaluate the level reached by their farm and to know the ways to improve by comparison with other farms;

3) *Score evolution*: the temporal comparison looks at the evolution of the results achieved by farms over time and those that are predictable in the future. This approach was not calculated in this research, because of the lack of farm data for different years, but it is still worth being cited because of its potential multiple utilizations. For example, farmers could use it to evaluate the trend of their own work, while policy-makers, comparing the performance of different farms systems over time, could obtain useful information for decision-making.

The assessment of social sustainability involves some relevant issues, considering that the current objectives of policy-makers are dedicated to the improvement of the environment and the landscape (e.g. Greening's conditions in the new Common Agricultural Policy) and the social value of agriculture (e.g. the Measures of the Rural Development Plan).

From a practical point of view, the social pillar often involves themes highly dependent on the farms' characteristics that still seem hard to measure because of the lack of a shared knowledge. Because of this, we believe we have shown an approach able to avoid the problems of qualitative assessments. Nevertheless, we point out the need for periodical revision of the method, in order to keep in step with the evolution of the social issues of the rural sector and improve the quantity and quality of information provided. For example, as the indicators used in the analysis relates specifically to social condition inside the farms, an interesting improvement may involve the direct assessment of social impact of farms' activities in their territories.

In addition, the study underlined that the integration of the method with assessments of environmental and economic performances could be a valid way to the validation of the social dimension. Because of its particular sensitivity to the local situation, a further approach might involve the comparison of the results of farms located in different geographical contexts.

ENDNOTES

[1] See Häni et al. (2003), Rasul and Thapa (2004), Van Cauwenbergh et al. (2007), Meul et al. (2008), Vilain et al. (2008), Zahm et al. (2008), Gómez-Limón and Fernandez (2010), Parachini et al. (2011), Bonneau et al. (2014).

APPENDIX

Table 3 Framework: sub-indicators, indicators and dimensions

Farm characteristic	Source	Code	Sub-indicator	Score range	Code	Indicator	Max score
Number of products involved in quality certified products	Questionnaire	S_1_a	Quality certified commodities	0–∞	S_1	Quality of the products	20
Number of quality certified products	Questionnaire	S_1_b	Quality certified food products	0–∞			
Number of vegetal organic products	Questionnaire	S_1_c	Vegetal organic production	0–∞			
Number of animal organic products	Questionnaire	S_1_d	Animal organic production	0–∞			
Aesthetics care of rural buildings	Questionnaire	S_2_a	Aesthetics of rural buildings	0–4	S_2	Rural buildings	12
Maintenance of the original use of the rural buildings	Questionnaire	S_2_b	Maintenance of the original use of the rural buildings	0–4			
Number of buildings built or renovate using "green techniques"	Questionnaire	S_2_c	Green building techniques	0–∞			
Maintenance of the farm green spaces	Questionnaire	S_3_a	Green maintenance	0–4	S_3	Landscape and territory	18
Presence of hedges, rows and wooded bands	Questionnaire	S_3_b	Hedges, rows and wooded bands	0–4			
Maintenance of the roads and paths	Questionnaire	S_3_c	Maintenance of the roads and paths	0–4			
Number of crop species/UAA (N/ha)	SIARL Database	S_3_d	Crop diversification	0–6			
Selling through direct sale (farm shop)	Questionnaire	S_4_a	Direct sales	0–3	S_4	Short food supply chain	30
Selling through online website, portals, etc.	Questionnaire	S_4_b	Online sales	0–1			
Income from short chain channels/income from conventional	Questionnaire	S_4_c	Direct sales relevance	0–20			

channels (%)							
Number of product sold through direct sale channels	Questionnaire	S_4_d	Product sold through direct sales	0–∞			
Sales through ethical purchasing groups	Questionnaire	S_4_e	Sales through ethical purchasing groups	0–3			
Sales through farmers' markets	Questionnaire	S_4_f	Sales through farmers' markets	0–2			
Sales to restaurant and shops	Questionnaire	S_4_g	Sales to restaurant and shops	0–2			
Sales to dining halls	Questionnaire	S_4_h	Sales to dining halls	0–3			
Social services (e.g. snow removal, composting, maintenance of private and public green areas)	Questionnaire	S_5_a	Social services	0–2	S_5	Related activities	20
Type of related activities (e.g. agritourism, B&B, restaurant)	Questionnaire	S_5_b	Type of related activities	0–10			
Educational farm, teaching	Questionnaire	S_5_c	Educational farm	0–5			
Presence of other social activities (e.g. disadvantages people, children, social inclusion)	Questionnaire	S_5_d	Social activities	0–5			
UAA/annual amount of work (ha/h)	SIARL Database, Questionnaire	S_6_a	Annual Work Unit (AWU)	0–8	S_6	Work	25
Number of new employments (in the last 5 years)	Questionnaire	S_6_b	New employments	0–∞			
Number of farm products processing	Questionnaire	S_6_c	Farm products processing	0–∞			
Number of workers who resided in the farm buildings	Questionnaire	S_7_a	Workers who resides in the farm buildings	0–2	S_7	Sustainability of the employment	15
Number of local workers/total number of	Questionnaire	S_7_b	Local workers	0–1			

employee (%)							
Number of female workers/total number of workers (%)	Questionnaire	S_7_c	Female workers	0–4			
Age of the entrepreneur	Questionnaire	S_7_d	Youth entrepreneurship	0–4			
Youth workers/total number of workers (%)	Questionnaire	S_7_e	Youth employment	0–4			
Training courses for workers formation	Questionnaire	S_8_a	Training courses	0–∞	S_8	Training	10
Presence of trainees from schools and universities	Questionnaire	S_8_b	Trainees	0–2			
Training activities unfold in farm	Questionnaire	S_8_c	Training activities unfold in the farm	0–2			
Employment of disadvantaged people among the workers	Questionnaire	S_8_d	Disadvantaged people hired	0–5			
Number of animal species bred	SIARL Database	S_9_a	Animal species	0–∞	S_9	Livestock management	25
S_9_b1 Type of stable S_9_b2 Management by physiological phases S_9_b3 Systems of ventilation or heat reduction S_9_b4 Quality/quantity control of the feed S_9_b5 Attendance at birth S_9_b6 Systems of cleaning	Questionnaire	S_9_b	Cattle	0–20			
S_9_c1 Type of flooring S_9_c2 Systems of ventilation or heat reduction S_9_c3 Handling	Questionnaire	S_9_c	Swine	0–20			

systems of animals S_9_c4 Presence of materials of environmental enrichment							
S_9_d1 Type of stable S_9_d2 Systems of ventilation or heat reduction S_9_d3 Presence of openings and/or windows S_9_d4 Quality/quantity control of the feed	Questionnaire	S_9_d	Poultry	0–20			
S_9_e1 Outdoor spaces and pasturage S_9_e2 Quality/quantity of the feed S_9_e3 Attendance at birth S_9_e4 Systems of cleaning	Questionnaire	S_9_e	Sheep and goats	0–20			
Participation in associations	Questionnaire	S_10_a	Association	0–5	S_10	Associations and social implications	15
Position of responsibility in associations	Questionnaire	S_10_b	Responsibility in the associations	0–5			
Participation in consortium	Questionnaire	S_10_c	Consortium	0–5			
The entrepreneur resides in farm	Questionnaire	S_10_d	Residence of the entrepreneur in farm	0–2			
Direct sales managed in cooperation with other farms and associations	Questionnaire	S_11_a	Direct sales managed in cooperation	0–2	S_11	Cooperation	10
Selling of other farm's products in the direct sale	Questionnaire	S_11_b	Selling of other farm's products	0–2			
Agritourism managed in cooperation	Questionnaire	S_11_c	Agritourism managed in cooperation	0–2			
Farm structures and/or machineries	Questionnaire	S_11_d	Farm structures and machineries	0–2			

managed in cooperation			managed in cooperation				
Workforce managed in cooperation	Questionnaire	S_11_e	Workers managed in cooperation	0–2			
Separate waste management	Questionnaire	S_12_a	Waste management	0–4	S_12	Waste management	15
Use of recyclable materials for the farm activities	Questionnaire	S_12_b	Recyclable materials for the farm activities	0–4			
Use of compost	Questionnaire	S_12_ç	Use of compost	0–3			
Mulching	Questionnaire	S_12_d	Mulching	0–4			
Presence of public farm spaces	Questionnaire	S_13_a	Public farm spaces	0–5	S_13	Accessibility to the farm spaces	10
Presence of recreational farm spaces	Questionnaire	S_13_b	Recreational farm spaces	0–5			
Self produced feed/Feed requirement (%)	Questionnaire	S_14_a	Self-produced feed	0–5	S_14	Sustainable use of materials	15
Self-produced fertilizer/fertilizer requirement (% of nitrogen)	Questionnaire	S_14_b	Self-produced fertilizers	-1–5			
Amount of livestock comeback	Questionnaire	S_14_c	Livestock comeback	0–4			
Use of methods for the water saving	Questionnaire	S_14_d	Saving water	0–2			
Self-produced seeds and plants	Questionnaire	S_14_e	Self-produced seeds and plants	0–2			
Schooling level of the farm workers	Questionnaire	S_15_a	Schooling level of the farm workers	0–10	S_15	Education	10

ACKNOWLEDGEMENTS

The authors would thank the Città Metropolitana di Milano, the Parco Agricolo Sud Milano and the Fondazione Cariplo for their support.

REFERENCES

1. Bessière J (1998) Local development and heritage: traditional food and cuisine as tourist attractions in rural areas. Sociol Rural 38(1):21–34
2. Bonneau M, Klauke TN, Gonzàlez J, Rydhmer L, Ilari-Antoine E, Dourmad JY, de Greef K, Houwers HWJ, Cinar MU, Fàbrega E, Zimmer C, Hviid M, van der Oever B, Edwards SA (2014) Evaluation of the sustainability of contrasted pig farming systems: integrated evaluation. Animal 8(12):2058–2068
3. Briquel V, Vilain L, Bourdais J-L, Girardin P, Mouchet C, Viaux P (2001) La méthode IDEA (indicateurs de durabilité des exploitations agricoles): une démarche pédagogique. Ingénieries EAT 25:29–39
4. Broom DM (2010) Animal welfare: an aspect of care, sustainability, and food quality required by the public. J Vet Med Educ 37(1):83–88
5. Chatzinikolaou P, Manos B (2012) Review of existing methodologies and tools for measuring sustainability in rural areas. http://www.feem-project.net/belpasso_2012/files/ studpapers/Paper_Chatzinikolaou.pdf
6. Chinnici G, D'Amico M, Pecorino B (2002) A multivariate statistical analysis on the consumers of organic products. Br Food J 104(3/4/5):187–199
7. Demartini E, Gaviglio A, Bertoni D (2015) Integrating agricultural sustainability into policy planning: a geo-referenced framework based on rough set theory. Environ Sci Pol 54:226–239
8. Dimara E, Skuras D (2003) Consumer evaluations of product certification, geographic association and traceability in Greece. Eur J Mark 37(5/6):690–705
9. Elfkih S, Guldara I, Mtimet N (2012) Are Tunisian olive growing farms sustainable? An adopted IDEA approach analysis. Span J Agric Res 10(4):877–889
10. Falguera V, Aliguer N, Falguera M (2012) An integrated approach to current trends in food consumption: moving toward functional and organic products? Food Control 26(2):274–281
11. Fortun-Lamothe L, Combes S, Gidenne T (2009) Contribution of intensive rabbit breeding to sustainable development, a semi-quantitative analysis of the production in France. World Rabbit Sci 17:79–85
12. Fotopoulos C, Krystallis A (2002) Organic product avoidance: reasons for rejection and potential buyers' identification in a countrywide survey. Br Food J 104(3/4/5):233–260
13. Gafsi M, Favreau JL (2010) Appropriate method to assess the sustainability of organic farming systems. In: Proceedings of the 9th European IFSA Symposium, Vienna, 4-7 July 2010: 912-921.
14. Gaviglio A, Pirani A (2015) Consumer perception of cured pork meats: the added value of the organic attribute food analysis. Czech J Food Sci 33(1):32–36
15. Gaviglio A, Pirani A, Bertocchi M (2014a) Development of the environmental, social and economic sustainability in the periurban agricultural areas: governance opportunities in the South Milan Agricultural Park. Adv Eng Forum 11:417–423
16. Gaviglio A, Bertocchi M, Silvestri S, Pirani A (2014b) The use of environmental, economic and social indicators as decision support for the advancement of farms towards Sustainability. In: Sustainability of the agri-food system: strategies and performances. Proceedings of the 50th SIDEA conference, Lecce, 26-28 September 2013: 215-225.
17. Gaviglio A, Demartini E, Pirani A, Marescotti ME, Bertocchi M (2015b) National brands versus private labels versus niche products: a graphical representation of consumers' perception. Paper prepared for presentation at the EAAE-AAEA Joint Seminar 'Consumer Behavior in a Changing World: Food, Culture, Society. March 25 to 27, 2015 Naples, Italy.
18. Gavrilescu C, Toma C, Turtoi C (2012) Assessment of the sustainability of agricultural holdings in macroregion 1 using the IDEA method. Bulletin UASVM Horticulture 69(2):122–130
19. Girardin P, Bockstaller C, van der Werf HMG (1999) Indicators: tools to evaluate the environmental impacts of farming systems. J Sustain Agric 13:5–21

20. Gómez-Limón JA, Riesgo L (2009) Alternative approaches to the construction of a composite indicator of agricultural sustainability: an application to irrigated agriculture in the Duero basin in Spain. J Environ Manag 90(11):3345–3362
21. Gómez-Limón JA, Sanchez-Fernandez G (2010) Empirical evaluation of agricultural sustainability using composite indicators. Ecol Econ 69(5):1062–1075
22. Hamm U, Gronefeld F (2004) The European market for organic food: revised and updated analysis, The European market for organic food: revised and updated analysis Hand MS, Martinez S (2010) Just what does local mean. Choices 25(1):13–18
23. Häni F, Braga F, Stämpfli A, Keller T, Fisher M, Porche H (2003) RISE, a tool for holistic sustainability assessment at the farm level. Int Food Agribus Man Rev 6(4):78–90
24. Hansen JW (1996) Is agricultural sustainability a useful concept? Agric Syst 50(2):117–143
25. INEA (2014) L'agricoltura Lombardia Conta 2014. http://dspace.inea.it/handle/inea/1037
26. Jacobs M (1995) Sustainable development—from broad rhetoric to local reality. Conference Proceedings from Agenda 21 in Cheshire. Cheshire County Council, Document No. 493.
27. Kihlberg I, Risvik E (2007) Consumers of organic foods—value segments and liking of bread. Food Qual Prefer 18(3):471–481
28. Littig B, Griessler E (2005) Social sustainability: a catchword between political pragmatism and social theory. Int J Sustain Dev 8(1/2):65–79
29. Lockie S (2009) Responsibility and agency within alternative food networks: assembling the "citizen consumer". Agric Hum Values 26(3):193–201
30. Magnusson MK, Arvola A, Hursti UKK, Åberg L, Sjödén PO (2003) Choice of organic foods is related to perceived consequences for human health and to environmentally friendly behaviour. Appetite 40(2):109–117
31. Makatouni A (2002) What motivates consumers to buy organic food in the UK? Results from a qualitative study. Br Food J 104(3/4/5):345–352
32. Mazzocchi C, Sali G, Corsi S (2013) Land use conversion in metropolitan areas and the performance of agriculture: Sensitivity Index of Agricultural Land (SIAL), a tool for territorial analysis. Land Use Policy 35:155–162
33. Meul M, Van Passel S, Nevens F, Dessein J, Rogge E, Mulier A, Van Hauwermeiren A (2008) MOTIFS: a monitoring tool for the integrated farm sustainability. Agron Sustain Dev 28:321–332
34. Migliorini P, Scaltriti B (2012) Evaluation of sustainability of the farms in the Agricultural Park of South Milan and their production chain. Special issue New Medit 4:53–56
35. Omann I, Spangenberg JH (2002) Assessing social sustainability. The social dimension of sustainability in a socioeconomic Scenario. Presented at the 7th Biennial Conference of the International Society for Ecological Economics" in Sousse (Tunisia), 6-9 March 2002.
36. Pacini C, Giesen G, Wossink A, Omodei-Zorini L, Huirne R (2004) The EU's Agenda 2000 reform and the sustainability of organic farming in Tuscany: ecological-economic modelling at field and farm level. Agr Syst 80(2):171–197
37. Paracchini ML, Pacini C, Jones MLM, Pérez-Soba M (2011) An aggregation framework to link indicators associated with multifunctional land use to the stakeholder evaluation of policy options. Ecol Indic 11(1):71–80
38. Paracchini ML, Bulgheroni C, Borreani G, Tabacco E, Banterle A, Bertoni D, Rossi G, Parolo G, Origgi R, De Paola C (2015) A diagnostic system to assess sustainability at a farm level: The SOSTARE model. Agric Syst 133:35–53
39. Pieniak Z, Aertsens J, Verbeke W (2010) Subjective and objective knowledge as determinants of organic vegetables consumption. Food Qual Prefer 21(6):581–588
40. Platania M, Privitera D (2006) Typical products and consumer preferences: the 'soppressata' case. Br Food J 108(5):385–395
41. Rasul G, Thapa GB (2004) Sustainability of ecological and conventional agricultural systems in Bangladesh: an assessment based on environmental, economic and social perspectives. Agric Syst 79:327–351

42. Reig-Martınez E, Gómez-Limón JA, Picazo-Tadeo AJ (2011) Ranking farms with a composite indicator of sustainability. Agric Econ 42:561–575
43. Rigby D, Woodhouse P, Young T, Burton M (2001) Constructing a farm level indicator of sustainable agricultural practice. Ecol Econ 39:463–478
44. Santini F, Paloma SG (2013) Short food supply chains and local food systems in the EU: a state of play of their socio-economic characteristics, Publications Office
45. Seyfang G (2006) Ecological citizenship and sustainable consumption: examining local organic food networks. J Rural Stud 22(4):383–395
46. Singh RK, Murty HR, Gupta SK, Dikshit AK (2009) An overview of sustainability assessment methodologies. Ecol Indic 9(2):189–212
47. Smith CS, McDonald GT (1998) Assessing the sustainability of agriculture at the planning stage. J Environ Manag 52:15–37
48. Thiollet-Scholtus M, Bockstaller C (2015) Using indicators to assess the environmental impacts of wine growing activity: The INDIGO® method. Eur J Agron 62:13–25
49. Van Cauwenbergh N, Biala K, Bielders C, Brouckaert V, Franchois L, Garcia Cidad V, Hermy M, Mathijs E, Muys B, Reijnders J, Sauvenier X, Valckx J, Vanclooster M, Van der Veken B, Wauters E, Peeters A (2007) SAFE—a hierarchical framework for assessing the sustainability of agricultural systems. Agric Ecosyst Environ 120(2-4):229–242
50. Van Der Werf HMG, Petit J (2002) Evaluation of the environmental impact of agriculture at the farm level: a comparison and analysis of 12 indicator based methods. Agric Ecosyst Environ 93(1):131–145
51. Van Ittersum K, Meulenberg MT, Van Trijp H, Candel MJ (2007) Consumers' appreciation of regional certification labels: a Pan-European study. J Agric Econ 58(1):1–23
52. Vanhonacker F, Lengard V, Hersleth M, Verbeke W (2010) Profiling European traditional food consumers. Br Food J 112(8):871–886
53. Verbeke W, Pieniak Z, Guerrero L, Hersleth M (2012) Consumers' awareness and attitudinal determinants of European Union quality label use on traditional foods. Bio-based Appl Econ 1(2):213–229
54. Verlegh PW, Steenkamp JBE (1999) A review and meta-analysis of country-of-origin research. J Econ Psychol 20(5):521–546
55. Viglizzo EF, Frank F, Bernardos J, Buschiazzo DE, Cabo S (2006) A rapid method for assessing the environmental performance of commercial farms in the Pampas of Argentina. Environ Monit Assess 117:109–134
56. Vilain L (2008) La méthode IDEA. Edicagri éditions, France
57. Von Wiren-Lehr S (2001) Sustainability in agriculture—an evaluation of principal goals-oriented concepts to close the gap between theory and practice. Agric Ecosyst Environ 93:131–145
58. Wells S (2001) Pandora's seed: the unforeseen cost of civilization. Random House, New York, New York, USA
59. Zahm F, Viaux P, Girardın P, Vilain L, Mouchet C (2008) Assessing farm sustainability with the IDEA method—from the concept of farm sustainability to case studies on French farms. Sustain Dev 16:271–281
60. Zanoli R, Naspetti S (2002) Consumer motivations in the purchase of organic food: a means end approach. Br Food J 104(8):643–653
61. Zimmerer KS, Bassett TJ (2003) Political ecology: an integrative approach to geography and environment-development studies. Guilford Press.

CHAPTER 14

Do the initial land endowments lift people out of chronic and transient poverty? Evidence from central and Northern Mozambique

Raul Pitoro

Department of Community Sustainability, College of Agriculture and Natural Resources, Michigan State University, East Lansing, MI-USA.

ABSTRACT

This study aims at assessing the most prevalent type of poverty in rural Mozambique, and if the causes of chronic and transitory poverty are similar to recommended policy, strategies to address each of them caused by different factors. This paper concludes that, most poverty is transient (66%) than chronic (34%), unlike in earlier studies where the determinants of chronic and transient poverty are not necessarily similar. However, it is recommended to prioritize addressing the chronic poverty given its damaging and long-lasting effects. Also, it is believed that by addressing chronic poverty some synergies can be generated, which allow tackling of the transient poverty as well. The most important set of variables for transient poverty are the household landholdings, head's age, family and hired labor, land quality, and livestock. All these covariates tend to increase the transient poverty, except the family labor which is likely to decrease it. The same variables are important to chronic poverty, in addition to education especially in men, where the number of members with self-employment and widowed household heads has a negative effect on chronic poverty. Policies aimed at reducing chronic poverty should concentrate more on improving household characteristics such as investing in education, agricultural reform that encourages landholding expansion and energy such as human power or animal traction for farming while reducing transient poverty would call for policies oriented at allowing economically active families to earn income for their livelihoods.

Keywords: Chronic, transient, poverty, Mozambique, panel data.

1. INTRODUCTION

While poverty is well documented in Mozambique, few studies have systematically used the existent panel data to make a distinction between chronic and transitory poverty which estimate their determinants (Pitoro, 2016).

A study of such nature is of interest to policymakers, development practitioners, and economists as it allows the decomposing poverty into two categories, which permit the developing intervention strategies to address the prevailing type of poverty.

Chronic poverty is defined as the living status in which households' incomes fall below the poverty line for extended period of time or contemporaneously, while transient poverty is when households' income move cyclically below and above the poverty line. In a sense, vulnerable households are more likely to be categorized as transient poor families.It is important to mention that, a transient poor household has much higher ability to bounce back after a severe shock suggesting that, strategies to mitigate these shocks are likely to have great impact in preventing them to be poor in future.

MPD (2010) reported that, one of the millennium development goals (MDGs) for Mozambique is to reach an absolute consumption poverty rate of 40% by 2015, from an estimated 80% in 1990. To reach this goal, the government has been implementing development projects aiming to meet three types of needs namely: primary needs, secondary needs, and other needs.

Despite these efforts, it is still not clear what type of poverty is being targeted. Even if one has identified the pathways out of poverty, one important aspect is to identify the prevalent type of poverty, to assess whether policies will have a similar effect on addressing chronic and transitory poverty and assess whether the determinants of chronic differ from those of transitory poverty. To address these issues, one needs to decompose the total poverty into chronic and transient.

As argued by Garza-Rodriguez et al. (2010), chronic poverty is more unfair and damaging than transient poverty. Understanding the causes and consequences of these types of poverty is very important in policy-making arena. In general, mixed results are found concerning the congruency between transient and chronic poverty determinants. For instance, in Latin America and Asia, studies have found that variables explaining transient poverty are different from those explaining chronic poverty (Garza-Rodriguez et al., 2010; Jalan and Ravalion, 2000); whereas in Eastern Africa, these differences are minimal, as per example from Kenya (Muyanga et al., 2007).

One important aspect of poverty study is whether the poverty reduction interventions will have a similar effect on addressing chronic and transitory poverty or whether the determinants of chronic differ from those of transitory poverty. To answer those questions, one needs to decompose poverty into chronic and transient components. As argued by Garza-Rodriguez et al. (2010), chronic poverty is more damaging than transient poverty; a chronically poor person is perceived to be in such state for so long.

As such, understanding the causes and consequences of these types of poverty is very important in policymaking arena. However, as presented by Cunagura (2008), distinguishing transitory from chronic poverty is an important methodological challenge in poverty analysis. This becomes much more challenging when using a two-period three-year panel to make this distinction with certainty that, large panel periods would be appropriate.

In fact, the distinction between transitory and chronic poverty is of great interest to policy makers, as it provides insights in what development strategies to pursue, based on the prevalent type of poverty. For policy interventions, to distinguish between chronic and transient poverty is not sufficient. One also needs to understand whether the determinants of chronic and transient poverty are different (Pitoro, 2016) and additionally assess, whether policies that address one are different from the other or whether policies that address the transient poverty would also be effective in addressing the chronic poverty.

This study aims to investigate the extent the initial landholdings endowments can address chronic and transient poverty, and the congruence of their determinants using the two-period panel data. The study empirically estimates the relationship between landholdings, transient and chronic poverty in rural Mozambique. It addresses two main questions: i what are the determinants of chronic and transient poverty? ii What is the role of the initial landholding endowments that affect transient and chronic poverty?

The results of this study give a guidance of whether the safety net or more activist policies used to remove poverty traps which are the focus (Dang et al., 2014) are appropriate. As noted by Jalan and Ravalion (1998), to design policies which harness poverty, it is necessary to know the severity of each component of the total poverty, in order to determine whether chronic or transient poverty is determined by the same or different factors.

Given that implementation of social assistance programs in developing country suffer two main problems: leakage (targeting not eligible beneficiaries) and under-coverage (missing the eligible beneficiaries), this study could assist in improving target, by identifying households based on their type of poverty, in order to identify the specific type of program to be implemented. For instance, in Ethiopia, Nega (2010) found that Food-to-Work decreased both total and transient poverty, while the food security package increased total, chronic, and transient poverty and the better-off households benefited the most.

2. METHODOLOGY

2.1 Data sources and collection techniques

The data for the analysis are drawn from a repeated regionally representative agricultural household survey, conducted by the Mozambique Ministry of Agriculture to understand the investments needed to guarantee a robust response, to the new rural environment resulting from the spike in food prices, in the domestic and international markets.

For that purpose, data covering the period before and after the food price crises, in 2008 and 2011, respectively, were collected. With financial assistance from USAID/Mozambique and technical assistance from Michigan State University, the survey sample size was 1,186 households in the Central and Northern regions in the five high agricultural potential provinces (Manica, Tete, Sofala, Nampula, and Zambezia).

This type of data has the advantage of allowing and controlling unobserved time-invariant household characteristics, which is one of the limitations of using cross-section data in empirical studies (Garrett and Ruel, 1999). Panel data were preferred for this research, as it allows an in-depth understanding of the size of the landholdings, poverty, and dynamics in rural Mozambique thus, contributing to more effective policy intervention design.

Several things are noteworthy with respect to survey design. First, although the panel is covering only five provinces out of ten originally interviewed in 2008, it uses the weights of the household Agricultural Survey conducted in 2008 (TIA, 2008). There was no random replacement in 2011, implying that the sample is the only representative of 2008 population.

Given that two sources of attrition were identified in 2011, the TIA2008's weights are used along with an attrition correction factor (Inverse Probability Weights), to control for the attrition bias. The first is that, the 2011 survey team did not go to all the TIA2008 districts in the center/north (for financial reasons). The second is that for the TIA2008 not all the households were re-interviewed during the revisit in 2011 (due to refusal or unavailability of the respondents and households have moved or dissolved).

Tests for attrition have shown the evidence of the presence of attrition bias, and as proposed by Wooldridge (2002), the appropriate inverse probability weights (IPW) were applied to the data. Donovan and Mather (2007) provide a detailed application of this method for previous TIA2002-2005 panel data in Mozambique.

Out of seventy-two districts sampled in 2008, a total of forty-two districts were not re-visited in 2011 in Zambezia, Manica, Tete, and Sofala provinces. These not represented at the provincial level in these provinces, but are representative of the areas surveyed in 2011. Nampula Province is the only one that did not drop a district, between the two survey years. All its data are a representative at province level.

Although since these data contain information on agricultural sector, some limitations are observed such as lack of information on physical wealth found in earlier research, as associated with low poverty and important to decreasing transient and chronic poverty (Jalan and Ravalion, 2000; Muyanga et al., 2007). The distance to markets, road, schools, and health posts, commonly used as measurements of infrastructures are also lacking in the data set.

As found by Muyanga et al. (2013), distance to markets is positively associated with total poverty while Adakhari et al. (2014) found that, in Nepal distance to primary schools and health posts is significant determinant of food consumption. For instance, they found that houses that are as twice as the one far away from a hospital tend to have 9% lower consumption, as compared to another house. With these earlier findings in mind, these omitted variables would have had an effect on the ability of the models to explain exit from poverty as well as vulnerability to chronic poverty.

2.2 Method of data analysis

The distinction between chronic and transitory poverty is based on the framework developed by Jalan and Ravalion (1998), which recognizes that chronic poverty is as a result of household characteristics that prevent people from meeting their basic needs, resulting from long periods of limited income while transitory poverty is caused by income and other shocks associated with household characteristics, that prevent them from meeting their basic needs temporarily.

Yaqub (2000) identifies two methods to distinguish chronic from transient poverty: the spells and components methods. The spells method defines that, chronically the poor depend on the number of times the household has been in poverty. In essence, the emphasis is on the time that an individual experiences deprivation (Harper et al., 2003), which is suggested by Hulme and Shepherd (2003) that, this can be five or more years. The components method identifies chronically poor, if the individual's permanent income is below the poverty line.

The components method has been adopted in this study due to its advantages over the spells method. The attractiveness of the components method resides on the fact that, it considers income transmission between periods (Aaberge and Mogstad, 2007) and depth of poverty (MacKay and Lawson, 2002) while the spells method does not.

Drawing from insights of the pertinent literature (Jalan and Ravallion, 1998; Dulcos et al., 2006; Garza-Rodriguez et al., 2010; Ribas and Machado, 2007; Panganiban, 2010; Muyanga et al., 2007), to distinguish between chronic and transitory poverty we consider the following decomposition in general terms: let (y_{i1}, y_{i2},.., y_{iT}) be the household *i*'s (positive) normalized income stream over T years. At any point in time *t*, a household *i*'s poverty is expressed as , P_{it}^{a}which is based on poverty measures developed by Foster et al. (1984). Jalan and Ravalion (1998) claim that using $\overline{y_i} = t^{-1}\sum_{t=1}^{T} y_{it}$ as an estimate of household *i*"s "permanent income," the chronic poverty is estimated by replacing household"s income y_{it} for all periods *t* by the estimated permanent income, which results in an assessment of chronic poverty for the household $i,\ P_{it}^{C\alpha} = t^{-1}\sum_{t=1}^{T}(1-\overline{y_i})^{\alpha}$ .Then, transitory poverty is the difference between total poverty and chronic poverty, which is thus given by: $P_i^{T\alpha} = P_{it}^{\alpha} - P_{it}^{C\alpha}$.

The regional-level estimates of chronic, transient, and total poverty are obtained by summing each type of poverty (excluding the non-poor), divided by all sample households irrespective of their poverty status.

For empirical estimation of determinants of transitory and chronic poverty, the squared poverty gap is used, and two models are estimated by a regression of each poverty measure on a set of explanatory variables. The initial characteristics are used as explanatory variables. The choice of squared poverty gap is because of the measure of poverty that meets several conditions for empirical work, including the convexity of the poverty function and the transfer of axiom as defined by Jalan and Ravalion (1998).

Models are the regressions of measures of the chronic and transient poverty on the same set of explanatory variables, as before. For the chronic poverty, the econometric model is expressed as

$$P_i^{C\alpha} = C_i = \begin{cases} C^* if\, C^* > 0, where\, C^* = x_i'\beta^C + +\varepsilon_i^C \\ 0\, otherwise \end{cases}$$

Where C^* is latent variable, Cl is the observed chronic poverty, βC a vector of estimable parameters, xi a set of explanatory variables and εi are the model error terms. Similarly, transient poverty model is estimated as:

$$P_i^{T\alpha} = Tr_i = \begin{cases} Tr^* if\, Tr^* > 0, where\, Tr^* = x_i'\beta^{Tr} + +\varepsilon_i^{Tr} \\ 0\, otherwise \end{cases}$$

Finally, a Censored Qunatile regression is used to deal with censored data issue, resulting from the fact that several observation on the dependent variables are non-poor (taking a value of zero). This strategy has been used by researchers such as Jalan and Ravalion (1998) to address the limitation of the usual Tobit model which is not robust to misspecification and the estimates are inconsistent and inefficiency in the presence of heteroscedasticity and non-normality in the distribution of error terms.

According to Panganiban (2010), Quantile regression has the advantage of being robust to distributional misspecification in the error terms and large outliers in the income data as the one used in this study. Following Garza-Rodriguez et al. (2010), to focus on the poorest, the 70th quantile is used for chronic poverty and 90th quantile for transient. The bootstrapping techniques were used to obtain the standard errors of the parameter estimates.

2.3 Description of variables and expected signs

To implement the conceptual framework adopted for this study, it is hypothesized that the ability of a household to move out or into poverty and to move out of transitory and chronic poverty is a function of its demographic characteristics (including education, age of the head and family labor), access to rural services, agricultural production, agricultural technologies, and the assets that the households have access to and are able to control as described below.

2.4 Household demography

The variation in household characteristics is believed to be highly correlated with poverty transition. For instance, the human capital theories argue that household earnings potential is a function of education attainments and age (Muller, 2002). Therefore, the initial household head's education and age are added in the model as continuous variables.

In fact, several studies have found that higher level of education of the household members decreases the likelihood of falling into poverty. Muyanga et al. (2007) found that household headed by educated heads experience more chronic than transient poverty. Garza-Rodriguez et al. (2010) found that illiteracy is inversely associated with transient poverty in Mexico, perhaps because illiterate people are more likely to earn lower income and can hardly aspire to earn high income in the course of their lifetime.

To concur with earlier findings that the success of education in poverty reduction hinges on students exceeding beyond secondary school education (Muyanga et al., 2007) and that a change of household head's education from no education to post-secondary education increases income by 34% in Kenya (Muyanga et al., 2013), the number of male household members with secondary school was included in the model. Male household members are used because they are most likely to influence the decision-making within the household compared to female members.

Jalan and Ravalion (1998) found some evidence of life cycle events being determinants of transient poverty but falling up to 45 years of age. In their study, they found that household characteristics such as household size, education levels of the head and the labor force are more important for chronic poverty than transient poverty. On the other hand, Muyanga et al. (2013) found that asset holding is an increasing function of household size and age of family head. As per this evidence, the age of the household head was included.

Furthermore, Gender of the household head is an important determinant of poverty. Garza-Rodriguez et al. (2010) found that male family head decreases both total and chronic poverty in Mexico while Muyanga et al. (2007) found that female-headed households tend to experience more chronic than transient poverty in Kenya. Therefore, the gender of the household head was hypothesized to influence poverty. The household gender dummy is also included to capture gender differences.

Household size has been found to be both a burden and a blessing in household livelihoods in the earlier studies. To test these relationships in this study, we added into the estimation model, the number of adult and active members to capture the potential of income generation, which is hypothesized to be more linked to transient poverty than chronic poverty as well as with the potential to move out of poverty. Muyanga et al. (2007) found that households with large dependence ratios (large familiy size) experienced chronic poverty as opposed to transient poverty, but Garza-Rodriguez et al. (2010) found an inverse relationship between the family size and transient poverty, that large households which have lower dependency ratio have greater number of people. This contributes to family income, making the household to cope with external issues to reduce income variability which is the leading cause of transient poverty.

2.5 Rural services

Rural services such as infrastructure, access to credit and self-employment were added to the models to capture their ability to reduce poverty in rural settings. For instance, Muyanga et al. (2007) found that Kenyan households that accessed credit either in cash or kind were less likely to be poor than those that did not.

On infrastructure, Muyanga et al. (2013) found that a one Kilometer decrease from homestead to the nearest motorable road increases ascenders' asset wealth by 15%. A lower effect is found on the distance to the health post, where a kilometer decrease leads to an income increase for poverty ascenders, descenders, and consistently non-poor by 2 and 3%, respectively. Given the lack of measures of distance in our data, we assessed the infrastructural effect through a dummy variable measuring remoteness of the village, defined as a village with public transport and roads travelable throughout the year assigned a value of zero and one otherwise.

2.6 Agricultural production and technologies

Higher agricultural output and use of improved agricultural production technologies are associated with the ability to exiting poverty and transient poverty than chronic poverty, thus, added to the models. Muyanga et al. (2007) found that households that adopted modern productivity-enhancing technologies such as fertilizer were less likely to face chronic poverty.

2.7 Household asset endowments

As indicated by Jayne et al. (2003), the initial asset endowments are essential for pro-poor growth. Jalan and Ravalion (1998) found that households with large cultivated areas in China are less vulnerable to chronic poverty. Large cultivated areas were found to be positively associated with high income in Nepal (Adhikari and Bjorndal, 2014). To account for these relationships, the cultivated land size and livestock possession measure as total household tropical livestock units were added into the models.

3. RESULTS AND DISCUSSION

The upsurge of poverty was further strengthened by an increase in poverty gap by an average of 2% points per year although, the squared poverty gap ratio decreased by 2% points in three years. Using the international poverty line of US$1.25 per capita per day based on parity purchase price (PPP), the poverty rate was much higher with similar trend although at a slower pace, with the exception that the difference between the two survey periods is not significant.

These poverty estimates are below those stated by MPD (2010) using consumption indicator and local poverty lines which in 2008/09 estimated 54.7% of people living in poverty in the entire country with the estimates at 49.6 and 56.9% in the Northern and Central Mozambique; respectively.

3.1 Characteristics of sampled households

The 2008 to 2011 panel permits an understanding of the short-term poverty dynamics over the three-year spell since 2008. The other features of the sample households are presented in Table 2. The average size of cultivated land per household did not increase over time estimated at 2.59 ha in 2008 and 2.37 ha in 2011 mainly due to limited access to manpower and alternative power sources for land expansion (Bolardo et al., 2014).

Table 1 presents poverty measure estimates. Results show that headcount ratio increased from 32% in 2008 to 40% in 2011. In only three years, the headcount ratio increased by 8% corresponding to 2.7 points decrease on average per year between 2008 and 2011.

Poverty	2008	2011	Total	% change	Significance
Local poverty lines					
Head count index	0.32	0.40	0.36	3.0	**
Poverty gap ratio	0.23	0.29	0.26	1.0	**
Squared poverty gap	0.43	0.41	0.42	-1.0	+
Poverty line =US$ 1.25/day based on Purchasing Power Parity (PPP)					
Head count index	0.78	0.80	0.79	2.0	
Poverty gap ratio	0.56	0.58	0.57	1.0	
Squared poverty gap	0.48	0.49	0.48	1.0	
Number of observations	1,172	1,172	2,344		

Singificance level: + at 10%; * at 5%; ** at 1% Source: Author's computation from TIA 2008 and Partial Panel 2011.

However, the aggregated agricultural production and access doubled over time from an average of about 1.6 tons of wheat equivalent units in 2008 and 2.7%, respectively. The size of inherited land represents about 52% of the cultivated land and follows the same pattern as the cultivated land size with no statistical change over time. Access to quality land observed 2.7 percentage points increase from 34 % of sampled households in 2008. The use of improved inputs is persistently low with the percentage of households, using chemical fertilizer and pesticides estimated at 6 and 2.5%, respectively. Similar results are reported by Mabiso et al., (2014) using earlier panel data (2002 to 2005).

During the study period, a gradual shift of the rural occupational structure was observed. The number of households residing in non-remote areas noted about 0.7 percentage points increase and the number of migrant workers increased over this period.

Results in Table 2a show that the adoption of improved agricultural technologies is low in rural Mozambique and did not vary over time. For instance, the use of chemical fertilizer is estimated at 6%, while the adoption of pesticides is estimated at not more than 2.5%. The structure of family labor occupation observed significant changes over time. Results in Table 2a and b show that an increase in agricultural workers more than doubled in 2011 from an average of 2 in every 10 household members reported in 2008. The local and international migration increased from 2008 to 2011, with domestic migration increasing from 0.32 members per household to 0.64 members in 2011.

Table 2a. Characteristics of sample households, 2008-2011 (continuous variables).

Characteristics	Total		Year				Operated land size/AE 08		Meandifferences(t-test)	
			2008 (1)		2011 (2)		quintiles			
	Mean	SD	Mean	SD	Mean	SD	Q1 (3)	Q5 (4)	(1-2)	(3-4)
Total land owned (ha)	2.94	3.74	3.09	4.06	2.79	3.35	1.58	5.83	-	**
Cultivated land size (ha)	2.48	3.41	2.59	3.79	2.37	2.95	1.16	5.10	-	**
Inherited land size (ha)	1.28	4.68	1.33	5.02	1.22	4.29	0.83	2.22	-	**
Head's education (years completed)	3.1	2.96	3.1	2.97	3.1	2.95	3.7	2.6	-	**
Males in secondary school (number)	0.22	0.55	0.21	0.54	0.22	0.55	0.40	0.13	-	**
Head's age (years)	42.1	13.28	42.3	13.32	41.9	13.24	43.5	41.5	-	-
HH size (number of members)	6.0	2.79	6.0	2.84	6.0	2.75	8.1	4.5	-	**
HH size (Adult equivalent)	4.5	2.0	4.45	2.0	4.4	2.1	6.0	3.5	-	**
Number of months with food reserves	7.3	4.27	7.3	4.20	7.2	4.34	6.8	7.9	-	-
Total Net HH income (in '000 2011 MZM)	40,17	252,9	47,08	339,7	32,95	101,24	65,05	27,19	-	*
HH's Maize production (Kgs)	792.8	1724.04	784.4	1811.21	801.7	1628.89	972.5	715.3	-	-
Average Maize yield (Kg/ha)	925.6	1010.11	939.0	103,2	911.2	986.49	1259.[illegible]	687.1	-	-
Mean Maize price (MZM/Kg)	5.30	2.94	5.09	2.96	5.52	2.92	5.0[illegible]	5.73	-	-
Aggregated production (in Wheat equivalent units)	2,577.2	48089.59	1,587.9	4377.77	3,610.6	68619.97	1,69[illegible].8	1,668.4	-	-
People aged 15-59 years (number	2.6	1.36	2.7	1.35	2.6	1.37	3.[illegible]	2.2	-	**
People with self-employment (number)	0.69	0.92	0.74	1.01	0.65	0.80	0.30	0.58	-	**
Total Tropical Livestock units per HH	0.94	2.72	0.98	2.93	0.91	2.47	1.11	1.01	-	-
HH used improved seeds for cereals	55.0	0.50	55.8	0.50	54.2	0.50	[illegible]8.6	55.0	-	-
Agricultural technologies used (number)	3.2	1.50	3.1	1.48	3.2	1.53	3.1	3.3	-	-
Number of agricultural workers	0.33	0.69	0.21	0.54	0.49	0.80	0.32	0.36	**	-
Number of non-agricultural workers	0.13	0.34	0.11	0.33	0.14	0.35	0.10	0.11	-	-
Number of domestic migrants per HH	0.46	0.75	0.32	0.60	0.64	0.87	0.42	0.47	**	-
Number of overseas migrants per HH	0.02	0.14	0.00	0.03	0.04	0.19	0.02	0.02	**	-
Number of observations	3,244	1,172	1,172	-	-	-	-	-	-	-

Singificance level: - at 10%; * at 5%; ** at 1%; SD is standard deviation; HH: household. Source: Author's computation from TIA 2008 and Partial Panel 2011.

Table 2b. Characteristics of sample households, 2008-2011 (discrete variables).

Characteristics	Percentage reporting			Operated land size/AE 08 quintiles		Group difference (chi² test)	
	Total	2008 (1)	2011 (2)	Q1 (3)	Q5 (4)	(1-2)	(3-4)
HH lives in non-remote village (1=Yes)	41.0	40.7	41.4	44.6	34.1	-	**
Female-headed HH (1=Yes)	18.2	18.1	18.3	15.3	20.4	+	-
Widowed HH head (1=Yes)	7.4	7.6	7.2	5.2	7.2	-	-
HH is food insecure (1=Yes)	24.3	25.3	23.0	25.7	15.4	-	**
HH have access to credit (1=Yes)	4.1	2.7	5.5	4.2	3.7	-	-
HH has good land quality (1=Yes)	35.3	34.0	36.7	51.4	19.5	-	**
HH used fertilizer (1=Yes)	6.0	6.1	6.0	2.6	8.6	-	**
HH used pesticide (1=Yes)	2.5	2.6	2.4	0.0	4.9	+	**
HH used manure (1=Yes)	4.3	4.3	4.3	3.6	3.3	-	+
HH used irrigation (1=Yes)	3.9	3.0	4.8	3.2	4.9	-	*
HH used improved seeds for cereals (1=Yes)	55.0	55.8	54.2	58.6	55.0	-	-
HH used improved seeds for beans (1=Yes)	28.5	28.1	28.9	26.3	31.3	-	-
HH used improved seeds for vegetables (1=Yes)	12.5	12.3	12.8	13.6	11.0	-	-
HH does crop rotation (1=Yes)	29.1	29.4	28.9	27.5	26.6	-	-
HH does intercropping (1=Yes)	76.6	75.1	78.2	73.9	81.3	-	-
HH does line sowing (1=Yes)	52.2	51.8	52.6	58.8	59.6	-	-
HH used permanent labor (1=Yes)	4.9	4.7	5.1	5.7	4.7	-	-
HH used seasonal labor (1=Yes)	28.3	27.0	29.7	30.1	25.3	-	-
HH used animal traction (1=Yes)	11.6	12.7	10.5	9.1	15.8	-	**
HH used at least one improved agricultural technology (1=Yes)	97.5	97.9	97.0	97.0	98.6	-	+
HH hired seasonal labor (1=Yes)	28.3	27.0	29.7	30.1	23.8	-	-
Number of observations	3,244	1,172	1,172	-	-	-	-

Singificance level: + at 10%; * at 5%; ** at 1%; SD is standard deviation. Source: Author's computation from TIA 2008 and Partia Pane 2011.

3.2 Determinants of total poverty (squared poverty gap)

Asargued by Walker et al. (2004) and Jalan and Ravalion (1998), squared poverty gap is preferred as it provides information on how far people are from the poverty line and satisfies two essential conditions: the convexity poverty function and the income transfer axiom. I then aim to understand the determinants of the total poverty (severity of income poverty) in rural Mozambique by estimating the predictors of the Squared Poverty Gap with varying the poverty lines.

The first two columns of Table 3 pertain to income poverty severity, the poverty gap and squared poverty gap estimated in Tobit models using the local poverty lines and $1.25/day PPP, respectively. An important estimation concern is an endogeneity of the explanatory variables. For this particular study, the endogeneity of landholdings is a major concern, as cultivated land size may result from income accumulation implying that households make decisions on their cultivated land sizes depending on the income they have or generate.

The inherited land size is used as an instrument for land size, assuming that the decision of transferring land from the predecessors is not made by the heirs who are being investigated, so it is an exogenous decision, which therefore, meets the required conditions for an instrumental variable. Given that the reliability of Instrumental variables estimates depends on the validity of an instrument, caution must be exercised while interpreting these results because of lack of other good instruments resulting in the estimation of just-identified IV-models with no further IV tests implemented.

In columns three and four, we estimate consistent parameters addressing the potential endogeneity of the initially cultivated land size/AE. Given that a value of one in the head count index is assigned to poor households, higher value in the severity measures indexes high severity, the negative signs of coefficients imply a reduction of poverty and positive signs of coefficients reveal an increase in poverty. As expected, many of the variables explaining variation in household income also explain the variation in poverty, although with an opposite sign.

Results in Table 3 indicate that keeping other factors constant, the Squared Poverty Gap in 2011 were both significantly higher than in 2008 (a year of high food prices) by 11. In fact, results show a poverty increase between the two survey years and they are consistent with the descriptive statistics in Table 1, highlighting the worsening of poverty in rural Mozambique between the two study years.

Table 3. Determinants of Squared Poverty Gap at local and $1.25/day PPP poverty lines

Variables	RE-estimation models		IV-estimation models	
	Local poverty lines	$1.25/day PPP	Local poverty lines	$1.25/day PPP
Year=2011	0.876+	-0.009	0.094**	0.017
	(1.68)	(-0.22)	(2.68)	(0.58)
HH lives in non-remote village (1=Yes)	-1.602**	-0.094+	-0.130**	-0.088**
	(-2.68)	(-1.91)	(-3.27)	(-2.96)
Log of cultivated land per AE	-0.765*	-0.132**	0.066	-0.344**
	(-2.29)	(-4.74)	(0.96)	(-6.56)
Male-headed HH (1=Yes)	-0.224	0.028	-0.144*	-0.013
	(-0.26)	(0.36)	(-2.54)	(-0.21)
Head's education (years completed)	-0.087	-0.011	-0.004	-0.005
	(-0.78)	(-1.19)	(-0.48)	(-0.86)
Males in secondary school (number)	-0.436	-0.013	0.016	-0.007
	(-0.71)	(-0.26)	(0.42)	(-0.27)
Head's age (years)	0.080**	0.006**	0.003*	0.001
	(3.85)	(3.47)	(2.39)	(0.86)
Males in secondary school (number)	-0.436	-0.013	0.016	-0.007
	(-0.71)	(-0.26)	(0.42)	(-0.27)
Head's age (years)	0.080**	0.006**	0.003*	0.001
	(3.85)	(3.47)	(2.39)	(0.86)
Widowed head (1=Yes)	-1.249	-0.077	-0.155*	-0.068
	(-1.03)	(-0.73)	(-1.99)	(-1.01)
People aged 15 59 years (number)	-0.660**	0.015	-0.017	-0.031*
	(-2.85)	(0.80)	(-0.80)	(-2.06)
People with self-employment (number)	-1.308**	-0.090**	-0.082**	-0.072**
	(-3.65)	(-3.48)	(-3.43)	(-3.96)
HH have access to credit (1=Yes)	-0.279	-0.095	0.076	-0.114
	(-0.20)	(-0.84)	(0.76)	(-1.62)
HH has good land quality (1=Yes)	-1.333*	-0.109*	-0.049	-0.155**
	(-2.30)	(-2.32)	(-1.11)	(-4.18)
HH used fertilizer (1=Yes)	-1.172	-0.072	-0.135+	-0.001
	(-0.90)	(-0.68)	(-1.82)	(-0.01)
HH used improved seeds (1=Yes)	-0.868	-0.024	-0.083*	-0.016
	(-1.54)	(-0.51)	(-2.11)	(-0.54)
HH used permanent labor (1=Yes)	-1.416	-0.006	-0.056	0.028
	(-0.98)	(-0.06)	(-0.69)	(0.43)
HH hired seasonal labor (1=Yes)	-2.052**	-0.155**	0.000	0.000
	(-3.06)	(-2.96)	(.)	(.)
HH used animal traction (1=Yes)	0.254	-0.012	-0.019	0.041
	(0.26)	(-0.15)	(-0.32)	(0.89)
Total Tropical Livestock units per HH	-0.136	-0.010	-0.009	-0.010
	(-1.15)	(-1.15)	(-1.22)	(-1.60)
Constant	2.325	1.099**	0.125	0.508**
	(1.30)	(6.57)	(1.09)	(5.41)
Observations	2.334	2.334	2.334	2.334
Exogeneity test (p-value)	-	-	0.443	0.000

Marginal effects; *t* statistics in parentheses; + at 10%; * at 5%; ** at 1%; District FE included, Source: Author's computation from TIA 2008 and Partial Panel 2011.

The determinants of poverty in Mozambique have been well documented using a variety of available nationally representative data (Boughton et al., 2005; Boughton et al., 2006; Cunguara, 2008; Datt et al., 2000; Jayne et al., 2003; MPD, 2010; Walker et al., 2004). A common finding in those studies is the positive effect of the size of cultivated land/AE in reducing poverty. To some extent, results in Table 3 confirm the welfare effect of cultivated land size in poverty reduction.

Although this relationship is true, the average farm size in rural Mozambique is very low, suggesting that creating conditions that encourage land expansion and utilization plays a major role in agricultural growth and poverty reduction. The cultivated land size increases significantly the total net household income and reduces the severity of poverty. Results show that, an additional percent of cultivated land/AE results in increased household income sufficient to shrinking the distance between income and poverty line (squared poverty gap index) by about 4%. It is noteworthy that the results are sensitive to the poverty line used. This effect vanishes when the size of cultivated land/AE is considered endogenous.

Under high poverty rates, especially when the international poverty line is used, the effect of the size of cultivated land/AE in reducing poverty is even higher, suggesting that increasing the size of cultivated land/AE is an important poverty reduction strategy under high poverty rates.

3.3 Determinants of transient and chronic poverty

For policy intervention targeting on poverty reduction, the distinction between chronic and transient poverty is made, the next step is to assess their determinants. Therefore, this study focuses on decomposing the total income poverty into chronic and transient poverty using the program "Distributive Analysis Stata Package" (DASP version 2.3), developed by Araar and Dulcos (2013), and estimates their determinants.

3.3.1 Poverty decomposition

Table 4 presents the decomposition of squared poverty gap index into transient and chronic poverty; both comprise total poverty. Without correcting for bias, the total poverty stands at 0.420, with transient poverty constituting 65% (0.273) of the total poverty. Correcting for biases brought about by using panel data of a small number of time periods, which is -0.027, the transient poverty now accounts for as much as 66% (0.198) of total poverty. In fact, the high share of transient poverty implies prevalent high cyclical income fluctuations in Northern and Central Mozambique, suggesting that much of the poor population can rise above the poverty line temporarily.

Table 4. Squared poverty gap index chronic and transient poverty decomposition

Total Tropical Livestock units per HH	-0.136 (-1.15)	-0.010 (-1.15)	-0.009 (-1.22)	-0.010 (-1.60)
Constant	2.325 (1.30)	1.099** (6.57)	0.125 (1.09)	0.508** (5.41)
Observations	2,334	2,334	2,334	2,334
Exogeneity test (p-value)	-	-	0.443	0.000

Marginal effects; *t* statistics in parentheses; * at 10%; * at 5%; ** at 1%; District FE included, Source: Author's computation from TIA 2008 and Partial Panel 2011.

3.3.2 Results from censored quantile regression

The research question addressed in this section is whether the determinants of chronic and transient poverty are congruent.Table 4 reports the parameter estimates for the two poverty measures to answer this question. The estimated parameters investigate the effect of the initial condition (using the initial landholdings as explanatory variables) on chronic and transient poverty.

Essentially, this estimation procedure aims to identify whether the determinants of poverty measures are persistent or not over time. The models predict chronic poverty fairly well than transient poverty as can be seen from the pseudo-R^2 (0.11 or 0.12). This echoes earlier studies by Jalan and Ravalion (1998) in China; Panganiban (2010) in the Philippines; and Garza-Rodriguez (2010) in Brazil.

As argued by Jalan and Ravallion, this could be due to lack of variation in the survey reflecting idiosyncratic shock to income. This argument appears to reflect the survey data of the present study where the standard deviation of transient poverty (0.20) is smaller than that of chronic poverty (1.60). The most interesting finding is that variables explaining chronic poverty are not the same that define transient poverty, suggesting that the determinants of chronic and transient poverty are not congruent; however, synergies are expected from policy interventions expected to tackle chronic poverty.

The estimates in Table 4 suggest that the determinants of chronic and transient poverty are not totally congruent; however, policies to address chronic poverty may as well tackle transient poverty, but the inverse does not hold. The most important variables for transient poverty are remoteness, head's age, family, fertilizer use, and livestock. All these variables are important even when the cultivated land is considered exogenous, except the remoteness and livestock that become insignificant.

All these covariates tend to decrease the transient poverty, except the remoteness, which is likely to decrease it. The most important variables for the chronic poverty are the cultivated land size, access to self-employment, use of fertilizer, improved seeds and hiring seasonal labor.

Nevertheless, when the cultivated land size is considered endogenous, only male headship, head's education and civil status of the head (widowed head) are important for chronic poverty (Table 5). All these variables tend to decrease chronic poverty except the widowed heads who tend to increase it. It is noteworthy that the important variables have opposite effect for each type of poverty. Results in Table 4 show that 1% increase in the initial cultivated land size is likely to decrease chronic poverty in about 0.4% under the current poverty lines.

3.4 Household's demographic characteristics

As one would expect, demographic characteristics (education, male heads, and widowed head) seem to be less important for transient than chronic poverty. Results in Table 4 indicate that only family labor and head's age are important for transient poverty. Education is an important factor for avoiding chronic poverty, perhaps because an educated person can easily aspire to have higher income over the course of their lifetime. The effect of the size of family labor seems to suggest that households with a greater number of people that can contribute to household income can help households to cope with external shocks leading to transient poverty.

3.5 Rural services

The improvement of infrastructures for transport is important for transient poverty. Results show that promoting self-employment opportunities in rural Northern and central Mozambique is likely to decrease significantly the likelihood of being chronically poor. Results show that an additional household member with accessing or engaging in self-employment is likely to reduce the likelihood of the household being chronically poor by a sizable amount, about 19%.

3.6 Productive assets

Increasing cultivated land sizes decrease the chance of being chronically poor. This result is consistent with findings by Jalan and Ravallion in China; households with higher cultivated land are less vulnerable to chronic poverty.

Table 5. The effect of initial endowments on transitory and chronic poverty.

Variables dependent variables: Log of chronic/transient poverty	Quartile regression		IV-quantile regression	
	Chronic (70th quantile)	Transient (90th quantile)	Chronic (70th quantile)	Transient (90th quantile)
HH lives in non-remote village (1=Yes)	0.105	0.294*	-0.000	0.145
	(0.59)	(1.73)	(-0.55)	(1.27)
Log of cultivated land per AE	-0.388*	0.222	-0.000	0.023
	(-1.80)	([illegible].23)	(-1.32)	(0.05)
Male-headed HH (1=Yes)	-0.181	-0.028	-0.000**	0.213
	(-0.85)	(-[illegible].19)	(-2.74)	(1.12)
Head's education (years completed)	-0.003	0.019	-0.000*	0.004
	(-0.10)	(0.80)	(-1.82)	(0.19)
Males in secondary school (number)	-0.094	0.[illegible]74	0.000	0.160
	(-0.93)	(0.53)	(0.99)	(1.35)
Head's age (years)	0.014*	-0.[illegible]07*	0.000	-0.009*
	(2.38)	(-2.11)	(0.57)	(-2.33)
Widowed head (1=Yes)	-0.067	0.032	0.188**	0.088
	(-0.16)	(0.2[illegible])	(1.69e+15)	(0.36)
People aged 15-59 years (number)	0.115	-0.119*	-0.000	-0.306**
	(1.56)	(-2.12)	(-1.30)	(-5.20)
People with self-employment (number)	-0.186**	-0.034	-0.000	0.021
	(-2.97)	(-0.6[illegible]	(-0.69)	(0.40)
HH have access to credit (1=Yes)	-0.196	-0.03[illegible]	0.000	0.051
	(-0.77)	(-0.09	(0.01)	(0.23)
HH has good land quality (1=Yes)	-0.227	0.060	-0.000	-0.024
	(-1.58)	(0.66)	(-1.20)	(-0.17)

HH used fertilizer (1=Yes)	-0.591*	-0.385*	-0.000	-0.489*
	(-2.16)	(-2.11)	(-0.57)	(-2.25)
HH used improved seeds (1=Yes)	-0.330*	0.031	-0.000	0.193
	(-2.17)	(0.28)	(-0.59)	(1.62)
HH used permanent labor (1=Yes)	-0.358	0.392	-0.000	0.362
	(-1.31)	(0.79)	(-0.53)	(1.62)
HH hired seasonal labor (1=Yes)	-0.328+	0.295	-0.000	0.109
	(-1.86)	(1.01)	(-0.15)	(0.88)
HH used animal traction (1=Yes)	0.021	0.347	0.000	-0.086
	(0.09)	(1.25)	(0.91)	(-0.50)
Total Tropical Livestock units per HH	-0.028	-0.036*	0.000	-0.014
	(-1.18)	(-2.07)	(0.80)	(-0.74)
Constant	3.176**	0.328	0.000*	-0.483
	(7.67)	(1.24)	(2.27)	(-1.21)
Observations	2,344	1,768	2,344	1,768
R-square	0.114	0.047		
Pseudo R-square	-	-	0.121	0.080

Marginal effects; *t* statistics in parentheses; Significance level: + at 10%; * at 5%; ** at 1%. Source: Author's computation from TIA 2008 and Partial Panel 2011.

An additional percentage of cultivated land decreases the likelihood of being chronically poor by about 0.4%. However, when the cultivated land size is considered endogenous, only male headship, head's education and civil status of the head (widowed head) are important for chronic poverty. This suggests that policies aimed at reducing chronic poverty should focus on agricultural reform that encourages landholding expansion given as it as a positive effect on income. However, it is noteworthy that land expansion will not be effective if complementary services are unavailable. It is acknowledged that small-scale agriculture is the main livelihood strategy for most rural households which accounts for a majority of the nation's agricultural production (85%) (Shapito et al., 2009) and that 80% of the area under cultivation in Mozambique is used for rain-fed production with limited use of improved inputs; efforts to increased production and productivity should be top priority in the government's development agenda when considering land reform.

3.7 Agricultural production and technologies

Similar to productive assets, the agricultural production and technologies are important for both types of poverty. It appears that hiring seasonal labor for agricultural productionand adopting of chemical fertilizers and improved seeds are more important for chronic than transient poverty as they reduce the chances of households being chronically poor by 60% and being transient poor by 39%.

Those adopting improved seeds and hiring seasonal labor are less likely to being chronically by 33% compared to those not using these inputs. The possession of livestock is more important for transient than for chronic poverty.

4. CONCLUSION AND RECOMMENDATIONS

Poverty decomposition shows that 66 percent of poverty is transient and 34 percent is chronic. Unlike most earlier studies, this study concludes that the determinants of chronic and transient poverty are not congruent, suggesting that the chronic and transient have different causes. However, multiple effects are expected from interventions aiming to promote agricultural growth and labor market.

Although the majority of poverty is transient, fighting chronic should be a priority given the fact that chronic poverty is more unfair and damaging than transient poverty. So, long-term interventions such as education and self-employment are recommended to tackle chronic poverty because an educated person can easily aspire to have higher income over the course of their lifetime. On the other hand, with the promotion of rural non-farm economy, households with a large size of family labor are likely to have higher income to cope with external shocks. The cultivated land size is more important for the poorest households under high poverty rates while education, labor market and infrastructure are important for the less poor under lower poverty rates.

The most important set of variables for transient poverty are the household landholdings, head's age, family and hired labor, land quality, and livestock. All these covariates tend to increase transient poverty, except the family labor which is likely to decrease it. The same variables are important for the chronic poverty in addition to education especially in men, the number of members with self-employment, and widowed household heads, which all tend to decrease chronic poverty. It is noteworthy that the important variables have opposite effect on each type of poverty. Policies aimed at reducing chronic poverty should concentrate more on improving household characteristics such as investing in education, agricultural reform that encourages landholding expansion and alternative power sources for agricultural production while reducing transient poverty would call for policies oriented at allowing economically active families to earn income for their livelihoods.

REFERENCES

1. Aaberge R, Mogstad M (2007). On the Definition and Measurement of Chronic Poverty. Institute for the Study of Labor IZA Discussion Paper N. 2659.
2. Adhikari CB, Bjorndal T (2014). Economic Relationship between Access to Land and Rural Poverty in Nepal. Appl. Econ. J. 21(1):20-41.
3. Araar A, Duclos J (2013). User Manual -DASP version 2.3. DASP: Distributive Analysis Stata Package. Université Laval: PEP, CIRPÉE and World Bank. June 2013.
4. Boughton D, Mather D, Tschirley D, Walker T, Cunguara B, Payongayong, E (2006). Changes in rural household income patterns in Mozambique 1996–2002 and implications for Agriculture"s contribution to poverty reduction. MINAG Working Paper No.61E. Maputo: Ministry of Agriculture.
5. Boughton D, Mather D, Tschirley D, Walker T, Payongayong E (2005). Pro-Poor Rural Economic Growth for Post-Civil War Recovery: Myth or Reality in Mozambique? A poster paper for the 26th International Association of Agricultural Economics Meeting, Brisbane, August 2006.
6. CANR (2016). Dissertation Defense - Raul Pitoro - 12/11/15: College of Agriculture and Natural resources (n.d.). Retrieved from http://www.canr.msu.edu/events/ dissertation_defense_raul_pitoro_12_11_15
7. Cunguara B (2008). Pathways Out of Poverty in Rural Mozambique. M.Sc. Thesis Submitted to Michigan State University.
8. Datt G, Simler K, Mukherjee S, Dava S (2000). Determinants of Poverty in Mozambique: 1996-97. Food Consumption and Nutrition Division Discussion Paper no. 78. International Food Policy Research Institute, Washington, DC.

9. Donovan C, Mather D (2007). Impacts of Prime-Age Adult Mortality on Rural Household Income, Assets, and Poverty in Mozambique: Analysis with the TIA Panel Data Set. No 55254, Food Security Collaborative Policy Briefs, Michigan State University, Department of Agricultural, Food, and Resource Economics.
10. Duclos J, Araar A, Giles J (2006). Chronic and Transient Poverty: Measurement and Estimation, with Evidence from China. CIRPEE Working Paper No. 06-11. Downloadable from SSRN: http://ssrn.com/abstract=895181
11. Foster J, Greer J, Thorbecke E (1984). Notes and Comments: A Class of Decomposable Poverty Measures. Econometrica 52:761-766.
12. Garret J, Ruel M (1999). Are Determinants of Rural and Urban and Nutritional Status Different? Some Insights from Mozambique. World Development. 27:1955-1975.
13. Garza-Rodriguez J, Gonzalez-Martinez M, Quiroga-Lozano M, Solis-Santoyo L, Yarto-Weber G (2010). Chronic and Transient Poverty in Mexico: 2002-2005. Econ. Bull. 30:3188-3200.
14. Harper C, Marcus R, Moore K (2003).Enduring poverty and the conditions of childhood: life course and intergenerational poverty transmissions. World Dev. 31:535-554.
15. Hulme D, Shepherd A (2003). „Conceptualizing Chronic Poverty. World Dev. 31:402-423.
16. Jalan J, Ravallion M (1998). Determinant of Transient and Chronic Poverty. Evidence from Rural China. World Bank. Policy Research Working Paper Series No. 1936. Washington DC.
17. Jayne T, Yamano T, Weber M, Tschirley D, Bentica R, Chapoto A, Zulu B (2003). Smallholder Income and land distribution in Africa: implications for poverty reduction strategies. Food Policy 28:253-275.
18. Mabiso A, Cunguara B, Benfica R (2014). Food (In) security and its drivers: insights from trends and opportunities in rural Mozambique. Food Sec. 6:649-670.
19. McKay A, Lawson D (2002). Chronic Poverty: A Review of Current Quantitative Evidence Chronic PovertyResearch centre working paper no 15.
20. (Ministry of Planning and Development) (2010). Poverty and Well-being in Mozambique: The Third National Assessment. National Directorate of Studies and Policy Analysis. October 2010.
21. Muller C (2002). Censored Quantile Regressions of Chronic and Transient Seasonal Poverty in Rwanda. Centre for Research in Economic Development and International Trade, School of Economics, University of Nottingham, University Park, Nottingham NG2 2RD.
22. Muyanga M, Ayieko M, Bundi M (2007). Transient and Chronic Rural Household Poverty: Evidence from Kenya. PMMA Working Paper No. 2007-20.
23. Muyanga M, Jayne TS, Burke WJ (2013). Pathways into and out of Poverty: A Study of Rural Household Wealth Dynamics in Kenya. J. Dev. Stud. 49:1358-1374.
24. Nega F (2010). Rural Poverty Dynamics and Impact of Intervention Programs upon Chronic and Transitory Poverty in Northern Ethiopia. Afr. Dev. Rev. 22:92-114.
25. Panganiban M (2010). Transient, Chronic, and Intergenerational Poverty: Evidence from the Cebu Longitudinal Health and Nutrition Survey. Unpublished MSc Thesis. A thesis submitted to the University of the Philippines School of Economics in partial fulfillment of the requirements for the degree of Master of Arts in Economics.
26. Pitoro R (2016). The effect of the initial land endowments on Transient and Chronic Poverty: Evidence from Rural Mozambique. Poster Presentation at the Annual World Bank Conference on Land and Poverty. Washington DC, March 14-18, 2016.
27. Ribas P, Machado A (2007). Distinguishing Chronic poverty from Transient poverty in Brazil. Developing model from Pseudo-Panel Data. International Poverty Centre. Working Paper No. 36.
28. Shapito A, Gage D, Rinehardt B (2009). Mozambique –Country Proposal Addressing Global Food Insecurity. Fall 2009 Capstone Workshop.
29. Walker T, Tschirley D, Low J, Tanque M, Boughton D, Payongayong E, Weber M (2004). Determinants of Rural Income in Mozambique in 2001-2002. Research Report No 57 E. May 2004 (CDIE Reference number PN-ADH-865).

30. Wooldridge J (2002). Econometric Analysis of Cross Section and Panel Data. Cambridge, MA: MIT Press.
31. Yaqub S (2000). Poverty Dynamics in Developing Countries. Sussex, England: institute of development studies.

CHAPTER 15

Impacts of Rural Labor Resource Change on the Technical Efficiency of Crop Production in China

Ning Yin and Yapeng Wang

College of Economics & Management, Huazhong Agricultural University, Wuhan 430070, China

ABSTRACT

This paper probes effects of the evolvement of labor resources on technical efficiency in crop production in rural China. Based on twelve years of data on crop production of 30 provinces in China, a stochastic frontier production function model is used to measure crop production efficiency in three crop-functional areas—the production area, the consumption area, and the balanced area. Then effects of both quantity and quality change in labor force on technical efficiency in different regions of China are analyzed. Results show that rural China generally has an increasing number of employees shifted to non-agricultural sectors and a decreasing trend of the stock of human capital. However, both these two changes in rural labor force have significantly positive effects on improving crop production efficiency. In addition, China's technical inefficiency is at an average of 22.2%. Dynamically, the technical efficiencies show a tendency to rise steadily throughout China and in three areas, while the consumption area possesses the highest technical efficiency. Those results may lend some experience for other countries that are currently experiencing rural labor force and economic transition.

Keywords: labor; rural migration; human capital; technical efficiency; crop production

1. INTRODUCTION

Economic growth of traditional agriculture depends largely on the initial endowments mainly including labor and land input, whereas the engine for modern agricultural growth has been proved to be factors such as capital input and well-qualified labor input [1,2]. The concept, human capital, has been employed to reveal the source of technical progress, which makes increasing contribution to output growth in modern agriculture. China has witnessed steady growth in crop production over the past ten years. Both accumulated productive factor inputs and advancing technical efficiency in crop production are contributors to this encouraging growth. Nevertheless, technical efficiency as a main embodiment of production efficiency makes more sustainable contribution to output growth than productive factors [3].

Several studies have focused on the analysis of technical efficiency, especially on its measurement and influential factors. Existing methods of measuring technical efficiency include stochastic frontier analysis (SFA) and data envelopment analysis (DEA) [4,5]; factors influencing technical efficiency cover human capital, family characteristics, agricultural public investment, transport infrastructure, and other social elements [6,7]. Among these factors, human capital shows a significantly positive impact on technical efficiency in many fields [8,9] and plays an important role in explaining regional differences of technical efficiency [10]. For the purpose of ensuring the long-term sustainable development of the Chinese economy, increasing the investment of human capital and improving the efficiency of technology application and crop production are of particular importance.

With the deepening industrialization and urbanization in China, traditional factors start to fail to satisfy the demands of modern agriculture development, so requirements for the conformation of new factor inputs have been put forward. Meanwhile, due to technical progress, rural labor resources have been experiencing changes in both quantity and quality. Many researchers contend that human capital has an increasingly pronounced effect in the development of society, economy, employment, income, and so on [11,12,13,14]. Specifically, the evolvement of rural labor resources, especially rural migration, has an increasingly important influence on grain output and economic development in the countryside

[15,16,17]. However, questions still remain to be discussed on the impacts of both quantity and quality change in labor resources in rural China on crop production efficiency, differences among sub-national areas of China, and the trend of China's crop production efficiency under this circumstance.

The objective of this study is to explore the relationship between the change in rural labor resources and crop production efficiency in China. The important part of this study is the consideration of the heterogeneity of rural laborers. The crop production efficiency is analyzed from both perspectives of the quantity and quality change in rural labor resources. Since rural labor resources denote physical as well as mental power in rural areas, in this study, the evolvement of rural labor resources is defined as the dual-effects of the decreasing population caused by rural migration and the quality improvement due to accumulated human capital stock. It is hoped that this study may lend some support for agricultural policy-making for countries also experiencing extensive rural migration and changes in the labor force. Additionally, most existing research starts on a macroeconomic, national level, while few have obtained a regional perspective. Among the few studies from a sub-national perspective, three areas—Eastern, Middle, and Western China are mostly explored, but this dividing standard overlooks the characteristics of targeted crop production. Therefore, in this study, the regions studied are divided according to resource endowments, comparative advantages, and consumption characteristics of each area for crop production, according to the dividing standard put forward in government documents.

2. EMPIRICAL SPECIFICATION

In this research, the following function will be applied to estimate technical efficiency according to the stochastic frontier production function model affected by technical inefficiency [18,19,20]:

$$Y_{it} = f(X_{it}, \beta)\exp(v_{it} - u_{it}) \ (i = 1, \ldots, N) \tag{1}$$

Then, the model can be turned into the following form by taking the logarithm:

$$\ln Y_{it} = \ln f(X_{it}, \beta) + v_{it} - u_{it} \tag{2}$$

where Y_{it} represents the grain output of province i in t period; X_{it} denotes the total factor input of province i in t period; β is an unknown parametric vector; $f(X_{it}, \beta)$ is the production frontier which shows the optimal production technology, with the definition "the maximum possible output gained within given factor inputs and their combinations." The error term (v_{it} - u_{it}) falls into two parts. $v_{it} \sim N(0, \sigma^2_v)$ is a systematic random error representing the uncontrollable factors such as disasters and measurement errors; u_{it} denotes the inefficiency which affects output, namely, the distance between real production and the frontier. u_{it} is subjected to the condition $u_{it} \geq 0$ and $u_{it} \sim N(M_{it}, \delta, \sigma^2_u)$, among which M_i is an exogenous variable related to efficiency.

The crop production efficiency is measured by the ratio of observed outputs to the corresponding stochastic outputs. This measuring method is the most frequently used output-oriented technical efficiency in previous studies [18]. The technical efficiency (TE) of crop production can be expressed as follows:

$$\mathrm{TE}_i = \frac{E[f(X_{it}, \beta)\exp(v_{it} - u_{it})]}{E[f(X_{it}, \beta)\exp(v_{it} - u_{it})|u_{it} = 0]} = \exp(-u_{it}) \tag{3}$$

This ratio is in a [0,1] interval due to the non-negativity of u_{it}, which means TE = 1 when there is no efficiency loss (u_{it} = 0). The model is estimated with a maximum likelihood method instead of an ordinary least squares (OLS) method because the error term in the stochastic frontier production function fails to be subject to classical assumptions in OLS. In addition, we make a distributional hypothesis to the two error terms: making σ^2_v and σ^2_u substituted by $\lambda = \sigma^2_u/(\sigma^2_v + \sigma^2_u)$, among which λ belongs to a [0,1] interval. In this way, the significance of inefficiency can be shown clearly when we estimate λ. When λ is close to 1, the error is mainly caused by inefficiency in technology use. On the contrary, λ being close to 0 means that random error is the primary reason for the error in the frontier production function.

Battese and Coelli [21] proposed the technical inefficiency model as an extension model for inefficiency effects in stochastic frontiers for the cross-sectional data. The inefficiency term u_{it} is defined as follows:

$$u_{it} = M_{it}\delta + \omega_{it} \tag{4}$$

where M_{it} denotes the exogenous variables affecting inefficiency; δ is the parameter to be estimated; random variable $\omega_{it}$is subject to a normal distribution $u_{it} \sim N(u_{it}, \sigma^2_u)$, which is truncated at $-M_{it}\delta$.

Based on the above discussion on model choosing, a stochastic frontier production function model is adopted in this paper to measure the technical efficiency in China's crop production. A function of technical inefficiency is also employed to analyze the implication of migration in different crop production areas on its technical efficiency. When it comes to production function choosing, several functions such as Leontief function, quadratic function, Cobb–Douglas production function, and so on can be applied [22]. Cobb–Douglas production function, a common and standard model in previous production researches, is employed in this study.

According to the literature and available dataset, the input variables included in the production frontier model are the classic production inputs: land, labor, and capital. Land denotes the total cultivated area, while Labor refers to the total labor used for crop production. Capital input is captured by two main factors, mechanical traction input and fertilizer use. These variables have different statistical standards, so we adjusted and converted these variables to maintain consistency with the statistical criteria of crop production. The modification is based on the following methods:

a) Labor inputs in crop production = employment in primary industry × (agricultural outputs/primary industry outputs) × (cultivated area of food/cultivated area of agricultural products);

b) The ratio "cultivated area of food/cultivated area of agricultural products" is used to modify and compute machine use, fertilizer, and pesticide use.

Therefore, referring to Equation (2), the function is specified as follows:

$$\ln Y_{it} = \beta_0 + \beta_1 \ln X_{1it} + \beta_2 \ln X_{2it} + \beta_3 \ln X_{3it} + \beta_4 \ln X_{4it} + v_{it} - u_{it} \tag{5}$$

where Y_{it} depicts the total grain output of province i in t period, the sum of rice, wheat and corn outputs; X_{mit} (m = 1,..., 4) represents four types of factor inputs of province i during t period, including labor input, cultivated area, mechanical traction input, and fertilizer input in crop production. β_m (m = 1,..., 4) is the coefficient representing the output elasticity of labor input, cultivated area, mechanical traction input, and fertilizer input, respectively, and u_{it} represents effects of technical inefficiency.

Despite the lack of accurate statistics on migration in China's official database, quite a few methods measuring migration volume have been applied in previous studies. One refers to the calculating method: the amount of migration = (urban employment − urban registered staffs) + (total rural employment − rural agricultural employment) [23]; while another method uses the difference between actual employees between agriculture and primary industry to show the amount of migration [24]. In this study, a new index to show migration scale, the ratio of employees in agriculture to employees in all industries, is adopted for two reasons. Firstly, the annual change of this index exactly reflects the rural-to-urban migration trend—the decline of this ratio indicates an increasing amount of migration. Secondly, the purpose of this study is to grasp the implication of migration on crop production efficiency from a macroeconomic perspective, so this new index appears more clear and direct.

The stock of human capital is also a very important variable apart from the migration variable. According to Schults' theory of human capital [25], among all factor inputs, human capital is the primary resource in economic growth. The process of rural-to-urban migration can be seen as farmers' decision-making process that helps them find an optimal price for human capital. An expanding body of literature has proven that the overall quality of farmers is dynamically improved due to the popular value on human capital investment, though the loss of rural human capital occurs during labor outflow [26,27,28]. To our knowledge, few studies, however, have probed the relationship between human capital and crop production efficiency with an empirical method. Admitting the basic fact of rural laborers' heterogeneity, this study integrates human capital as a variable into the migration index to reflect the influence of rural human capital change on crop production efficiency. The level of human capital is represented by a variable H_i, which shows the scalable stock of human capital [29]. The computing function is $H_i = e^{\varphi(Ei)R}L_i = h_iL$, where $\varphi(E_i)$, under the condition $\varphi(0) = 0$, is the production efficiency of laborers who have had education for E years. Different from previous studies that use years of education to measure the quality of labor force, this method highlights that different education stages contributing differently to production efficiency. The education attainment of rural laborers in each province is calculated annually, referring to the method of counting education span and return rate in a study by Li [30]. The variable L_i represents the human capital of an ordinary laborer, so the volume of agricultural employees is used.

Therefore, referring to Equation (4), migration and the stock of human capital considered, the inefficiency model in this study is as follows:

$$u_{it} = \delta_0 + \delta_1 M_{1it} + \delta_2 M_{2it} + \omega_{it} \tag{6}$$

where explanatory variable M_1 is the ratio of agricultural employees to employees in all industries, representing the migration scale; M_2 captures the stock of human capital; i presents the cross-section unit of dataset; t reflects the 12 years of time series from 2001 to 2012. Parameters δ_1 and δ_2 represent the influence of labor outflow (proportion of non-agricultural industries) and rural human capital on crop production inefficiency, respectively. A positive numerical value of δ means a positive influence on inefficiency, namely a negative impact on technical efficiency and otherwise.

3. MATERIAL AND METHODS

In this study, we use a panel dataset of 30 provinces in China from 2001 to 2012, excluding 4 provinces (Hong Kong, Macao, Taiwan, and Tibet) because these areas possess particular resource endowment and productivity. The panel data are drawn from *China Rural Statistical Yearbook* and *China Statistical Yearbook* published by the National Bureau of Statistics (NBS) [31,32]. Since these yearbooks cover data from previous years, we used yearbooks from 2002–2013. Unfortunately, data on our key variable, rural employee volume in the primary industry in 2013, have not yet been included. The most updated data we were able to use was that from 2012. This may be one of the shortcomings of this study, and more information is needed for further research.

To capture a better understanding of disparities in different regions in China, we divided these 30 provinces into three crop-functional areas: the production area, the consumption area, and the balanced area (Table 1), according to the dividing standard put forward in government documents considering regional resource endowment, comparative advantages, and consumption characteristics. In the crop production area, the total volume of crop production is larger than the total volume of crop that has been consumed. On the contrary, in the consumption area, the amount of crop consumed is larger than that of produced crop. Compared to the balanced area, the two volumes are basically equal. Compared with studies dividing China into eastern, middle, and western areas [33,34], this regionalism appears more targeted when crop production is the focus of research.

Table 1. Provinces that are included in three crop-functional areas.

Crop Functional Area	Provinces
production area	Liaoning, Hei Longjiang, Jilin, Hebei, Inner Mongolia, Shandong, Jiangsu, Anhui, Jiangxi, Henan, Hubei, Hunan and Sichuan
consumption area	Beijing, Tianjin, Shanghai, Zhejiang, Fujian, Guangdong and Hainan
balanced area	Shanxi, Chongqing, Guangxi, Ningxia, Xinjiang, Qinghai, Shaanxi, Yunnan, Guizhou and Gansu

The descriptive statistics of variables (Table 2) reveal some interesting patterns. For instance, the production area has an absolute advantage in factor input volume; however, when labor resources are focused on, we find that the migration scale in the production area and the balanced area are roughly close, while the stock of human capital varies in each area. The stock of human capital in the production area is approximately twice the stock volume in the balanced area. This suggests an evolvement disparity of human capital in these two different areas.

Table 2. Descriptive statistics of output and inputs in three crop-functional areas.

		Total Grain Output (10^4 tons)	Labor Input (10^4 person)	Cultivated Area (10^3 hectare)	Mechanical Traction (10^4 kw)	Fertilizer Use (10^4 tons)	Employee Proportion (%)	Stock of Human Capital (10^4 persons)
N	Mean	1629.12	330.79	3422.83	1662.19	107.33	0.59	1285.60
	S.D.	1303.62	266.69	2604.85	1732.70	89.96	0.16	1015.03
	Max	5761.49	1477.72	11,519.54	8231.54	479.18	0.88	4748.33
	Min	58.03	8.15	141.34	39.69	2.54	0.16	50.49
PA	Mean	2881.60	483.00	5761.65	2940.29	181.41	0.60	1902.82
	S.D.	995.71	279.90	2097.77	1984.18	85.87	0.11	1051.21
	Max	5761.49	1477.72	11,519.54	8231.54	479.18	0.82	4748.33
	Min	1239.10	178.08	2743.27	591.15	60.87	0.30	675.63
CA	Mean	490.56	119.21	934.43	529.50	41.66	0.43	664.66
	S.D.	457.97	134.23	871.44	437.72	40.02	0.17	711.75
	Max	1600.10	456.84	3125.60	1369.88	134.64	0.78	2215.08
	Min	58.03	8.15	141.34	39.69	5.29	0.16	50.49
BA	Mean	873.45	285.54	2242.30	872.51	61.57	0.68	951.31
	S.D.	481.88	197.36	1280.38	571.47	42.04	0.10	689.85
	Max	1749.10	757.90	4399.57	2641.48	176.95	0.88	2285.21
	Min	86.80	24.33	169.43	106.27	2.54	0.43	54.40

Note: N stands for the nationwide data; PA stands for the production area; CA is the consumption area, BA is the balanced area.

The larger stock of human capital in the production area may derive from two aspects: the preliminary large quantity of labor force and the improvement of education attainment in this area.

The consumption area experiences the lowest rate of agricultural employees, which is reasonable because, in the consumption area, the first industry does not dominate in economic development. Therefore, redundant rural laborers have more initiations and opportunities to participate in off-farm employment. Theories on migration motivation have demonstrated that the economic development gap has been an important contributor to migration [35,36,37].

4. RESULTS AND DISCUSSIONS

4.1. Migration and the Stock of Human Resource Change

Table 3 shows the change in the average proportion of employees in the primary industry to employees in all three industries, as well as the trend of average stock of human capital. Since the annual change in employee proportion measures the ratio of employees in the first industry, a decreasing trend, as shown in the results, indicates that people taking jobs in the first industry is declining and the amount of laborers taking off-farm employment is increasing. This reflects an increasing migration scale to the non-agricultural sectors.

Table 3. The average migration scale and stock of human resource change in China.

Year	Employee Proportion	Stock of Human Capital
2001	1374.82	0.68
2002	1359.63	0.67
2003	1334.16	0.65
2004	1310.26	0.63
2005	1303.03	0.61
2006	1287.62	0.59
2007	1263.32	0.57
2008	1259.72	0.56
2009	1256.72	0.55
2010	1247.14	0.54
2011	1219.13	0.53
2012	1211.66	0.52

As the migration is increasing, rural China experiences a descending stock of human capital. Both quantity and quality change in rural labor force have contributed to this decrease. With more rural labor force migrating to non-agricultural sectors, the remaining laborers in rural areas tend to be the elderly, children, and women. Studies have shown that young and more educated laborers are more likely to migrate [38,39]. Consequently, problems including rural aging laborers, female farmers, and unattended children have been increasingly prominent in recent China [38,40].

4.2. Crop Production Efficiency

The crop production efficiency in China and the three crop-functional areas from 2001 to 2012 is measured respectively with Equation (3). Results show that the average technical efficiency displays a marked difference in the three areas (Table 4). The highest technical efficiency goes to the main consumption area and then the production area. A possible explanation for this result may be that the economic power of the consumption area ensures its advantage in agricultural technology application, which is exactly what the technical efficiency represents. In all of China, technical inefficiency is at an average of 22.2%, possibly deriving from two facts. Firstly, the technical frontier is gradually advanced because of the rapid technical progress, so the measuring standard of technical efficiency is pushed forward constantly; secondly, technical efficiency loss occurs when heterogeneous farmers fail to fully utilize new technology because they are confined to household conditions and endowments.

Table 4. The crop production efficiency in China and three crop-functional areas.

Year	Nationwide	Production Area	Consumption Area	Balanced Area
2001	0.752	0.764	0.860	0.717
2002	0.765	0.783	0.859	0.734
2003	0.750	0.738	0.853	0.739
2004	0.774	0.786	0.873	0.750
2005	0.767	0.790	0.874	0.748
2006	0.780	0.810	0.903	0.731
2007	0.779	0.792	0.908	0.747
2008	0.791	0.826	0.916	0.754
2009	0.779	0.805	0.914	0.756
2010	0.786	0.827	0.910	0.762
2011	0.799	0.861	0.923	0.756
2012	0.814	0.875	0.922	0.791
Average	0.778	0.805	0.893	0.749

Seen from a dynamic perspective, the technical efficiency in China and three functional areas is ascending constantly except for the general slump in 2003. In 2003, the main production area and the main consumption area witnessed the lowest technical efficiency: 0.738 and 0.853, respectively. This is in relation to three years of grain output reduction from 2000 to 2002 when farmers' grain planting initiatives diminished. Because of the low prices, the enthusiasm of farmers to produce grain crops tended to descend. At the same time, the arable land in China had declined in these three years, which also contributed to the reduction of grain output. However, the subsidies on crop production and agricultural tax exemption policy were published in 2004 to encourage farmers to grow grain products, and prove to be effective. The technical efficiency of the whole China, the main production area and the main consumption area increased by 2.4%, 5.2%, and 2.0%, respectively, from 2003 to 2004.

4.3. Output Elasticity of Input Factors

Equations (5) and (6) are estimated with the method of three-stage maximum likelihood estimation. Firstly, the output elasticity and other parameters are estimated with the OLS method. We then attain the variance ratio through a two-phase grid search and use this ratio to adjust output elasticity and other parameters. At last, the maximum likelihood estimated value is obtained with numerical method using these parameters as initial values [21]. The estimation result (Table 5) shows that all coefficients are highly significant at a 1% or 5% level, except for the coefficient for mechanical traction in the production and consumption areas and that for labor input in the main consumption area.

Output elasticity of labor input in crop production is negative from the nationwide perspective, which means that there is surplus labor in China's crop production. This is accordant with results of previous studies [6]. When regional results are considered, only the consumption area has a positive, but not significant, output elasticity of labor input, while the production area and the balanced area have negative ones. This robust result implies that there is surplus labor in rural China, and increasing the quantity of labor input in crop production alone cannot necessarily lead to output growth, but studies have suggested a positive effect of ameliorating the quality of labor resources on yields growth [34].

Output elasticity of cultivated area appears to be positive and significant at a 1% level in four models. The robust result means that enlarging cultivated areas of crops can improve grain output, which conforms to the neoclassic economic theory. Although the model agricultural production increasingly depends less on the enlargement of land resource because of the scarcity, ensuring the fundamental levels of farming land is necessary when taking food security into consideration.

Table 5. Estimation results of stochastic frontier production function model.

	Variable	Parameter	Coefficient	Std. Err.
Nationwide Scope	constant	β_0	−0.086	0.124
	labor	β_1	−0.175 ***	0.025
	land	β_2	0.957 ***	0.031
	machine	β_3	0.046 **	0.018
	fertilizer	β_4	0.127 ***	0.028
Production Area	constant	β_0	2.548	0.002
	labor	β_1	−0.121 ***	0.013
	land	β_2	0.571 ***	0.011
	machine	β_3	0.002	0.017
	fertilizer	β_4	0.279 ***	0.024
Consumption Area	constant	β_0	−0.503	1.288
	labor	β_1	0.015	0.082
	land	β_2	1.083 ***	0.333
	machine	β_3	−0.008	0.168
	fertilizer	β_4	−0.176 ***	0.044
Balanced Area	constant	β_0	2.725 ***	3.40×10^{-8}
	labor	β_1	−0.213 ***	7.76×10^{-9}
	land	β_2	0.617 ***	1.11×10^{-8}
	machine	β_3	−0.049 ***	3.76×10^{-9}
	fertilizer	β_4	0.261 ***	3.24×10^{-9}

Notes: ** and *** indicate estimated coefficients are statistically significant at 5% and 1%, respectively.

Capital inputs have shown significantly positive impacts on grain output growth nationwide, which implies that more investments on mechanization and fertilizer can contribute to crop production improvement. This result strongly supports opinions of neoclassic economics that a capital factor plays an important role in production function formation. It is also clear that China's agricultural development at the present stage can still be identified as neoclassic agricultural development. This developing pattern is an accompanying process with China's industrialization and reflects the positive impact of industrialization on agricultural development.

Among the capital inputs, fertilizer input is significantly positive in the production and balanced areas, while a significantly negative coefficient is shown in the consumption area. This supports the findings of previous studies that proper chemical input can contribute to crop production, but excessive use of fertilizer does occur [41]. Additionally, with the development of agricultural mechanization, the parameter of mechanical traction in crop production is negative, except for the production area where the mechanization level still needs to be improved.

Results from perspectives of different functional areas suggest that different aims should be directed at improving crop production in different areas. To achieve more efficient and targeted policies, investment in capital input should have a distinct emphasis on different crop-functional areas.

4.4. Technical Inefficiency

The estimation results of Equation (6) show the technical inefficiency nationwide and in three functional areas (Table 6). The estimation results of λ are significant and close to 1, which indicates that technical inefficiency consequently leads to most errors when input factors are under control.

Table 6. Estimates of technical inefficiency model.

	Variables	Parameter	Coefficient	Std. Err.
Nationwide Scope	constant	δ_0	0.005	0.107
	migration scale	δ_1	0.605 ***	0.132
	the stock of rural human capital	δ_2	−0.0001 ***	0.00003
		λ	0.953 ***	0.020
	logarithm of likelihood function		156.136	
Production Area	constant	δ_0	0.256 ***	0.065
	migration scale	δ_1	0.247 **	0.104
	the stock of human capital	δ_2	−0.0001 ***	0.00002
		λ	0.994 ***	0.016
	logarithm of likelihood function		123.647	
Consumption Area	constant	δ_0	−0.060	0.549
	migration scale	δ_1	1.167	0.0001
	the stock of human capital	δ_2	−0.002	0.001
		λ	0.644 ***	0.233
	logarithm of likelihood function		94.359	
Balanced Area	constant	δ_0	1.347 ***	0.199
	migration scale	δ_1	−0.697**	0.277
	the stock of human capital	δ_2	−0.001 ***	0.0001
		λ	0.999 ***	0.019
	logarithm of likelihood function		100.132	

Notes: ** and *** indicate estimated coefficients are statistically significant at 5% and 1%, respectively. Migration scale is defined as the proportion of employees in the agricultural sector to that in all industries.

The estimation results of δ_1 (Table 6) show that the migration scale in a region has a pronounced impact on crop production efficiency. Notably, the interpretation of signs of coefficients needs to be clarified. As regards the coefficient of δ_1 in the nationwide model, since migration scale is interpreted by the ratio of employees in the agricultural sector to employees in all industries, a larger ratio means more people stay in the agricultural sector and indicates a lower migration degree. The positive parameter of the migration scale indicates that, when more people are employed in the agricultural sector, i.e., the migration scale is smaller, there is more technical inefficiency. To make it simple, when employees in the agricultural sector comprise a larger proportion, crop production is less affected. Therefore, after laborers migrate from agricultural sectors, this proportion decreases, and crop production efficiency increases. The estimation results of δ_1 show that rural migration helps to improve crop production efficiency in rural China, as well as in the production and balanced areas. For all of China, a 1% drop in employment in the agricultural sector corresponds to a 0.6% rise in technical efficiency. In the consumption area where the migration level is higher, labor needs to increase in the agricultural sector.

The estimation results of δ_2 show that the stock of rural human capital poses a significant impact on technical inefficiency of crop production. The negative values of δ_2, all of which are significant at the 99% level, indicate a positive traction of human capital to technical efficiency. The balanced area, compared to the other two areas, experiences improved effects of the stock of rural human capital on the improvement of technical efficiency. However, the positive effects of human capital are limited because no δ_2 coefficients, nationwide or in the three areas, are larger than 0.02. This does not necessarily mean that the stock of human capital is not important for technical efficiency improvement. One reason could be that the stock of human capital is decreasing annually, and the positive effects can hardly contribute to the increase in technical efficiency.

Generally, the evolvement of rural labor resources in China has a positively significant impact on crop production efficiency in China. Government efforts on the improvement of crop production efficiency in China should increase efforts to achieve the quality amelioration of rural laborers and the advancement of rural migration. On the other hand, the regional disparity of the impact shows that reinforcing education and migration plays a more essential role in the balanced area for the purpose of improving technical efficiency of crop production.

5. CONCLUSIONS

The extensive off-farm migration is a transforming process of resource allocation with a traditional agricultural pattern, since the number of rural laborers descends during this process. Meanwhile, the off-farm migration displays its incentive effect on the demand growth for rural education and investments on human capital. Therefore, the migration has posed an impact on both quantity and quality of agricultural labor. It is under this background that this study focuses on impacts of human capital change on crop production efficiency.

It is found that there exists an increasing migration scale to the non-agricultural sectors and decreasing trend of the stock of human capital in rural China. Additionally, China's technical inefficiency is at an average of 22.2%, and the average technical efficiency displays a marked difference in the three areas. The highest technical efficiency goes to the consumption area and then the production area. Dynamically, the technical efficiencies show a tendency to rise steadily throughout China and in the three crop-functional areas.

One of the major findings is that the decrease in employees in the primary industry is beneficial for improving crop production efficiency. It is not assumed that the migration will make labor a "bottleneck" resource for crop production; on the contrary, redundant labor in production is released by migration, and the resource allocation efficiency can be thus improved. In addition, the off-farm migration can contribute to the scale expanse and specialization of production because the patterns and organizations of crop production are reconstructed during migration. Additionally, off-farm employment helps to increase family income and this can help to improve the potential application of new agricultural technology.

The quality of rural labor and the stock of human capital has revealed positively significant but faint effects on technical efficiency improvement. The low marginal effect of labor quality derives from the qualitative surplus of the labor force in rural China. It also reveals that the technical progress in China's agricultural production hardly depends on skill-based technical progress. However, human capital is of paramount importance in agricultural growth.

Agriculture in China is experiencing a transformation from a traditional to modern pattern. During this period, human capital appears to be a crucial explanation for the growth of agricultural outputs because the increase in human capital stock brings higher technical efficiency. This conclusion accords with the New Economic Growth Theory and indicates that the agricultural pattern in China is transforming towards endogenous development. Although this developing pattern mostly exists in developed economies, it will be the orientation and destination of future agriculture in China. Additionally, this can lend experience to other countries that are currently experiencing rural labor force and economic transitions.

Future research on this topic is needed. First, more efforts on the stock of human capital are necessary since it has played an important role in the growth of agricultural economics. Second, obtaining more specific data on the grain production and rural human capital over a longer time period is necessary. Third, future research that focuses on the impact of labor change in the agricultural sector on crop production can unfold from the perspectives of farm households and provide more empirical evidence in other areas in the world.

ACKNOWLEDGMENTS

This article is funded by the National Social Science Foundation of China (13AGL004).

AUTHOR CONTRIBUTIONS

Ning Yin designed the research topic, completed data analysis, and drafted the manuscript; Yapeng Wang provided suggestions to construct and edit the paper.

REFERENCES

1. Xu, X.; Jeffrey, S.R. Efficiency and technical progress in traditional and modern agriculture: Evidence from rice production in China. *Agric. Econ.* **1998**, *3*, 157–165.
2. Yang, M.; Bai, R.; Liu, M. Development of Agricultural Mechanization and Construction of Modern Agriculture. *Trans. Chin. Soc. Agric. Mach.* **2005**, *36*, 68–72.
3. Huang, J.; Zhou, X. Technical Efficiency and Growth of Total Factor Productivity of Crop production in China, 1978–2008. *South China J. Econ.* **2010**, *9*, 40–52.
4. Liu, N. Evaluation on the Grain Production Capacity of Major Grain Production Area Based on Super-efficiency Output-DEA Model. *Soft Sci.* **2011**, *25*, 79–83.
5. Li, G.; Feng, Z.; Fan, L. An Empirical Analysis about the Effect of Household Endowments on the Technical Efficiency of Farmer's Household Management—Evidence from the Farmers of Hubei Province. *Stat. Res.* **2008**, *25*, 35–42.
6. Wang, X.; Jiang, T. Research on China Agricultural Technical Efficiency based on Public Investment on Agricultural. *Chin. Rural Econ.* **2009**, *5*, 79–86.
7. Li, Z.; Li, D. Empirical Research on the Effect of Transportation Infrastructure Construction on Agricultural Technical Efficiency. *Forum Sci. Technol. China* **2012**, *2*, 127–133.
8. Yan, M.; Wang, W. Empirical Test on Technical Efficiency of Human Capital Structure in China Based on Stochastic Frontie. *J. Math. Pract. Theory* **2012**, *42*, 11–18.
9. Fu, X.; Wu, L. Technical Efficiency, Capital Deepening and Regional Disparity. *Econ. Res. J.* **2006**, *10*, 52–61.
10. Fu, Q.; Jin, N. Empirical Test on Efficiency of Human Capital and R & D Investment in China Based on Stochastic Frontier Production Function. *Tech. Econ.* **2009**, *28*, 5–10.
11. Plummer, P.; Taylor, M. Entrepreneurship and human capital, distilling models of local economic growth to inform policy. *J. Small Bus. Enterp. Dev.* **2004**, *11*, 427–439.
12. Hong, P.Y.; Pandey, S. Human capital as structural vulnerability of US poverty. *Equal Oppor. Int.* **2007**, *26*, 18–43.
13. Khorasgani, M.F. Higher education development and economic growth in Iran. *Educ. Bus. Soc.* **2008**, *1*, 162–174.

14. Jiang, Y.; Shi, X.; Zhang, S.; Ji, J. The threshold effect of high-level human capital investment on China's urban-rural income gap. *China Agric. Econ. Rev.* **2011**, *3*, 297–320.
15. Fan, D.; Zhu, Y. The Outflow of Agricultural Labor Force on Grain Production Affection in the Context of the Dual Economy. *J. Hebei Univ. Econ. Bus.* **2012**, *33*, 39–43.
16. Zhang, X.; Shen, L. The Outflow of Agricultural Labor Force and Economic Development. *J. Guangdong Coll. Financ. Econ.* **2009**, *8*, 80–84.
17. Xu, C. Influence the Outflow of Agricultural Labor Force on Family Supporting. *Occup. Circle* **2007**, *2*, 74–76.
18. Xu, H.; Ayachit, M.; Reddyreddy, A. Formal Modeling and Analysis of XML Firewall for Service-Oriented Systems. *Int. J. Secur. Netw.* **2008**, *3*, 147–160.
19. Aigner, D.J.; Lovell, C.A.K.; Schmidt, P. Formulation and estimation of stochastic frontier production function models. *J. Econom.* **1977**, *6*, 21–37.
20. Meeusen, W.; van Den Broeck, J. Estimation from Cobb-Douglas production functions with composed error. *Int. Econ. Rev.* **1977**, *18*, 435–444.
21. Battese, G.E.; Coelli, T.J. A model for technical inefficiency effects in a stochastic frontier production function for panel data. *Empir. Econ.* **1995**, *20*, 325–332.
22. Griffin, R.C.; Montgomery, J.M.; Rister, M.E. Selecting functional form in production function analysis. *Western J. Agric. Econ.* **1987**, *12*, 216–227.
23. Guo, J., Li, Z. Agricultural Output Increase under the Context of Selective Labor Transfer, A Empirical Test of Hypothesis of Non-Skill-Biased Technical Change and Quality-Surplus Labor Force. *Jilin Univ. J. Soc. Sci. Ed.* **2011**, *51*, 100–110.
24. He, J.; Shu, H.; Tian, Y. Study on Calculation of Rural Labor Force Transfer in China and Influential Factors. *China Popul. Resour. Environ.* **2011**, *21*, 148–152.
25. Schultz, T.W. Investment in human capital. *Am. Econ. Rev.* **1961**, *51*, 1–17.
26. Li, Z.; Guo, J. Selective Labor Transfer and Human Capital, the Theory and the empirical. *Thinking* **2010**, *36*, 112–117.
27. Bai, N.; Li, J. China's Urbanization and Rural Labor Migration. *Chin. J. Popul. Sci.* **2008**, *4*, 1–10.
28. Sheng, L. Analysis of the Determinants of Rural Labor Migration in China. *China Rural Surv.* **2007**, *3*, 2–15.
29. Hall, R.; Jones, C. Why do Some Countries Produces Much More Output per Worker than Others? *Q. J. Econ.* **1999**, *114*, 83–116.
30. Li, G. Human Capital and TFP Growth of Regional Agriculture in China, Empirical Study Based on Data Envelopment Analysis. *J. Financ. Econ.* **2009**, *35*, 115–128.
31. National Bureau of Statistics of China (NBSC). *China Rural Statistical Yearbook 2002–2011*; NBSC: Beijing, China, 2014.
32. National Bureau of Statistics of China (NBSC). *China Statistical Yearbook 2002–2011*; NBSC: Beijing, China, 2014.
33. Hu, J.; Wang, S.; Yeh, F. Total-factor water efficiency of regions in China. *Resour. Policy* **2006**, *31*, 217–230.
34. Ma, J.; Ma, Y.; Bai, Y.; Xia, B. Study on the Technical Efficiency of Creative Human Capital in China by Three-Stage Data Envelopment Analysis Model. *Discret. Dyn. Nat. Soc.* **2014**, *2014*, 964275.
35. Stark, Q. Rural-to-urban migration in less developed countries. *Econ. Dev. Cult. Chang.* **1984**, *32*, 475–486.
36. De Brauw, A.; Rozelle, S.; Taylor, E. Migration remittances and agricultural productivity in China. *Am. Econ. Rev.***1999**, *89*, 287–291.
37. De Brauw, A. Seasonal migration and agricultural production in Vietnam. *J. Dev. Stud.* **2010**, *46*, 114–139.
38. Heenan, D. Social capital and older people in farming communities. *J. Aging Stud.* **2010**, *24*, 40–46.
39. Benjamin, D.; Brandt, L.; Rozelle, S. Aging, wellbeing and social security in rural northern China. *Popul. Dev. Rev.***2000**, *26*, 89–116.
40. Zhou, Z.; Sun, X.; Liu, Y. Psychological Development and Education Problems of Children Left in Rural Areas. *J. Beijing Norm. Univ.* **2005**, *1*, 71–79.

41. Verhoeven, J.; Arheimer, B.; Yin, C.; Hefting, M. Regional and global concerns over wetlands and water quality. *Trends Ecol. Evol.* **2006**, *21*, 96–103.

CHAPTER 16

Effects of local institutions on the adoption of agroforestry innovations: Evidence of farmer managed natural regeneration and its implications for rural livelihoods in the Sahel

Joachim N. Binam[1], Frank Place[2], Arinloye A. Djalal[1] and Antoine Kalinganire[1]

[1]*World Agroforestry Centre, (ICRAF)-West and Central Africa Regional Office-Sahel node, P.O. Box E5118, Bamako, Mali*

[2]*International Food Policy Research Institute (IFPRI), Institution and Market, Washington 20006, USA.*

ABSTRACT

The present study aims at (1) assessing how the existing local formal and informal institutions affect farmer managed natural regeneration (FMNR) practices and, (2) evaluating the benefits of such practices on livelihoods. The propensity score with continuous treatments was used to assess the effects of a set of covariates on FMNR as well as the impacts of that practice on income, cereal production and caloric intake using data collected from 1,080 rural households in Burkina Faso, Mali, Niger, and Senegal. This study demonstrated that regeneration of trees on farms, whereby farmers play an active role in the types of trees and their densities, is important as a practice and safety-net by providing cash income, caloric intake and diet, and crops supplements throughout dryland areas of West Africa. Overall, FMNR cannot be excluded as a recommendation in any geographical region. In addition, the study concludes that the effects of institutions in fostering FMNR practices in the Sahel are mixed. In areas with well-structured formal and informal institutions, populations seem to have adopted a better collaboration attitude with the local government by developing plans for a good management and protection of natural resource including FMNR practices. However, in areas where these commissions are being assimilated to governmental institutions, the willingness to raise incentives towards a better management of natural resources is less perceived. While recognizing the benefits of trees and tree products on caloric intake and diet, there is a need to explore in much more details, the FMNR-food nexus in future researches by going beyond what was covered from this study.

Keywords: Agroforestry, Livelihoods, Local conventions, Natural regeneration Sahel

1. BACKGROUND

The dry areas of the developing world occupy about 3 billion hectares and are home to 2.5 billion people: 41% of the Earth's land area and more than one-third of its population. About 16% of this population lives in chronic poverty (Solh and Saxena 2011). Drylands have limited natural resources. They already face serious environmental constraints, which are likely to worsen as a result of climate change. The West African Sahel and dry savannas is one of the five developing-country regions where dryland agriculture faces serious challenges.

Dryland agro-ecosystems include a diverse mix of food, fodder and fiber crops; vegetables, rangeland, and pasture species; fruit and fuel-wood trees; medicinal plants; livestock, and fish. Dryland systems are characterized by persistent frequent droughts, water scarcity, rapid population growth, high climatic variability, land degradation and desertification, and widespread poverty.

Two of the biggest challenges facing Sahelian countries today are reducing poverty, especially among rural households, and preventing climate shocks by protecting the ecosystems, which provide essential services for a growing population whose survival is dependent to a large extent on a combination of subsistence rain-fed crop farming, and extensive livestock rearing, supplemented with the gathering of agroforestry tree products

(AFTPs) including wood, grasses, fruits, leafy vegetables, nuts, and condiments.

Connecting the valuation of ecosystem services with the rural poor in the identified countries could make a significant contribution to both these challenges. Most of the population in Sahel consists of smallholder subsistence farmers who produce their own food in less than one hectare (Place et al. 2013). Indeed, as indicated by Sanogo et al. (2016), those farmers critically depend upon local ecosystems for survival and are directly affected by changes in availability of ecosystem goods and services. Thus, the loss of ecosystem services, important for livelihoods and many other ecological and environmental functions, can be devastating for the rural poor in Sahel. A more productive use of natural resources contributes to increasing food security and raising incomes among the rural poor in the Sahel. For example, deforestation has contributed to soil erosion, loss of agricultural productivity, and the scarcity of fuelwood. The loss of wetlands has threatened the availability of water leading to a need to invest in ecosystem services in order to improve people's resilience (Polak, 2008).

By using their land properly, collectively and individually, smallholder farmers can provide valuable services, such as carbon sequestration, water flow, or biodiversity protection. For example, it has been shown according to Weston et al. (2015) that Sahelian parklands resulting from the conversion of forest to agricultural landscapes are important sources of benefits both for rural populations and the landscapes. Indeed, trees from agroforestry parklands can reduce wind speed while increasing soil fertility and air humidity and reduce diseases like fungal attacks (Bayala et al. 2014). Regarding soil properties, recent studies of Sahelian agroforestry parklands have revealed a decrease in soil bulk density and as a consequence, soil under trees displayed higher porosity compared to adjacent open areas (Sanou et al., 2010; quoted by Bayala et al. 2014). Beyond the above mentioned supportive services, trees in the parklands also contribute to the reduction of carbon in the atmosphere by accumulating biomass via photosynthesis. As asserted by Bayala et al. (2014), this process is important for improving soil properties when accumulated biomass is stored in the below-ground compartment as soil carbon. Nevertheless, the improvement in soil fertility parameters by trees has been a controversial issue because trees may have simply grown in spots of higher fertility. Ajayi et al. (2007) and Place (2009) claim that, incentive policies such as subsidies and institutional support for certain soil fertility management options may have a considerable indirect influence in shaping farmers' decisions on soil fertility replenishment and other sustainable management of common resource strategies.

In addition, two decades of research into the management of what economists call common-pool resources suggests that, under the right conditions, local communities can manage shared resources sustainably and successfully. These findings challenge the long-held belief in the "tragedy of commons." In a recent study, Allen et al. (2012) found that, tragedy is not inevitable when a shared resource is at stake, provided that people communicate. In many places, communities have come together for the sake of the environment and their own long-term well-being.

As such, institutions that support ecosystem services can be used as tools to help protect ecosystems and improve rural livelihoods by allowing smallholder farmers to generate income through providing valuable public goods. The systems that support the provision of ecosystem services range from a neighborhood, to a watershed, to a country, to a continent. Managing them effectively requires a set of nested institutions that can encompass some of these multiple scales and interact with each other to create a proper management. To enhance the capacity of farmers in sustaining tree-based production systems, an enabling institutional, technical, and policy environment needs to be promoted.

In a recent paper on land tenure and agricultural productivity in Africa, Place (2009) indicates that the adoption of agroforestry is influenced by various factors. Some of them (including climate conditions, household and farm characteristics, and attributes of the particular agroforestry technology) have relatively little to do with policy, while others are directly linked to the existing policy (local formal and informal institutions). Forest policies can inhibit tree growing on farms by regulating harvesting, cutting or sale of tree products and certain tree species. Such protective policies discourage farmers from planting and protecting new trees that emerge.

A number of studies provided a description of the functions of some of the major species found in Sahelian parklands indicating that agroforestry tree products (AFTPs) can play three main functions in the household economy of rural communities living in or adjacent to the parkland (Kouyaté 2005; Kalinganire et al. 2007). Firstly, they help to fulfill households' subsistence and consumption needs in terms of energy and nutrition as well as medical and construction purposes. Secondly, they serve as a safety-net in times of crises (e.g., income shortages due to crop failure) and thirdly, some AFTPs provide regular cash income (Kouyaté 2005; Kalinganire et al. 2007).

However, despite the importance of famer managed natural regeneration (FMNR) practices for farmers' livelihoods the benefits from FMNR have not been studied extensively in the Sahel region. What is mainly missing is an economic quantification and livelihood effects (Haglund et al. 2011; Nyemeck et al. 2015). Furthermore, there has been no systematic study of how such benefits vary across the landscape according to the parklands, or countries. Our paper aims to fill this gap by (1) assessing how the existing local formal and informal institutions affect farmer FMNR practices and, (2) evaluating the benefits of such practices on food production, rural income and caloric intakes and diet.

2. OVERVIEW OF RESEARCH ON FARMER MANAGED NATURAL REGENERATION

Farmer managed natural regeneration involves regeneration of trees on farms, where farmers are actively involved in manipulating the natural biological regeneration into one that suits them. The types of activities that farmers carry out under FMNR include: thinning of unwanted emergent trees, protection of desired emergent trees from grazing through micro structures or fencing, managing water for young trees, taking action against insects and disease, retention of mature trees so that the rootstock may regenerate more young trees, ploughing practices that preserve emergent trees, and annual care of the regenerated trees. In the drylands, it can be argued that virtually all farmers practice FMNR to some degree (e.g., all farmers will actively thin trees from their fields). It is rather the degree to which it is done which differs considerably. The technical potential for FMNR in the drylands is high throughout almost all dryland areas because no input system is required—just the in situ germplasm (seed and roots) in the soil.

Evidence of the active practice of FMNR is most available for the Sahel in Niger, Mali and Burkina Faso (Reij et al 2009). But there is also evidence from Ethiopia, Kenya, Uganda, Malawi, Zambia, and Zimbabwe. In two lakes bordering districts in Malawi, it was recently found that although faidherbia seeds are readily available, 70% of all plants established between 2008 and 2010 was through regeneration (some being mediated through livestock) (Glenn 2012).

The technical potential for FMNR is related to environmental conditions: in the drier environments, rainfall limitations make it more difficult to regenerate trees in the absence of supplemental watering. The other major limiting factor is the presence of germplasm in the soil. It is well known that the diversity of soil borne tree germplasm varies across sites, although there is no mapping of its variation. Indicators of poor soil borne tree germplasm would be the lack of trees in the landscape and heavily eroded soils.

Besides environmental conditions, other factors may limit the technical potential of FMNR. At the household level, adoption analysis was done on the Sahel with a sample of 1000 households involved in the study (Place et al. 2013). Very few factors were found to be significant in explaining the number of new trees regenerating on the farm. The most consistent finding was that FMNR was positively related to the number of mature trees on the plot, which supports the biological explanation of the dependence on existing soil-borne germplasm. It also suggests that farmers with a higher number of mature trees may see the benefits of trees more clearly and thus be keen to establish more.

3. OVERVIEW OF INSTITUTIONS GOVERNING NATURAL RESOURCE MANAGEMENT IN THE SAHEL

Institutions can broadly be defined as structures and practices of formal and informal rules that regulate social behaviors (Jessop and Nelson 2003, Geraldi 2007). Analyses of socio-ecological factors, such as tree tenure, land and tree property and access rights, as well as benefits sharing from tree resources, provide insight into what Howard and Nabanoga (2007) qualified as the decision-making and power loci of a community. Indeed, as reported by Toulmin et al. (2002), if the institution that creates and enforces tenure regulations is weak, then the regulations no matter how well formulated, may not be enforced or abided by. Hilhorst (2008) argued that these institutions work best when they are adaptive, when rules are clear, enforced and nuanced, when they are legitimized by external organizations, and are faced with slow exogenous change. Yatich et al. (2014) argued further that collaboration between informal and formal institutions is essential for effective resource management. Therefore, an understanding of governance institutions in relation to natural resources is important to better formulate effective pro-ecosystem service policies.

3.1 Institutions governing farmland in the Sahel

Generally, control over access to farmland is in the hands of the lineage that started farming first in the village, personified in the male-head or lineage. The head of the family can grant strangers temporary access to land (secondary right). Pastoralists tend to have a host in the community who will help them to secure their secondary rights to pasture and crop residues as well as rights of passage (Mikulcak, 2011; Leach et al. 2011) (Table 1).

Table 1 Land tenure arrangements in the Sahel

Tenure arrangement	Rights
Inheritance	Full rights (access, withdrawal, management, exclusion, alienation). After the death of the family founder), the field commonly used by the family is passed to the children. The land is divided on the basis of the existing law. Each married male heir becomes a head of his household and of the share of land he inherited.
Purchase	Full rights (access, withdrawal, management, exclusion, alienation). All 'bundles of rights' to the land, including exclusion and alienation, are sold from the land owner to the buyer. The buyer becomes owner of the land.
Lease	Access, limited withdrawal and management rights (in accordance with Owner). The land is transferred in exchange for either money or any other 'security deposit'. It remains at any time the property of the initial owner who holds exclusive alienation rights. The person leasing has certain management and usage rights. This tenure arrangement is valid as long as the deposit is not repaid
Loan (temporary borrowing)	The land is loaned for a temporary or undefined period, without any security deposit or monetary transaction. Borrowed land stays at any time the property of the initial owner who holds exclusive alienation rights. The land may at any time be resumed by the owner.

Source: Adapted from Mikulcak, 2011

Private rights apply on intensively managed lands, but farmers may allow communal access to some tree products. Permanent landholders usually reserve for themselves exclusive tree planting and felling rights, but may encourage tree pruning and gathering of tree products by other community members. Restrictions occur depending on the value of tree species and quantity harvested. Because planting in some extent confers ownership, permanent landholders generally do not authorize tree planting to those with secondary rights.

3.2 Institutions governing common lands in the Sahel

In the semi-arid zone of West Africa, forests, pastures, and fallows are resources that are used by multiple groups (herding cattle, cutting wood, gathering, hunting, and bee keeping) with competing interests. Since colonial times, central government has sought to control access and use of forest lands declared as public lands (Yatich et al. 2014). Some forests were even classified and thus protected. Herd mobility and secured access to strategic resources, such as water and dry season grazing, are critical for pastoralist production systems. Recently, most Sahelian countries have revised their forestry legislation but many basic provisions remain (see Table 2).

Table 2 Forestry policies in selected countries

Country	**General principles**	**Standard of Management**	**Measures of conservation/protection**	**Institutions of management**
Mali	• Land and ownership is vested on the state or local council notably under the decentralization process • Access, use and management of indigenous trees is controlled	• Bushfires regulated but not completely disallowed • Disconnections between provision of law and practices • Bylaws and local conventions not formally recognized and an mainly oral	• Special protection of 11 indigenous tree species • Existence of classified forest and parklands recognized • Necessity for implementation of texts and better communication • Some legalized bylaws are being considered as effective instrument in the governance of natural resources	• Forest service manages classified forests as well as on-farm protected tree species • Forest services are devolved at the regional level and manage forest jointly with local government/territories and communities
Niger	• Forest and natural resources are considered	• Agroforestry is being considered as type of	• Special protection of 15 indigenous tree species • The existence of	• Use of forest service and local communities also involved to manage forest

	as national patrimonies • Promotion of a secured and fair access of all to the natural resources • Forest considered as a safe investment	land use • The practice of bushfire is regulated • Integrated vision regarding natural resources manageme nt and multifuncti onal use of the space	private and public forest is recognized and legalized bylaws are being considered as effective instrument in the governance of natural resources • Necessity of the implementation of texts and better communication among actors • The law is however not well known and not fully implemented	resources • Presence of land administration services nationwide

Source: Compiled from Yatich et al. (2014)

In principle, the new forest codes go some way towards recognizing customary rights and, in some cases, devolving management of certain forest resources to local populations. In practice, even if farmland technically falls outside forest domain, because rural landholdings are often non-registered, they continue to fall under the state control. Consequently, many restrictions which were originally intended to protect trees are also applied to trees on farms and in fallows, with the result that farmers are prevented from carrying out basic management activities which are crucial in optimizing their land use systems.

3.3 Local governance institutions managing access and use of natural resources in Sahel

Many Sahelian countries have undergone an institutional shift within the last decades. Communities in rural areas display a wide diversity of cultures and type of livelihoods pursued. As noted by Hilhorst (2008) there is a juxtaposition of various formal and informal authority structures and laws. The most significant institutions that prevail in the selected countries pertain to customary authorities, village land management commissions, and local conventions (Ouedraogo 2007; Djire and Dicko 2007). Table 3 provides an overview of these institutions and the roles they can play in promoting, managing, and regulating access and use of natural resources.

Table 3 Local governance institutions managing access and use of natural resources in selected Sahelian countries of West Africa drylands

Type of institution	Description and role	Country			
		Burki na F.	Mali	Niger	Sene gal
Customary authorities	Organization structure that reflects socio-cultural and economic diversity headed by authorities chosen through customary decision-making process and expected to act as custodians. They also intervene to prevent, mediate or manage conflicts including those related to natural resources use	Yes	Yes	Yes	Yes
Village land management commissions	More formal governance institutions expected to contribute to reconcile the various, and sometimes contradictory land tenure regimes and eliminate tenure-related obstacles to socio-economic development. They play a significant role in natural resource management and even land administration.	Yes	Yes	Yes	Yes
Local conventions	More or less formalized negotiated agreement designed and formulated to regulate access and use of common natural resources. The defined rules and regulations are formulated through a process of stakeholder consultation and dialogue to address issues related to bushfire surveillance brigades, marking out livestock tracts, fixing periods for harvesting wild fruits or entering grazing lands,	Yes	Yes	Yes	Yes

	quotas for resource use, and protection of regenerating forest among others.				

Source: Adapted from Leach et al. (2011)

4. DATA AND METHODOLOGICAL FRAMEWORK

4.1 Site selection and data collection

Data used in this study were collected in 2012 from a survey of 1080 households and focus group discussions in four Sahelian countries, namely Burkina Faso, Mali, Niger, and Senegal.

Two research questions guided this study: (1) how do the existing local institutions (formal and informal) affect the practice of farmer managed natural regeneration? (2) How does that practice impact on rural livelihoods?

Villages were randomly selected in the countries, and within each village, a two-step procedure was used to sample households. Through key informants, all households in a village were listed and identified as having a high or low density of trees on their farms (which proxies for adoption of FMNR). Then a random sample of 10 households in each strata was selected giving sample sizes of 240 for Burkina Faso and Mali; 480 and 120 for the Republic of Niger and Senegal, respectively. Convenience sampling was used to invite respondents into the focus group discussions, basing the selection on the respondents' knowledge and practice of agroforestry. Local extension agents and enumerators helped to identify appropriate individuals with sufficient knowledge of agroforestry.

The household survey collected quantitative data about FMNR practices, including: tree species, number and age; sourcing of tree products (e.g., fruits and fuelwood) from farm and non-farm landscapes; all sales of tree products for the 2011–12 agricultural year; crop and livestock production and sales; income from other activities; and characteristics of households and their land. Household members were also asked qualitative questions to understand their perception of a broad set of benefits, costs, and risks associated with FMNR. One major interest was whether households perceive FMNR to have any additional environmental services, such as improved soil fertility, improved water management; if the regenerated vegetation is perceived as a buffer against shocks; and whether the integration of FMNR helps to reduce overall risks (e.g., variation of production and income).

The focus group discussions questionnaire aimed to identify the major benefits from FMNR and how they are distributed across individuals, households, and locations as well as to understand the main constraints in adopting FMNR. In addition, focus group discussions and exchanges with key respondents and experts' opinion captured the existing institutions governing natural resources in the different villages. These were used to ascertain how they affect the practice of farmer managed natural regeneration.

4.2 Methodological framework

Much of the work of propensity score analysis has focused on cases where the treatment variable is binary. In practice, more than two conditions may be compared. This often happens when we want to estimate the impact of dosage analysis with treatment dosage (Bia and Mattei 2008). Hirano and Imbens (2004) developed an extension to the propensity score method in a setting with a continuous treatment as in the case of this study where the treatment variable (FMNR practice) is a continuous treatment. The analysis of treatment dosage with propensity scores may be generalized in two directions. In the first, one estimates a single scalar propensity score using ordered logistic regression, matches on the scalar propensity score and, proceed in a two-treatment group situation (Joffe and Rosenbaum 1999). In the second direction, one estimates propensity score for each level of treatment dosage (i.e., if there are five treatment conditions defined by differential doses, one estimate five propensity scores for each participant). Propensity scores derived in this fashion are called *generalized propensity scores* (GPS). This method, originally developed by Imbens (2000), uses the inverse of a particular estimated propensity score as a sampling weight to conduct a multivariate analysis of outcomes.

4.3 The generalized propensity score estimator

Building on Imbens (2000), Hirano and Imbens (2004) developed a generalization of the binary treatment propensity score and labeled the method a GPS estimator or propensity score with continuous treatments. Thereafter, Bia and Mattei (2008) developed a software program in *Stata* called *gpscore* to implement the GPS estimator. The basic idea of the GPS according to Bia and Mattei (2008) is to assume a random sample of size N, indexed by $i = 1, 2, \ldots N$. For each unit of i in the sample, we observe a $p \times 1$ vector of pretreatment covariates, X_i; the treatment received, T_i; and the value of the outcome variable associated with treatment, Y_i. Using the counterfactual framework (Rubin 1997; Holland 1986), unit i has a set of potential outcomes, defined as, $\{Y_i(t)\}_{t \in \Gamma}$, i = 1,....., N, where τ is a continuous set of potential treatment values. Under this definition, Hirano and Imbens

(2004) refer to $\{Y_i(t)\}_{t\in\Gamma}$ as the unit-level dose-response function, and a dosage analysis is interested in the average dose-response function, $\mu(t) = E\{Y_i(t)\}$. According to Hirano and Imbens (2004), $\{Y_i(t)\}_{t\in\Gamma}$, T_i, and X_i, $i = 1,....,N$, are defined on the common probability space; T_i is continuously distributed with respect to the Lebesgue measure on Γ; and $Y_i = Y_i(T_i)$ is a well-defined random variable. With the preceding, the GPS is defined as the conditional density of the treatment T given the covariates, or the GPS is $R = r(T, X)$, where $r(t, x) = f_{T|X}(t|x)$. The GPS has a balancing property similar to the propensity score under the setting of binary treatment. Hirano and Imbens (2004) proved two theorems with respect to the balancing property of GPS: (1) weak un-confoundedness given the GPS and (2) bias removal with GPS.

The implementation of the GPS method consists of three steps.

***Step 1: modeling the conditional distribution of the treatment given covariates:* r(t, x)**

Practically, this is the step in which researchers estimate the GPS at a given level of treatment and observed covariates X and then perform the balancing check. Hirano and Imbens (2004) used a flexible parametric approach to estimate the GPS. That is, they assume a normal distribution for the treatment given the covariates:

$$g(T_i)|X_i \sim N(\beta_0+\beta' X_i,\sigma^2)$$

where $g(T_i)$ is a suitable transformation of the treatment variable. While the parametric model assumes the normal distribution, the actual distribution of treatment dosage T_i in the sample may not be normally distributed. To correct for normality, one can do a transformation of T_i by taking the logarithm of T_i or by applying other transformations. The formula for estimating GPS for each observation based on the estimated regression model is:

$$\hat{R}_i = \frac{1}{\sqrt{2\pi\hat{\sigma}^2}} exp\left[-\frac{1}{\sqrt{2\hat{\sigma}^2}}\{g(T)_i - \hat{B}_0 - \hat{B}' X_i\}^2\right]$$

Next, users need to test the balancing property and whether the GPS balances observed covariates. The balance check observes six steps (Bia and Mattei 2008).

Step 2: estimating the conditional expectation of the outcome given the treatment and GPS

In this step, the conditional expectation of the outcome, Y_i, given T_i and R_i, is modeled as a flexible function of its two arguments. In practice, one can use a quadratic approximation or polynomial approximations of order not higher than three. The quadratic approximation is

$$E[Y_i|T_i,R_i]=\delta_0+\delta_1 T_i+\delta_2 T_i^{\,2}+\delta_3 R_i+\delta_4 R_i^{\,2}+\delta_5 T_i R_i.$$

According to Hirano and Imbens (2004), there is no direct meaning to the estimated coefficients in the selected model, except that testing whether all coefficients involving the GPS are equal to zero can be interpreted as a test of whether the covariates introduce any bias.

Step 3: estimating the dose response function to discern treatment effects as well as their 95% confidence bands

This last step consists of averaging the estimated regression function over the score function evaluated at the desired level of the treatment. Given the estimated parameters in step 2, researchers can estimate the average potential outcome at treatment level *t*, which is also known as the dose-response function. This is the final statistic that shows the outcome differences associated with the treatment dosage. The dose-response function is given by:

$$\hat{E}[Y_t]=\frac{1}{N}\left(\hat{\delta}_0+\hat{\delta}_1 t+\hat{\delta}_2 t^2+\hat{\delta}_3\hat{r}(t,X_i)+\hat{\delta}_4\hat{r}(t,X_i)^2+\hat{\delta}_5 t\hat{r}(t,X_i)\right)$$

At this stage, one can estimate the dose response functions for user treatment values and use bootstrapping to form standard errors and confidence intervals. The *Stata dose response* programme (Bia and Mattei 2008) can draw two types of plots to show the final results: one plot shows the dose response function, and the other plot shows the estimated treatment effect function (also known as estimated derivatives).

4.4 Empirical model

To ascertain the effects of formal and informal institutions on investment in agroforestry through FMNR, the data collected in 2012 from 1080 households in Burkina Faso, Mali, Niger, and Senegal, are used. The dependent variable, practice of FMNR, is measured by the density of trees kept and managed by the household on its farmlands. In many observational impact studies involving technology adoption, there is generally a cohort using the technology compared against a control group that did not use the technology. However, FMNR is a complicated technology that does not necessarily fit into two categories. First, virtually all trees in the Sahel are regenerated naturally and it is not always easy to identify the degree to which the regeneration was facilitated by farmers' practices. Thus, adoption is defined more as the degree of natural regeneration with or without farmer management/assistance.

Based on field experience and observations, the household questionnaire was designed to capture four aspects related to two dimensions: (1) knowledge continuum including farmers' knowledge of FMNR practice and the diversity of the species managed on the farmland as compared to the main species found in the area; and (2) compliance continuum including ownership of at least one farm plot and the number and size of trees kept and managed in the farm. The "degree" of the practice can then be reflected in the tree density (i.e., number per hectare) and age distribution of the trees. Thus, farmers with a high tree density of young trees reflect active FMNR practitioners.

Older trees can often be more than 100 years old. Therefore, a good number of the mature trees may have been inherited by the current farmer and do not result from previous FMNR practiced by him or her. Trees may be kept on farm because of an appreciation for the benefits or an inherited way of farming from parents. There may also be unobserved variables at household or plot level that make certain sites more amenable to tree growth. Moreover, divergence between households in inherited and regenerated trees can also result from factors at the community or landscape level.

There are several institutional factors that have been identified as limiting the potential for FMNR, such as fire setting, free grazing, and rights and regulations over trees. The use of fire and free grazing systems generate benefits in terms of grass regeneration, clearing of debris, catching wild rodents for food and in the case of free grazing, offering a cheap mechanism for feeding livestock. Thus, it is not easy to find institutional reforms that can accommodate the interests of FMNR with others. However, practices such as controlled fires, rotational grazing (the flip side being rotational exclosure areas), and the promotion of livestock corridors are all options that have been successfully implemented in the drylands. These factors are generally regulated through governing formal and informal structures defined by codes of conduct, norms of behavior and conventions. In this study, we are interested in local conventions (LOCONV) and formal village land management commissions (COFO/CVGT).

In addition, a number of covariates are specified to reflect the potential effects of other variables on FMNR. These additional factors that may influence a household's decision about agroforestry practices are classified into two constructs: household structure and household endowment (access to assets and information). *Household structure variables* include household size and the number of active members in the household.

Access to assets provides households with leverage to invest in productive activities and generate more output, which increases their probability of participating in market transactions. Access to assets is an indication of endowment and wealth. In general, well-endowed households tend to experience lower transaction costs and have more flexibility in allocation of resources to market activities (Siziba et al., 2013). Included in this category are *production assets* including size of arable land, number of livestock owned; *transportation assets,* captured by ownership of transport equipment such as cart. The total land area may have indirect positive impacts on investment in trees by enabling farmers to generate production surpluses, to overcome credit constraints, where land can be used as collateral for credit (Kabore et al. 1997), and allow them to adopt improved agroforestry technologies that increase productivity of tree products. In other words, the more arable land a household has, the higher the level of output and thus the higher the propensity to practice FMNR.

Access to information tends to improve decision-making skills and affects the know-how of farmers. The construct proxy representing access to information consists of level of education of the household head, ownership of a cell phone, distance to the main market, and membership of farmer groups. The variable measuring proximity to the main market reflects how far farmers have to travel to reach important sources of information that are located in the nearest town where there are government offices and markets. It is expected that longer distances increase travel time and travel costs, which impact negatively on adoption of agroforestry innovation. Participation in farmer groups increases a household's awareness of the type of information needed for production and marketing decisions. Many farmer groups also engage in group marketing as well as credit provision to their members. Table 4 gives definitions and summary statistics of the variables used in the analysis.

Table 4 Summary of variables used in the analysis

Variable description	**Countries**			
	Burkina Faso (*N* = 240)	**Mali** (*N* = 240)	**Rep. of Niger** (*N* = 480)	**Senegal** (*N* = 120)
Continuous treatment variable: Number of young trees kept and managed in the farmlands	40 (50.9)	56 (54)	110 (135)	48 (59)
Outcome variables				

Income per capita (ln $)	4.09 (4.9)	7.58 (4.03)	7.04 (3.79)	5.98 (5.14)
Value of tree products (FCFA)	30,000 (38,480)	34,400 (68,030)	5525 (5500)	17,670 (19,765)
Crop production (ln kg)	7.2 (2.33)	7.31 (2.4)	5.5 (2.28)	6.87 (1.53)
Food consumption score (FCS)	22.52 (15.6)	23.87 (8.01)	29.3 (19.4)	31.7 (11.9)
Covariates				
Household structure				
Number of active members in the household	3 (1.72)	9 (4.8)	4 (1.5)	6 (2.0)
Household size	8 (4.0)	16 (10.0)	12(6.0)	12 (6.0)
Household endowment				
Production assets				
Size of arable land (ha)	5.96 (5.53)	9.09 (9.05)	2.17 (1.9)	8.3 (5.8)
Number of livestock units owned	10.48 (12.10)	17.2 (22.6)	9.9 (5.01)	5.34 (5.05)
Transportation assets				
If the household owns a cart (1 = yes)	29.2%	92.1%	43.5%	64.2%
Information assets				
If the household owns a TV (1 = yes)	56.7%	18.3%	1.5%	27.5%
If the household owns a cell phone (1 = yes)	61.3%	71.7%	25.2%	54.2%
If the highest level of education of the household head is primary (1 = yes)	84.5%	74.5%	10.04%	79.2%
If the highest level of education of the household head is secondary (1 = yes)	50%	38.3%	66.7%	61.7%
Average distance from the main markets (km)	20.07 (20.03)	5.3 (3.7)	7.4 (5.6)	7.1 (5.3)
If the household is a member of at least one	57%	28%	18%	15%

CBO (1 = yes)				
Institutional factors				
Proportion of villages with local conventions (%)	33	71	19	17
Proportion of villages with village land management commission (%)	98	0	29	17
If the household has access to credit (1 = yes)	70%	24%	19%	35%

Source: data compiled from the household surveys (values in parenthesis are standard deviations)

5. RESULTS AND DISCUSSION

Table 4 presents summary statistics for the sampled households. The average number of trees kept and managed on the farmland is reported in the top of the table. As is clear, there is significant variation among the countries. Niger is clearly an exception: the average number of trees recorded is more than double of that reported in Burkina Faso and Senegal.

A significant proportion of villages in Mali and Burkina Faso have formal and/or informal natural resource institutions that govern access and the use of natural resources. These can be identified as local conventions or a village land management commission. Of the sampled villages in Mali, 71% have a local convention. In Burkina Faso, 98% of the sampled villages have a local land management commission. Surprisingly, Mali is the only country where no land management commission was found in the sampled villages.

5.1 Factors influencing FMNR on the farmland

In general, the uptake of agroforestry technologies is more complicated than that of annual crops (Mercer, 2004) because of the multi-components and multi-years through which awareness, acceptability and evaluation take place (Ajayi et al. 2006). Several studies have been carried out to gain insights into the adoption of agroforestry and the factors that drive the adoption of agroforestry (Phiri et al., 2004; Place et al., 2002; Ajayi et al. 2006). In the process of estimating the conditional expectation of the selected outcomes, the study ran a maximum likelihood regression to ascertain the effects of certain of the covariates on the treatment and to create the GPS. The results of the maximum likelihood estimations are reported in Table 5. The covariates are ordered in blocks corresponding to the categories: household structure, household endowment (physical and information assets), and institutional covariates. As indicated by the chi-squared statistic, overall the model is valid in explaining factors that influence the investment in trees through FMNR in the selected Sahelian countries.

Table 5 Factors influencing investment on trees in the farmland among farmers in West Africa Sahel

Variable description	**Countries**				
	Burkina Faso (*N* = 240)	**Mali** (*N* = 240)	**Rep. of Niger** (*N* = 480)	**Senegal** (*N* = 120)	**Sahel** (*N* = 1,080)
	Coefficient (S.E)	**Coefficient (S.E)**	**Coefficient (S.E)**	**Coefficient (S.E)**	**Coefficient (S.E)**
Constant	0.677 (0.03)***	4.118 (0.227)***	1.206 (0.038)***	1.073 (0.069)***	2.351 (0.505)***
Covariates					
Household structure					
Number of active members in the household	−0.045 (0.037)	−0.004 (0.009)	−0.023 (0.038)	0.012 (0.054)	−0.29 (0.025)
Household size	−0.004(0.016)	0.005 (0.004)	0.026 (0.017)	0.032 (0.033)	0.023 (0.015)
Household endowment					
Production assets					
Size of arable land (ha)	0.042 (0.012)***	0.176 (0.003)***	0.199 (0.030)***	−0.017 (0.022)	0.064 (0.009)***
Number of livestock units owned	−0.009 (0.004)**	−0.001 (0.002)	−0.006 (0.006)	−0.002 (0.020)	0.004 (0.006)
Transportation assets					
If the household owns a cart (1 = yes)	−0.136 (0.146)	0.113 (0.183)	0.320 (0.126)**	−0.200 (0.225)	0.279 (0.168)*
Information assets					
If the household owns a cell phone	0.211	0.194	0.317	0.501	0.464 (0.157)*

(1 = yes)	(0.092)**	(0.111)*	(0.138)**	(0.241)**	**
If the highest level of education of the household head is primary/functional (1 = yes)	−0.047 (0.130)	0.027 (0.124)	0.550 (0.560)	0.285 (0.274)	0.143 (0.038)***
Average distance from the main markets	0.000 (0.001)	−0.017 (0.011)	0.045 (0.011)***	0.069 (0.018)***	0.003 (0.001)**
If the household is a member of at least one CBO (1 = yes)	0.174 (0.092)**	0.169 (0.097)*	0.063 (0.154)	−0.340 (0.287)	0.444 (0.172)**
Institutional factors					
If there is an operational local convention within the community (1 = yes)	0.156 (0.140)	0.167 (0.067)**	0.833 (0.254)***	0.060 (0.073)	0.678 (0.220)***
If there is existing land management commission within the village (1 = yes)	0.025 (0.012)**	–	−0.230(0.156)	−0.559 (0.276)**	−0.787 (0.215)***
If there is both local convention and land management commission within the village (1 = yes)	0.085 (0.072)	–	−0.908 (0.320)**	−0.116 (0.075)	−0.155 (0.374)
If the household has access to credit (1 = yes)	−0.351 (0.167)**	0.162 (0.120)	0.212 (0.149)	0.069 (0.017)***	0.353 (0.184)*
CHI- SQ	109.96	51.92	85.30	52.65	145.61
$P(\chi^2)$	0.000	0.000	0.000	0.000	0.000

Source: authors' estimations from household surveys data; ****P* < 1%; ***P* < 5%, and **P* < 10%

Consistent with our expectations, access to information, size of available land, access to markets and formal norms and rules are important factors influencing the FMNR practice. The important role of information in influencing FMNR is highlighted by the positive and significant sign of two variables: the ownership of a cell phone and membership of at least one community-based organization (CBO). In fact, agroforestry technologies as compared with conventional agricultural practices are new phenomenon (Ajayi et al. 2006). The promotion of tree-based systems requires skills in terms of management of trees. Capacity of doing this needs to be built and requires a multidimensional approach including field demonstrations as well as a proper dissemination of information using different information and communication technology tools. There is a growing agreement in developing countries that the use of cell phones is an effective and efficient way to communicate timely information on agroforestry innovations to farmers.

In addition, farmers can be ingenious in problem solving and if they pick-up information about FMNR from friends or other acquaintances they may well innovate and adapt the method to their own conditions. Because of its flexibility and multiple options, FMNR is a practice that can trigger the innovative creativity of farmers. The positive and significant sign of the variable CBO confirms the assertion that farmers tend to believe their trusted peers more than formal advisers when discussing the advantages and disadvantages of a new technique, approach or tool. It is therefore easier and helpful for them to interchange ideas and experiences on the technique through formal or informal CBOs.

Another important factor in explaining the practice of FMNR is the size of arable land. This variable has a positive and significant effect meaning that ownership of more arable land increase the practice of FMNR in the Sahel. In some countries including Niger for example, the habit of leaving and managing trees in crop fields is a standard farming practice. Typically, access to more arable land will encourage farmers to maintain and manage more valued trees on their farms (Pye-Smith 2013).

Effective demand in the market and the supply chain beyond production is also important in ensuring that farmers will receive attractive returns for their efforts in using sustainable agricultural practices such as FMNR. As such, policies and institutions that encourage and enable the marketing of indigenous tree products, permit FMNR practitioners to be aware of the existing opportunities of FMNR technology to provide immediate, medium and long term individual and public benefits simultaneously.

The institutional factors portrayed in this study are the formal and informal governance institutions established to play a role in natural resource management and land administration. These are captured by two dummy variables representing the village land management commissions and the local convention. According to the results reported in Table 5, those variables can exert (1) a positive influence on the FMNR practices, (2) negative impacts, (3) ambiguous or no direct effects.

Local conventions seem to have a positive and significant effect on FMNR practice and this result is particularly strong in Mali, Niger, and Sahel as the whole. Indeed, there is a growing interest in strengthening community-based natural resource management and participation in those countries (Yatich et al. 2014). Traditional West African tree tenure systems are extremely complex and further complicated by the current movements towards decentralization in many of West African Sahelian countries (Alinon and Kalinganire 2008). In fact, as noted by Howard and Nabanoga (2007), tenure is more than just legal ownership, it creates beneficiaries and victims based on how ownership is divided and how power relations accompany that ownership. Von Benda-Beckmann quoted by Leach et al. (2011) suggests that this includes not only the many rights that one owner may have in tree and other resources but also the ways in which these rights are expressed through ideology, legal institutions and power dynamics. Because the local conventions are more inclusive and their territorial coverage in many cases more important (in some cases, the facilitation of those conventions is increasingly focused on the more complex situations involving several local government territories or the district level, such as livestock corridors), they appear to be considered as a pro-agroforestry institution. In addition, they provide a more secured frame within which farmers in different communities are sensitized to invest in trees. Due to the community consensus plan materialized within the local conventions, farmers perceive the population of valued local species to be increasing, and they are eager to continue that trend by planting or protecting more trees each year.

The effect of the variable "village land management commission" is mixed, i.e., positive and significant in Burkina Faso and negative in Senegal. In areas where the village land management commissions work well because of their flexibility such as in Burkina Faso, there is a lot of buy-in and populations collaborate better with the local government in developing sustainable natural resource management plans and practices such like FMNR. However, in some countries like Senegal where these commissions are being assimilated to governmental institutions, local communities are less compliant to the established rules, norms and recommendations related to sustainable management of natural resource. In addition, forest codes are often poorly understood by rural people. Faced with lack of resources, most forest services are unable to enforce regulations properly. Moreover, because of some misbehavior from individual forest agents, these institutions are sometimes seen as disincentives towards FMNR practice. Therefore, the local management of resources will need to be accompanied by institutional change within forest services with a greater emphasis given to capacity building in participatory approaches and a formal acknowledgement of the sustainable nature of traditional management practices.

The results also indicate that there is a need to develop a more integrated institutional framework and find ways to enhance effective participation across institutions. Indeed, the results indicate that in villages where populations are regulated by both local conventions and a village land management commission, farmers seem to be discouraged to invest in trees on their farmlands.

5.2 Assessing the effects of FMNR practice on livelihoods and caloric intakes

5.2.1 Effects on income

With the *gpscore programme*, the study first took a logarithmic transformation of the treatment variable. Results of this transformation are not desirable: the Kolmogorov-Smirnov test of the normality assumption shows that the log-transformed treatment variable is not normally distributed. The analysis then chose the zero-skewness log transformation of the treatment variable; the Kolmogorov-Smirnov test shows that the assumption regarding normality under this transformation is statistically satisfied at the 0.01 level. Using this transformation, the study ran the maximum likelihood regression to create the GPS model for each country. Using the medians of the transformed treatment variable for the different sub-groups, the study estimated five sets of GPSs for all units. Results show that the GPSs improve covariate balances.

The study then moved to estimate the conditional expectation of the outcomes, given the treatment and GPS, and employed the conditional expectation to estimate the dose-response functions or treatment effects at the different selected treatment levels (i.e., percentile density of trees kept and managed on the farmland at the following levels: 25, 50, 75, 100%) and their 95% confidence bands. Results are presented in two types of plots: one plot shows the dose-response function and the other plot shows the estimated treatment effect function also known as estimated derivatives (Fig. 1).

Taking the dose-response function results, generally speaking in the Sahel, it appears that there is a sharp increase of income received from marketing of tree products when the number of trees kept and managed on the farmland ranges from 10 to 40/ha. After the threshold of 22000 CFA benefit, the total amount earned from selling tree products harvested on farm decreases to 20000 CFA when the number of trees ranges from 45 to 100/ha. The treatment effect functions show the marginal effect of maintaining and managing trees on income.

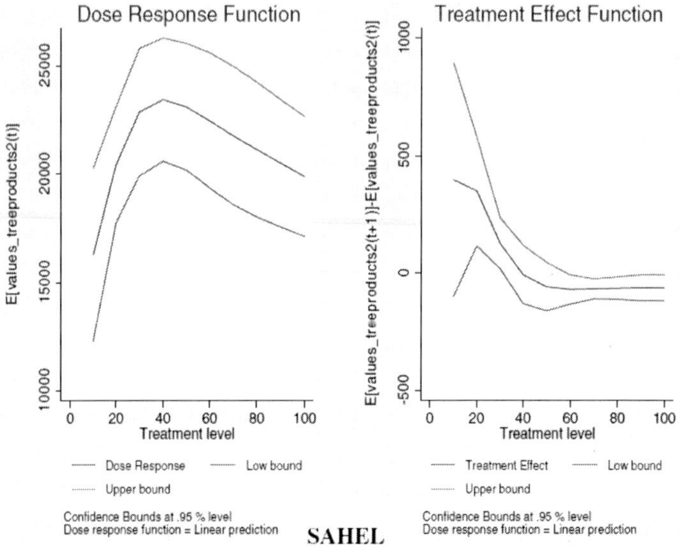

SAHEL

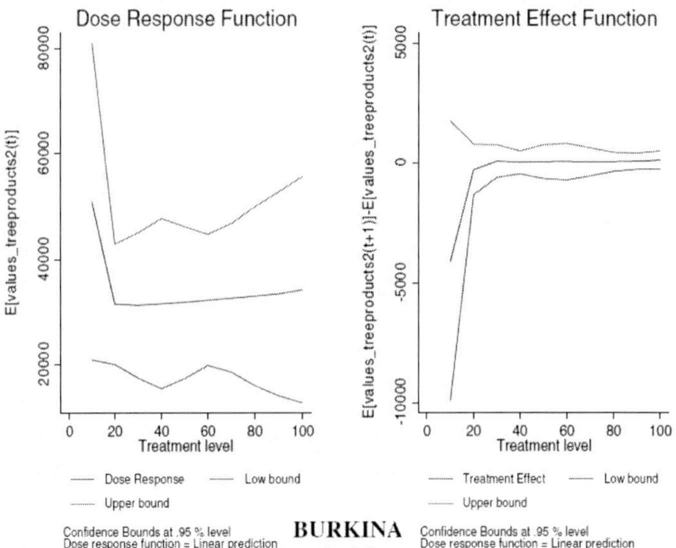

BURKINA FASO

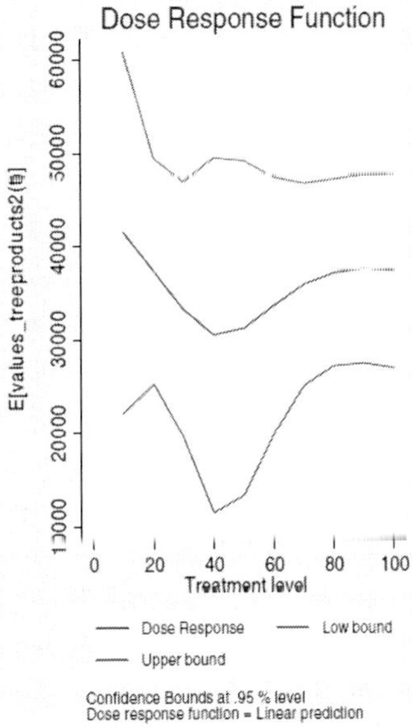
Dose Response Function
E[values_treeproducts2(t)]
Treatment level
Dose Response
Low bound
Upper bound
Confidence Bounds at .95 % level
Dose response function = Linear prediction

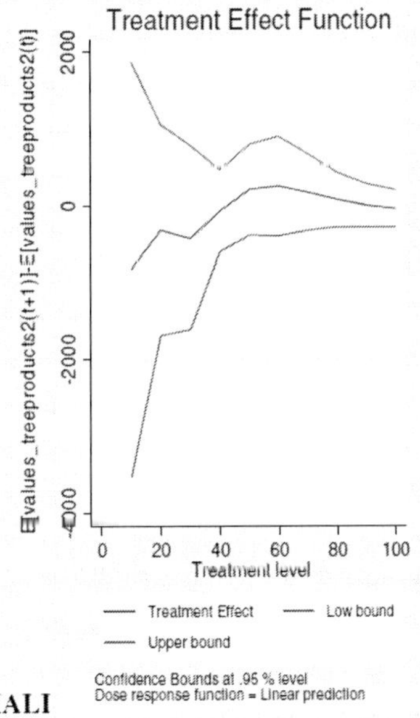
Treatment Effect Function
E[values_treeproducts2(t+1)]-E[values_treeproducts2(t)]
Treatment level
Treatment Effect
Low bound
Upper bound
Confidence Bounds at .95 % level
Dose response function = Linear prediction

MALI

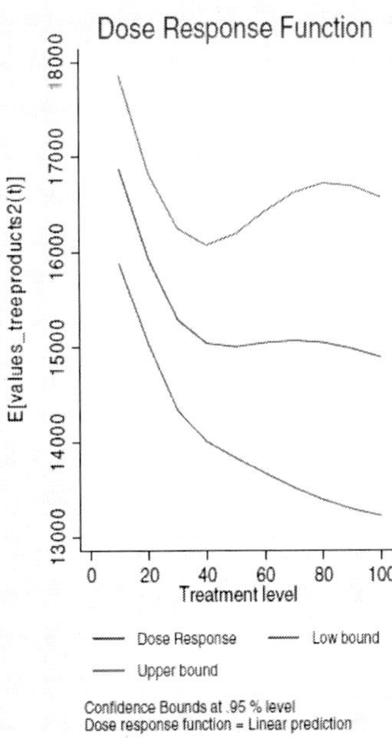
Dose Response Function
E[values_treeproducts2(t)]
Treatment level
Dose Response
Low bound
Upper bound
Confidence Bounds at .95 % level
Dose response function = Linear prediction

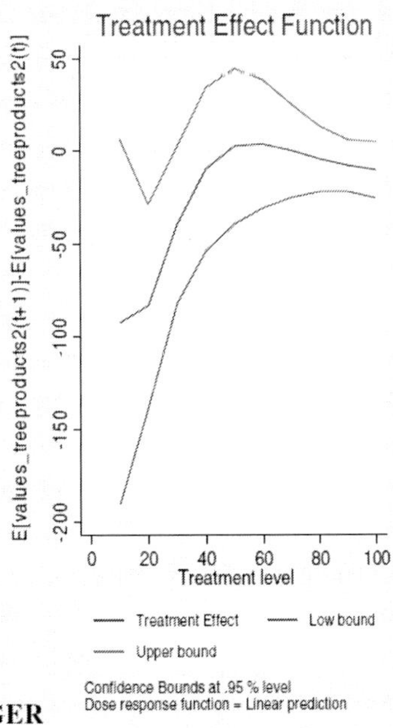
Treatment Effect Function
E[values_treeproducts2(t+1)]-E[values_treeproducts2(t)]
Treatment level
Treatment Effect
Low bound
Upper bound
Confidence Bounds at .95 % level
Dose response function = Linear prediction

NIGER

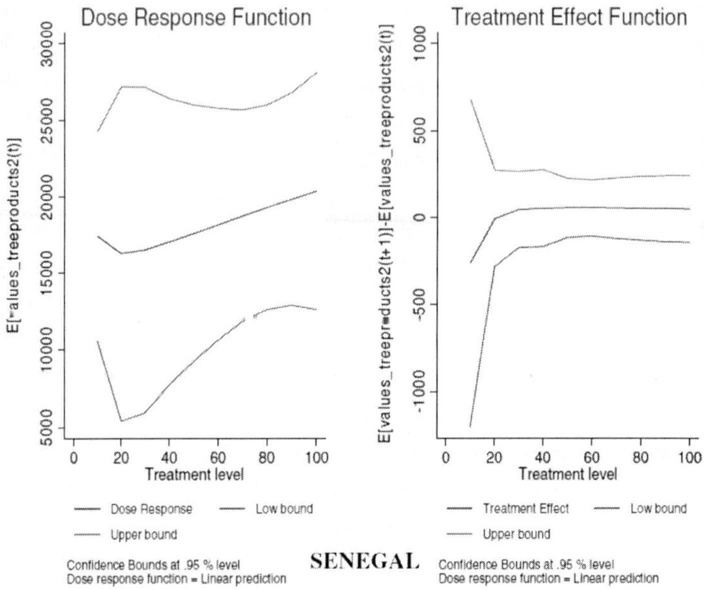

Figure. 1 Estimated effects of investment on trees on the household income-dose-response function results, derivative, and 95% confidence bands

Globally the plot suggests that in the Sahel, for those who have less than 45 trees/ha on their farmland, every one percentage-point increase in the number of trees kept and managed on the farm decreases the amount of income received from tree products. The marginal effects again show that farmers who are less involved in the practice of FMNR (i.e., farm with less than 40 trees/ha) are more sensitive to income change than those with more trees (i.e., higher than 40/ha). The findings may suggest that in order to earn more income from the FMNR practices, there is an optimal number of trees needed to be kept and managed on the farm.

Figure 2 shows the disaggregated results per countries. The results from the dose-response functions show that there is a sharp decline of income received from tree products in Burkina Faso when the density of trees on the farmland is less than 20/ha. The revenue received from tree products tends to be stable at 30,000 CFA/ha when the density of trees is higher than 20/ha. The maximum amount expected to be received by selling tree products is 40,000 CFA/ha with a farm of 15/ha density in Burkina Faso. In Mali, FMNR practitioners with 20 trees/ha are better off in terms of revenues received from tree products (42000 CFA). This amount sharply declines when the density ranges from 21 to 40 trees/ha. After the 40 trees/ha threshold, the revenue received from tree products gradually increases with every unit of tree increase during the practice of FMNR to be stabilized around 40,000 CFA. By contrast, even if the figures in terms of trend and density look alike in many cases across countries, Niger and Senegal exhibit the lowest amount of incomes that can be expected from tree products (15,000 to 17,000 CFA in Niger) and (15,000 to 20,000 CFA) in Senegal. These findings also suggest that the species favoured by farmers vary from place to place; so does the density of the tree. Some projects have advised farmers to keep 20–40 trees/ha, but densities of over 150 are not unusual. In addition, while there are common parkland species throughout the Sahelian/Sudanian ecozones, the composition of species differs across space due to rainfall conditions, topographical location, soils and farmers' preferences for tree species and functions (Faye et al. 2011).

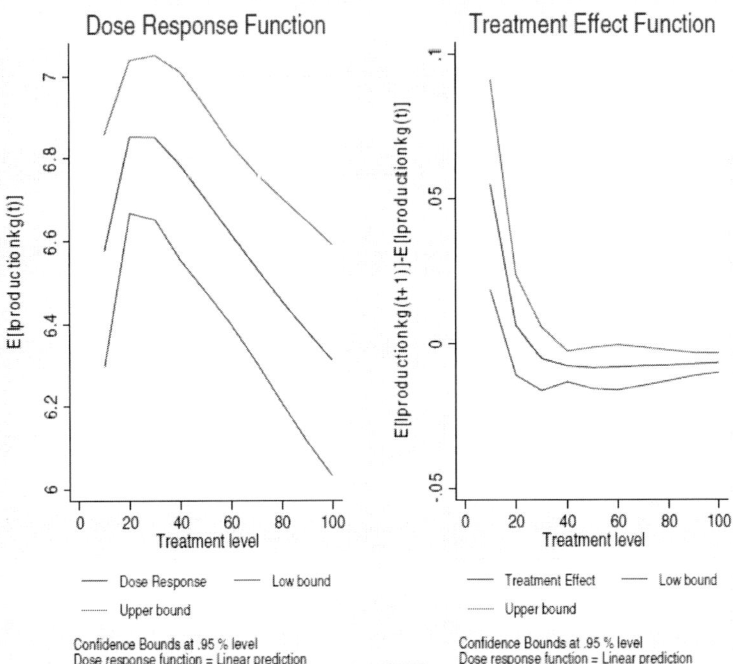

SAHEL

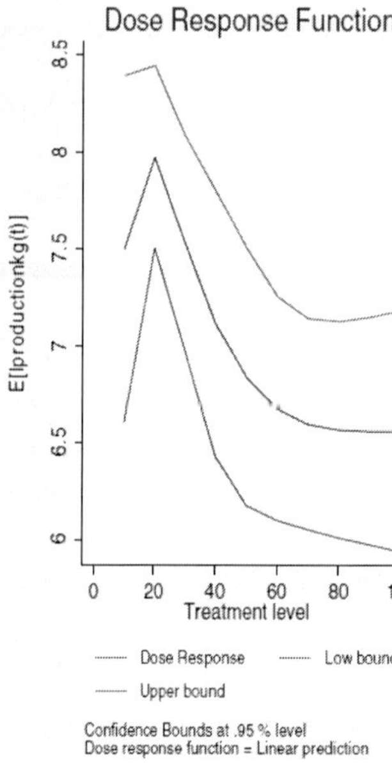

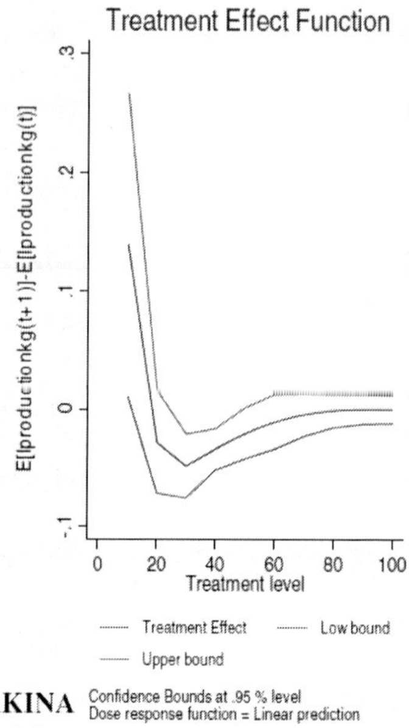

BURKINA FASO

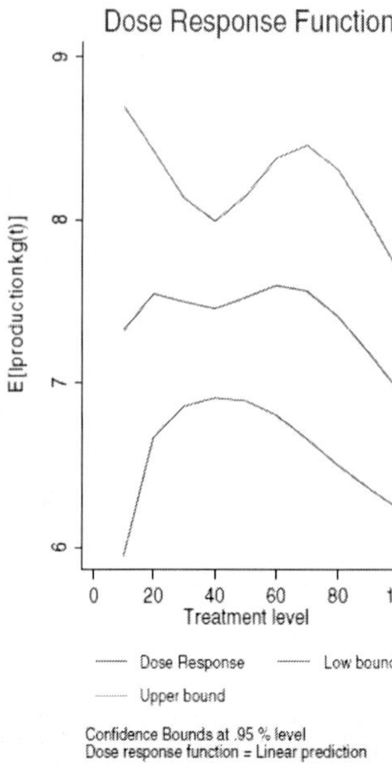

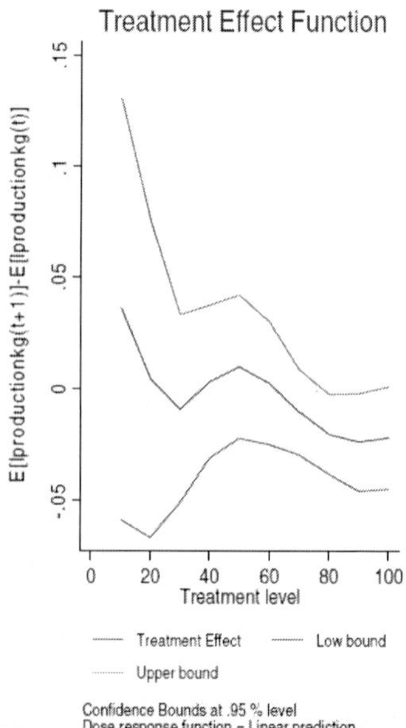

MALI

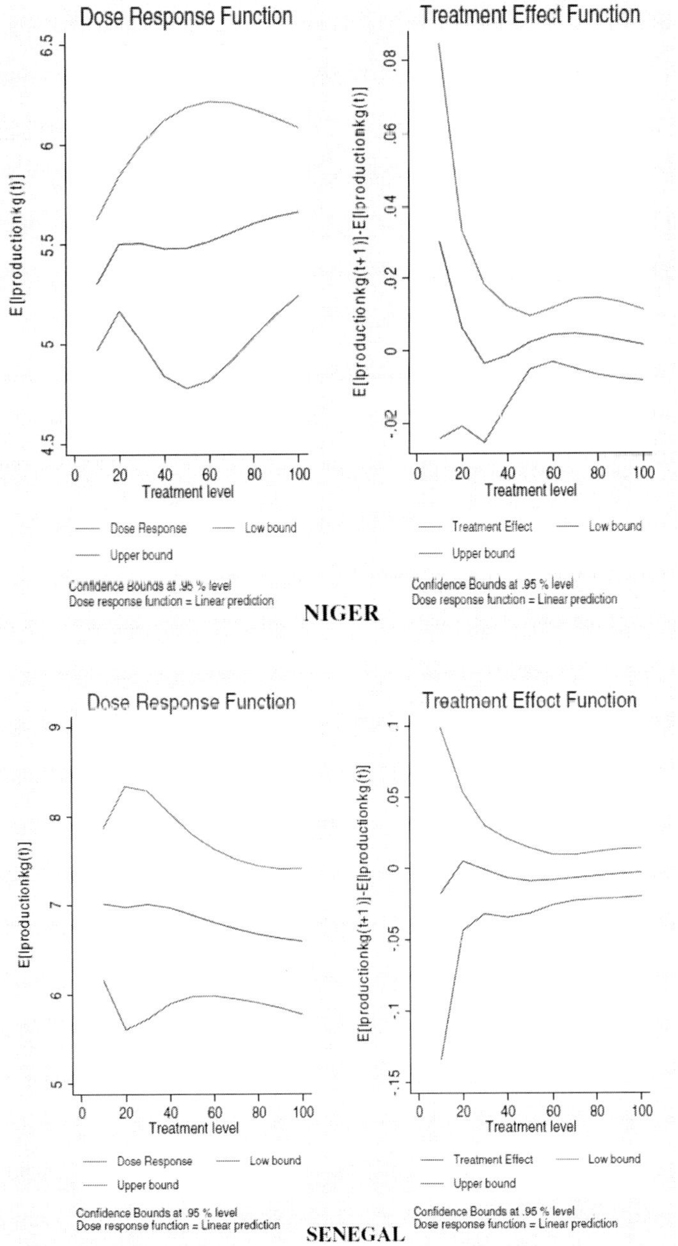

Figure. 2 Estimated effects of investment on trees on crop production-dose-response function results, derivative, and 95% confidence bands

5.2.2 Effects on crop production

Cereals and vegetables are the most common staple crops associated with trees in the Sahel. Crop yield improvement is another major benefit pathway of trees. The pathway to increase yields is due to micro climate regulation (from all trees), improved soil fertility in its broadest sense of physical, chemical and biological aspects (more so from certain species than others), and potential impacts on water availability for crops-under certain contexts only Bayala et al. (2014).

This study also tried to assess the impact of FMNR on crop production. Considering the dose-response results in Fig. 2, there is a sharp increase in crop production when the density of trees ranges between 15 to 20 trees/ha. When the density ranges from 20–40/ha there is an increase with a constant slope (in other words, the effects on crop production tend to be constant); above the density level of 40 trees/ha, the crop production decreases with increasing density levels. The dose-response plots show that, the maximum effect is reached when the density of trees ranges from 20 to 40/ha, indicating an increase of crop production by 915.985 kg ($=e^{6.82}$) or 274.25 kg/ha (=915.985/3.34) in the Sahel. The same calculations can be done for each of the countries indicating an increase of the expected quantity of cereals produced by 401.38 kg/ha in Burkina Faso; 198.9 kg/ha in Mali; 112.76 kg/ha in Niger and; 132.12 in Senegal.

The marginal propensity to benefit from increased crop production as indicated by the treatment effect functions shows that FMNR practitioners with less than 40 trees/ha are more sensitive to the changes in crop production than those with a higher number of trees/ha. Indeed, the plot suggests that for FMNR practitioners who have less than 40 trees/ha, every one percentage-point increase in keeping and managing trees on their farmland decreases the effects on crop production, and the stronger effect was associated with the management of about 20 trees/ha (+5% in the Sahel; +14% in Burkina Faso; +4% in Mali; +3% in Niger and about +1% in Senegal).

These findings again suggest that the effects of trees on crop production might be affected by both the density and quality of tree species. Indeed, many tree species also assist staple crops through soil-fertility improvement. This was demonstrated in an analysis of more than 90 peer-reviewed studies on the planting of nitrogen-fixing green fertilizers, including trees and shrubs, which found consistent evidence of benefits to cereals in semi-arid countries, although the level of response varied by soil type and by the technology used (Sileshi et al. 2008; Bayala et al. 2012). In fact, past studies on tree–crop interactions have clearly shown that trees have highly varying effects on the associated crops when comparing the yields of associated crops in the influence zone of trees with that of a treeless monoculture control plot (Oginosako et al (2006); Baggnian et al. (2013); Bayala et al. 2012, 2014).

5.2.3 Effects on caloric intake and diet

Most definitions of food security vary around that proposed by the World Bank (1986); major components of the most common definitions are summed up by Maxwell and Frankenberger as "secure access at all times to sufficient food for a healthy life" (Maxwell 1996). In an emergency food security assessment (EFSA), three key sets of indicators are used to estimate the dimensions of the food security problem (World Food Programme WFP 2009):

"Mortality rates give an indication of risks at the population level".

"Nutrition indicators are used to estimate nutrition status at the individual level".

"Food security indicators focus on assessing access to food and food consumption at the household level".

In our analysis, we employ the most commonly used food consumption indicator, i.e., the "food consumption score (FCS)". This is a proxy indicator that represents the dietary diversity, energy and macro and micro (content) value of the food that people eat. It is based on dietary diversity—the number of food groups a household consumes over a reference period; food frequency—the number of days on which a particular food group is consumed over a reference period, usually measured in days; and the relative nutritional importance of different food groups (World Food Programme WFP 2009). Households are asked to recall the foods that they consumed in the previous 7 days. Each item is given a score of 0 to 7, depending on the number of days on which it was consumed during a week. For example:

- If potatoes were eaten on three of the last seven days, they are given a frequency score of 3;
- If potatoes were eaten on three of the last seven days, even if they were eaten twice on each of those days, at two meals, they are still given a frequency score of 3.

In the analysis, food items were listed according to food groups and the frequencies of all the food items surveyed in each food group were summed. Any summed food group frequency value over 7 was recorded as 7. Each food group is assigned a weight reflecting its nutrient density. For example:

- Beans, peas, groundnuts and cashew nuts are given a weight of 3, reflecting the high protein content of beans and peas and the high fat content of nuts;
- Sugar is given a weight of 0.5, reflecting its absence of micronutrients and the fact that it is usually eaten in relatively small quantities.

The household FCS was calculated for each household by multiplying each food group frequency by each food group weight, and then summing these scores into one composite score.

Solving the problem of food and nutritional security requires a range of interconnected agricultural approaches, including improvements in the productivity of staple crops, the bio-fortification of staple foods, and the cultivation and/or management of a wider variety of edible trees that provide fruits, nuts and vegetables for more diverse diets (Frison et al. 2011). Exotic and indigenous fruits cultivated and managed in agroforestry systems are important in Sahelian countries. As well as directly providing edible products, trees in agroforestry systems support food production by giving shade and support to nutritious vegetable crops (Yamba and Sambo 2012; Maliki et al. 2012; Garrity et al. 2010; Susila et al. 2003). The results from the sampled households showed that overall, 39.78, 36.97, and 23.25% FMNR households respectively fall under poor, borderline and acceptable quality diet. In addition, those under borderline and acceptable quality diet are those managing on average a high and diversified number of trees in their farmlands.

Figure 3 exhibits the two-type plots reporting the effects of FMNR practice on caloric intake and diet.

According to the treatment effects plots, the contribution of trees on caloric intake and diet tends to decrease when the density ranges from 15-20 trees/ha, after the 20 trees/ha threshold, an increase in the number of trees kept and managed on the farmland increases the food consumption score gradually with every unit increase in the number of trees managed by the farmers to a maximum score of 30 points increase in Niger and Sahel. The score tends to stabilize at 22 points increase in Burkina Faso and Mali, and 30 points increase in Senegal, when the minimum density is 20 trees/ha. Summarized results of the regressions analysis are reported in Table 6.

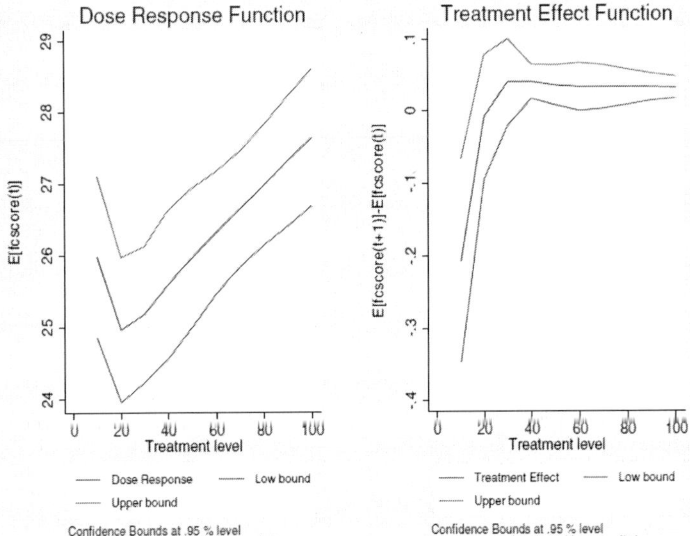

SAHEL

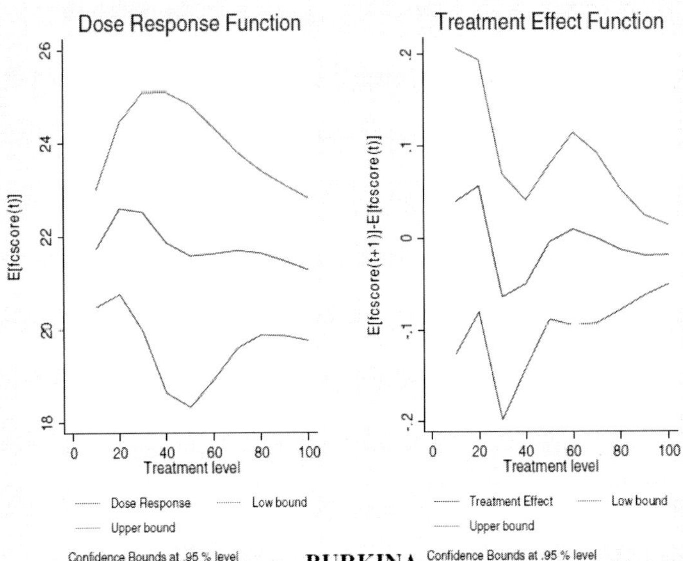

BURKINA FASO

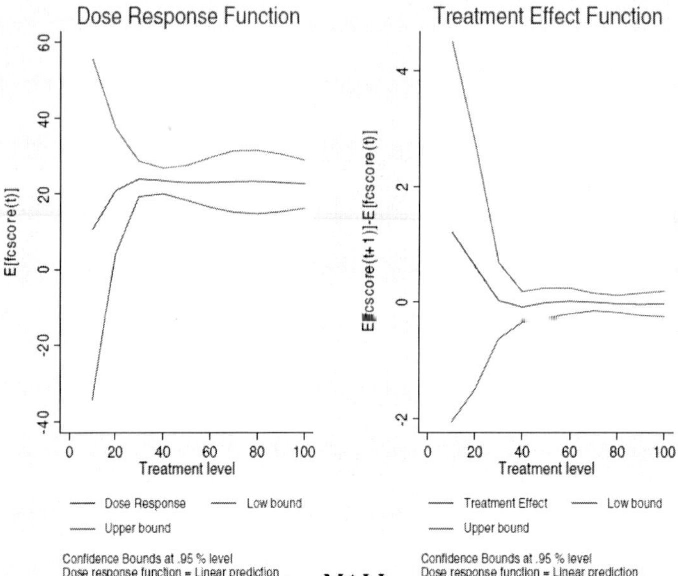
Dose Response Function
E[fcscore(t)]
Treatment level
Dose Response
Low bound
Upper bound
Confidence Bounds at .95 % level
Dose response function = Linear prediction
Treatment Effect Function
E[fcscore(t+1)]-E[fcscore(t)]
Treatment level
Treatment Effect
Low bound
Upper bound
Confidence Bounds at .95 % level
Dose response function = Linear prediction

MALI

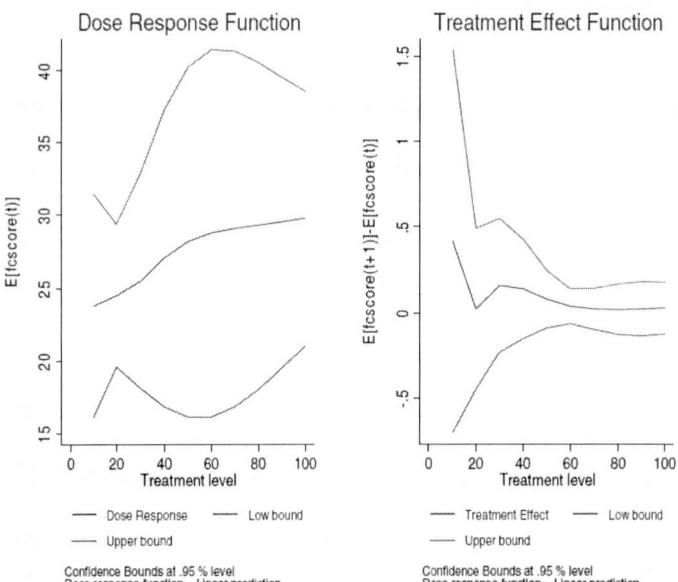
Dose Response Function
E[fcscore(t)]
Treatment level
Dose Response
Low bound
Upper bound
Confidence Bounds at .95 % level
Dose response function = Linear prediction
Treatment Effect Function
E[fcscore(t+1)]-E[fcscore(t)]
Treatment level
Treatment Effect
Low bound
Upper bound
Confidence Bounds at .95 % level
Dose response function = Linear prediction

NIGER

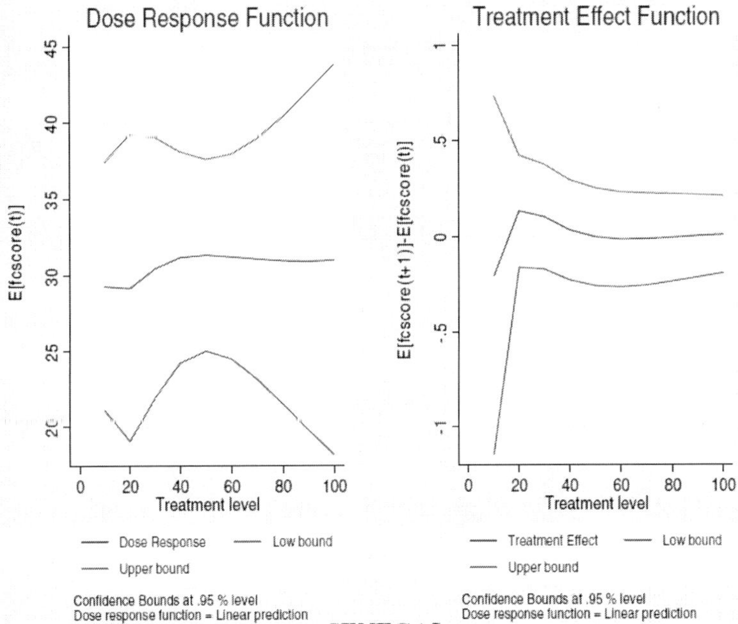

SENEGAL

Figure. 3 Estimated effects of investment on trees on caloric intakes and diet-dose-response function results, derivative, and 95% confidence bands

Table 6 Summary of key findings (regression analysis)

		Country				
		Burkina Faso	**Mali**	**Niger**	**Senegal**	**Sahel**
Factors affecting FMNR	Three categories of factors observed here					
	Pre-disposing factors featured by: • Landholding size	Positive influence	Positive influence	Positive influence	No significant influence	Positive influence
	Facilitating factors featured by access and use of information through: • Ownership of cell phone • Being member	Positive influence Positive influence Negative influence	Positive influence Positive influence No significant influence	Positive influence Positive influence No significa	Positive influence Positive influence Positive influence	Positive influence Positive influence Positive influence

	of community-based organization (CBO) Access to credit			nt influenc e		
	Re-enforcing factors featured by: • Average distance to the main market as a proxy of the market attractiveness of AFTPs • Existence of an operational local convention • Existence of an operational land management commission	No significant influence No significant influence Positive influence	No significant influence Positive influence	Positive influenc e Positive influenc e No significa nt influenc e	Positive influence No significant influence Negative influence	Positive influence Positive influence Negative influence
Effects on liveliho ods	*Effects on income* *Dose response function* indicating an increase of income given a range level of trees being kept in the farmland as a result of FMNR practice *Treatment effects function* that shows the marginal effect of FMNR on income	The revenue received from tree products tend to be stable at 30,000 CFA/ha (US$60) when the density of tree is higher than 20/ha. The maximum amount expected to be received by selling tree products is 40,000 CFA/ha (US$80) with a farm of 15/ha	FMNR practitioners with 20 trees/ha are better off in terms of revenues received from tree products (42,000 CFA or US$84). This amount sharply decline when the density ranges from 21 to 40 trees/ha. After the 40 trees/ha threshold, the revenue perceived from tree products gradually increase with every unit of tree increase during the practice of FMNR to be stabilized around 40000 CFA (US$80).	Exhibit with Senegal the lowest amount of incomes that can be expecte d from tree product s (15,000 to 17,000 CFA i.e., $30 to $34)	Exhibit with Niger the lowest amount of incomes that can be expected from tree products (15,000 to 20,000 CFA i.e., $30 to $40)	Indicates a sharp increase of income (up to a 22,000 CFA or US$44 threshol d) from trees products when the number of trees being managed in the farmland ranges from 10 to 40/ha Farmers who are less involved in the practice of FMNR (i.e., farm with less than 40 trees/ha) are more sensitive to

						income change than those with more trees (i.e., higher than 40/ha) as the effects of changes in income tend to be stable as it appeared in the plot.
	Effects on crop production *Dose response function* indicating an increase of income given a range level of trees being kept in the farmland as a result of FMNR practice	The maximum effect is noted when the density of trees ranges from 20 to 40/ha, indicating:				
		An increase of the expected quantity of cereals produced by 401.38 kg/ha	198.9 kg/ha increase observed among the FMNR practitioners	112.76 kg/ha maximum observed in Niger	132.12 kg/ha increase observed in Senegal	General increase of crop producti on by 274.25 kg/ha in the Sahel as the whole
	Treatment effects function that shows the marginal effect of FMNR on income	The plot suggests that for FMNR practitioners who have less than 40 trees/ha, every one percentage-point increase in keeping and managing trees in their farmland decreased the effects on crop production, and the stronger effect was associated with the management of about 20 trees/ha with a:				
		+14% increase in Burkina Faso	+4% increase in Mali	+3% increase in Niger	Only +1% increase in Senegal	+5% general increase in the Sahel as the whole
	Effects on calorie intakes	According to the treatment effects plots, the contribution of trees on food security tends to decrease when the density ranges from 15-20 trees/ha, after the 20 trees/ha threshold, and increase in the number of				

	Dose response function indicating an increase of income given a range level of trees being kept in the farmland as a result of FMNR practice *Treatment effects function* that shows the marginal effect of FMNR on income	trees kept and managed in the farmland increase the food consumption score gradually with every unit increase in the number of trees managed by the farmers to a maximum threshold level of FCS:				
		The score tends to be constant at 22 points increased when the minimum density is 20 trees/ha	Same as Burkina Faso	30 points increased in observed in Niger	The effects tend also to be constant at 30 points increased of the food consumption score in Senegal	30 points increase d is also observed in the Sahel as the whole

Source: authors' summary

CONCLUSIONS

This study demonstrated that regeneration of trees on farms, whereby farmers play an active role is important as a practice and tree products can serve as a safety-net in times of crises (e.g., income shortages from other income sources, e.g., crop failure) as they may be ready for harvest to serve as emergency food or to be processed and sold. Often, some of these species provide opportunities to marginalized members of the communities such as women and the poor, since they are a valuable source of income and help to mitigate risks during crisis by providing alternative food and economic resources. In addition, deficiency of iron and vitamin A is prevalent in most parts of dryland countries including Mali. However, fruit trees which are popularly used by the local communities contain appreciable amounts of nutrients and energy and thus useful food supplements. Pulp from the baobab fruit has been reported to contain an average of 283 mg/100 g of Vitamin C. The vitamin is important as an antioxidant, hence the prevention of certain types of diseases, ranging from pulmonary, to cancer prevention (Chadare, et al., 2008). For instance, a child could cover 100 percent of its vitamin C requirement by eating only about 10 g of baobab pulp a day. Moreover, the fruit and the kernel of karite have high nutritional value. From the results reported for Burkina Faso (Lamien et al. 1996), the pulp provides protein amounting to 1.9 g per 100 g of matter and about 4.8 Kcal per 100 g, while the kernel provides protein of 6.8 g per 100 g of matter and 579 Kcal per 100 g. The fruit is a source of vitamins and energy. The kernels undergo an elaborate preparation process to produce oil and butter for home consumption, and increasingly for selling. The kernels are dried, cracked, pounded to form a paste, boiled and the butter is then skimmed off. Every woman makes sure that she has enough shea butter in the home for cooking to last at least until the following harvest. In addition, in some places where the markets of fruit tree products are functioning well, fruit trees cultivation offers great potential for income generation.

The analysis of factors affecting FMNR in the Sahel did not find many household's characteristics constraining significantly the FMNR practice in more than one country. However, several institutional factors have been identified as promoting or limiting the potential for FMNR, such as rights and regulation rules over trees and natural resources in general. For example in the Sahel, the use of fire and free grazing systems generates benefits in terms of grass regeneration, clearing of debris, catching wild rodents for food and in the case of free grazing, offering a cheap mechanism for feeding livestock. Thus, it is not easy to find institutional reforms that can accommodate the interests of FMNR with others. However, practices such as controlled fires, rotational grazing and the promotion of livestock corridors through local conventions are all options that have been successfully implemented in the drylands.

Markets for tree products are other factors that impacts on incentives to manage trees. For FMNR in particular, market development may have different effects. On the one hand, markets surely do influence collection and marketing behavior of farmers as the case of shea in Burkina and Mali has demonstrated. In general, as tree product markets develop, there is more incentive to maintain trees on farms. There may be further incentives concerning the selection of species to retain based on market signals, but only if market signals persist for a long period of time since changes in tree species composition is a long term operation in the drylands.

RECOMMENDATION DOMAINS

1. Overall, FMNR cannot be excluded as a recommendation in any of the dryland zones. FMNR will continue to support the largest number of established trees on farms in the drylands. Its importance at a landscape level is likely to increase as well, the more that agricultural expands into woodland and bush lands. The other tree related alternatives to FMNR require the planting of trees. Opportunities for planting will remain very limited in arid drylands and limited in semi-arid drylands. A major exception will be low-lying areas of the landscape, but even there, other enterprises may be preferred over tree planting, thus reaffirming the importance of FMNR for trees. Within a particular dryland zone, there may be further nuances on recommendations for how to practice FMNR. For example, certain institutional arrangements such as improved grazing management may be an important complementary action in some places while not in others.

2. The issue of forest regulations which create disincentives for farmers is one that is widespread in the developing world. These include the banning of felling or cutting of a number of species without obtaining a prior permit, at a fee. Violation of such regulations entails a hefty fine, and so farmers will often remove young trees from their land to avoid having to deal with such regulations in the future. Among such regulations, the adverse effects of the Sahelian forest codes have long been recognized (e.g., McLain (1992)) and there have been many dialogues in the region to try and move forward. Although not backed by formal policy change, the recent regreening in Niger has been attributed to a significant extent by the relaxation of enforcement of such policies (Reij et al 2009). A recent analysis of the forest codes and recommendations for action are in Yatich et al (2014). *National policies need to be complemented by institutional support at the local level to reduce current constraints of property rights and other institutional constraints affecting the promotion of FMNR.* In addition, *there is a need to develop a more integrated institutional framework and find ways to enhance effective participation across institutions.*

3. Access to information tends to improve decision-making skills, which in turn affects the know-how of farmers. Indeed, agroforestry technologies as compared with the conventional agricultural practices that farmers have known, have been used to and have received training for a much longer period, are generally incipient and relatively new phenomenon (Ajayi et al. 2006). *The promotion of tree-based systems requires skills in terms of management of trees. Capacity of doing this needs to be built at all level and requires a multidimensional approach including field demonstrations as well as a proper dissemination of information using different information and communication technology tools. In addition, be aware of the existing opportunities of FMNR technology to provide immediate, medium and long term individual and public benefits simultaneously need to be well-communicated to many stakeholders.*

LIMITATIONS OF THE STUDY

1. While recognizing the benefits of trees and tree products on caloric intake and diet, the results from this study establishing a correlation between FMNR and caloric intakes should be taken with cautious given that, "a snapshot of one week consumption cannot provide enough information to claim that FMNR household are food secured or not. In fact, the FCS does not capture seasonal changes, quantify the food gap, capture intra-household food consumption etc..." In addition, while the evidence base for the role of forests and tree-based systems for food security and nutrition is growing (Sunderland et al. 2013; Bhaskar et al. 2015), there

remain many gaps in our understanding of this relationship and its potential contribution to reducing global hunger. There is a need to explore much more details the FMNR-food nexus in future researches by going beyond what was covered from this study, particularly in "relation to the integrated management of multi-functional landscapes, and the multi-scalar and *cross-sectoral governance* approaches that are required for the equitable delivery of these benefits".

2. The types of trees that will be desirable for farmers to retain, as well as the densities of trees, may also differ across locations. For example, where fertilizer use is extremely low, promoting regeneration of trees which have known positive soil fertility properties will be more important. Similarly, the promotion of higher tree canopies in locations with high current or expected temperatures would be recommended if current crop choices are to be maintained. Unfortunately, these types of nuances cannot be neatly mapped out as they require information that was not readily available. However, such information can be collected quickly and cheaply by programmes promoting FMNR.

3. Land tenure would has been a constraining factor, but in the study area, over 85% of all land plots were inherited in each of the countries. This still may be an issue for women and the plots that they manage, but this could not be assessed with data at hand.

ACKNOWLEDGEMENTS

The authors gratefully acknowledge the International Funds for Agricultural Development (IFAD), the United States Agency for International Development (USAID) who provided financial support for this study. Andre Bationo Babou from the National Institute of Environment and Agricultural Research (INERA) in Burkina Faso, Joseph Dakouo from Institute of Rural Economy (IER) in Mali, Abasse Tougiani from the National Institute of Agricultural Research (INRAN) in Niger and, Diaminatou Sanogo from Institute of Agricultural Research (ISRA) in Senegal, were invaluable resource persons during the field work. Finally, we are indebted to the thousands of households from rural communities in Burkina Faso, Mali, Niger Republic, and Senegal whose experience, knowledge, perseverance, and friendship provided the foundation for our understanding and interpretation. We are very indebted to the anonymous two referees and the editor in chief for their insightful comments and suggestions that helped to significantly improve the quality of this manuscript.

REFERENCES

1. Ajayi OC, Akinnifesi FK, Mullila-Mitti J, DeWolf JJ, Matakala W (2006) Adoption of agroforestry technologies in Zambia: synthesis and implications for policy. World agroforestry centre-mimeo
2. Ajayi OC, Akinnifesi FK, Sileshi G, Chakeredza S (2007) Adoption of renewable soil fertility replenishment technologies in the southern African region: Lessons learnt and the way forward. Nat Resour Forum 31:306–317
3. Alinon K, Kalinganire A (2008) Effectiveness of bylaws in the management of natural resources. The West African experience, CAPRI Working Paper 93. CGIAR, Washington DC
4. Allen JH, DuVander J, Kubiszewski I and Ostrom E (2012) Institutions for managing ecosystem services. Solut J Sustainable Desirable Future 2(6):44-49
5. Baggnian I, Adamou MM, Toudou A, Ali M (2013) Impact des modes de gestion de la Régénération Naturelle Assistée des ligneux (RNA) sur la résilience des écosystèmes dans le Centre-Sud du Niger. J App Biosci 71:5742–5752
6. Bayala J, Sileshi GW, Coe R, Kalinganire A, Tchoundjeu Z, Sinclair F, Garrity D (2012) Cereal yield response to conservation agriculture practices in drylands of West Africa: a quantitative synthesis. J Arid Environ 78:13–25
7. Bayala J, Sanou J, Teklehaimanot Z, Kalinganire A, Ouedraogo SJ (2014) Parklands for buffering climate risk and sustaining agricultural production in the Sahel of West Africa. Environ Sustain 6(2):28–34
8. Bhaskar V, Wildburger C, Mansourian S (eds) (2015) Forests, trees and landscapes for food security and nutrition: a global assessment report. International Union of Forest Research Organizations (IUFRO), Vienna
9. Bia M, Mattei A (2008) A Stata package for the estimation of the dose-response function through adjustment for generalized propensity score. Stata J 8(3):354–373
10. Chadare FJ, Hounhouigan JD, Linnemann AR, Nout MJR, van Boekel MAJS, Chadare FJ, Hounhouigan JD, Linnemann AR, Nout MJR, van Boekel MAJS (2008) Indigenous Knowledge and Processing of L. Food Products in Benin. Ecol Food Nutr 47 (4):338-362
11. Djire M, Dicko AK (2007) Les conventions locales face aux enjeux de la decentralization au Mali. Karthala, Paris
12. Faye M D, Weber J C, Abasse T A, Boureima M, Larwanou M, Bationo A B, Sonogo D D (2011) Farmers' preferences for tree functions and species in the West African Sahel. For. Trees and Livelihoods 20(2-3):113-136
13. Frison EA, Cherfas J, Hodgkin T (2011) Agricultural biodiversity is essential for a sustainable improvement in food and nutrition security. Sustainability 3(1):238–253
14. Garrity DP, Akinnifesi FK, Ajayi OC, Weldesemayat SG, Mowo JG, Kalinganire A, Larwanou M, Bayala J (2010) Evergreen agriculture: a robust approach to sustainable food security in Africa. Food Sec 2:197–214
15. Geraldi JG (2007) New institutional economics. MIP, Universitat Siegen
16. Glenn JV (2012) Economic assessment of landowner incentives: analyses in north Carolina and Malawi. MSc Thesis, Department of Natural Resources, North Carolina State University, Raleigh
17. Haglund E, Ndjeunga J, Snook L, Pasternak D (2011) Dry land tree management for improving household livelihoods: farmer managed natural regeneration in Niger. J Environ Manag 92(7):1696–1705
18. Hilhorst T (2008) Local governance institutions for sustainable natural resource management in Mali, Burkina Faso and
19. Niger. KIT Working Paper Series G1. Royal Tropical Institute
20. Hirano K, Imbens GW (2004) The propensity score with continuous treatment. In: Gelman A, Meng X-L (eds) Applied
21. Bayesian modeling and causal inference from incomplete-data perspectives. Wiley Interscience, West Sussex, pp 73–84

22. Holland P (1986) Statistics and causal inference (with discussion). J Am Stat Assoc 81:945–970
23. Howard PL, Nabanoga G (2007) Are there customary rights to plants? An inquiry among the Baganda (Uganda), with special attention to gender. World Dev 35(9):1542–1563
24. Imbens GW (2000) The role of the propensity score in estimating dose-response functions. Biometrika 87(3):706–710
25. Jessop B, Nelson K (2003) Institutions and rules, Research Paper 11/03. Network institutional Theory
26. Joffe MM, Rosenbaum PR (1999) Invited commentary: propensity scores. Am J Epidemiol 150:327–333
27. Kabore PD, Kabore PS, Maatman A, Ouedraogo AA, Ruijs AA, Sawadogo H, Schweigman C (1997) Analysis of farmers' strategies in the central and north-west regions of Burkina Faso: approach and some results. In: Asenso-Okyere WK, Benneh EY, Tims W (eds) Sustainable food security in west Africa. Klewer Academic Publisher, Boston, pp 80–128
28. Kalinganire A, Weber J, Uwamariya A, Kone B (2007) Improving rural livelihoods through domestication of indigenous fruit trees in the parklands of the Sahel in Indigenous Fruit Trees in the Tropics: Domestication, Utilization and Commercialization (eds F.K. Akinnifesi et?al.)., pp 186–203
29. Kouyaté AM (2005) Enquête ethnobotanique sur Detarium microcarpum Gulll. et Perr. Au sud du Mali. In: Kalinganire A, Niang A, Kone B (eds) Domestication des espèces agroforestières au Sahel: Situation actuelle et perspectives. ICRAF Working Paper No. 5. World Agroforestry Centre, Nairobi
30. Lamien N, Sidibe A, Bayala J (1996) The joy of cooking: recipes for the success of the shea tree. Agrofor Today 1996:11
31. Leach HB, Van der Stege C, Vogl CR (2011) Baobab (adansonia digitat L.) and tamarind (tamarindus indica L.) management strategies in the midst of conflict and change: a dogon case study from Mali. Hum Ecol 39(5):597–612
32. Maliki R, Cornet D, Floquet A, Sinsin B (2012) Agronomic and economic performance of yam-based systems with shrubby and herbaceous legumes adapted by smallholders. Outlook Agric 41(3):171–178
33. Maxwell D (1996) Measuring food insecurity: the frequency and severity of "coping strategies". Food Policy 21(3): 291–303
34. McLain R (1992) Recommendations for a New Malian Forest Code. Observations from the Land Tenure Center's Study of Land and Tree Tenure in Mali's Fifth Region. Madison, WI University of Wisconsin, Land Tenure Center, LTC Research Paper 109
35. Mercer DE (2004) Adoption of agroforestry innovations in the tropics: a review. Agroforestry Syst 61(1): 311- 328
36. Mikulcak F (2011) The Implications of Formal and Informal Institutions on the Conservation of On-farm Trees: An analysis from the Department of Mirriah, Republic of Niger. Master thesis in geography, Department of Human Geography, Stockholm University
37. Nyemeck BJ, Place F, Kalinganire et al (2015) Effects of farmer managed natural regeneration on livelihoods in semi-arid West Africa. Environ Econ Policy Stud 17(4):543–575
38. Oginosako Z, Simitu P, Orwa C, Mathenge S (2006) Are they competing or compensating on farm? Status of indigenous and exotic tree species in a wide range of agro-ecological zones of Eastern and Central Kenya, surrounding Mt. Kenya, ICRAF Working Paper no. 16. World Agroforestry Centre, Nairobi
39. Ouedraogo M (2007) La communalisation en milieu rural: les élus locaux peuvent-ils relever le défi ? l'Harmattan, Paris, p 98
40. Phiri D, Franzel S, Mafongoya P, Jere I, Katanga R, Phiri S (2004) Who is using the new technology? The association of wealth status and gender with the planting of improved tree fallows in eastern Zambia. Agroforestry Syst 79(2): 131–144

41. Place F (2009) Land tenure and agricultural productivity in Africa: a comparative analysis of the economics literature and recent policy strategies and reforms. World Dev 37(8):1326–1337
42. Place F, Franzel S, DeWolf J, Rommelse R, Kwesiga F, Niang A and Jama B (2002) Agroforestry for soil fertility replenishment: evidence on adoption processes in Kenya and Zambia. In: Barrett CB, Place F and Aboud AA (eds), Natural Resources Management in African Agriculture: Understanding and Improving Current Practices. CAB International, Wallingford, pp. 155–168
43. Place F, Nyemeck BJ, Kalinganire A (2013) Economic impacts of farmer-managed natural regeneration in the Sahel: end-of-project technical report for free university Amsterdam and IFAD. World Agroforestry Center, Nairobi
44. Polak P (2008) Out of Poverty: What Works when Traditional Approaches Fail. Berrett-Koehler Publishers, San Francisco
45. Pye-Smith C (2013) The quiet revolution: how Niger's farmers are re-greening the parkland of the Sahel. ICRAF Tree for Change no 12:48
46. Reij C, Tappan G, Smale M (2009) Agroenvironmental transformation in the Sahel-another kind of 'Green revolution', International Food Policy Research Institute (IFPRI) Discussion Paper 00914
47. Rubin DB (1977) Assignment to treatment group on the basis of a covariate. J Educ Stat 2(1):1–26
48. Sanogo K, Binam J, Bayala J, Villamor GB, Kalinganire A, Dodiomon S (2016) Farmers' perceptions of climate change impacts on ecosystem services delivery of parklands in southern Mali. Agr Syst. doi:10.1007/s10457-016-9933-z
49. Sanou J, Zougmoré R, Bayala J and Teklehaimanot Z (2010) Soil infiltrability and water content as affected by Baobab
50. (Adansonia digitata L.) and Néré (Parkia biglobosa (Jacq.) Benth.) trees in farmed parklands of West Africa. Soil Use Manage 26(1):75-81
51. Sileshi G, Akinnifesi FK, Ajayi OC, Place F (2008) Meta-analysis of maize yield response to planted fallow and green manure legumes in sub-Saharan Africa. Plant Soil 307(1-2):1–19
52. Siziba S, Nyikahadzoi K, Nyemeck BJ, Diagne A, Adewale A, Oluwole F (2013) Estimating the impact of innovation systems on maize yield: the case of Iar4d in southern Africa. Agric Econ Res Policy and Pract S Afr 52(3):83–100
53. Solh M, Saxena MC (eds) (2011) Food security and climate change in dry areas: proceedings of an International Conference, 1-4 February 2010, Amman, Jordan. PO Box 5466. International Center for Agricultural Research in the Dry Areas (ICARDA), Aleppo, p viii + 369
54. Sunderland TCH, Powell B, Ickowitz A, Foli S, Pinedo-Vasquez M, Nasi R, Padoch C (2013) Food security and nutrition: the role of forests. CIFOR Discussion Paper. Centre for International Forestry Research, Bogor
55. Susila AD, Purwoko BS, Roshetko JM, Palada MC, Kartika JG, Dahlia L, Wijaya K, Thieba D (2003) Quelle question des ressources naturelles pour le milieu rural? Les lecons de l'experience de la mise en oeuvre de la reforme agraire et fonciere et les defis de la decentralisation, enjeux et viabilite des communes rurales au Burkina Faso. Groupe de Recherche Action sur le Foncier (GRAF), Amsterdam, KIT
56. Toulmin C, Lavigne, Deville P, Traore S (2002) The dynamics of resource tenure in West Africa. James Cury, Oxford
57. Weston P, Hong R, Kaboré C, Kul AC (2015) Farmer-managed natural regeneration enhances rural livelihoods in dryland west Africa. Environ Manage 55(6):1402–141
58. World Bank (1986) Poverty and Hunger. Issues and options for food security in Developing countries, The World Bank, Washington
59. World Food Programme (WFP) (2009) Emergency food security assessment handbook, 2nd edn. WFP, Rome-Italy, p 297

60. Yamba B, Sambo M (2012) La Régénération Naturelle Assistée et la sécurité alimentaire des ménages de 5 terroirs villageois des départements de Kantché et Mirriah (Région de Zinder). Rapport pour le Fonds International pour le Développement Agricole
61. Yatich T, Kalinganire A, Weber JC, Koffi A, Dakouo JM, Samake O, Sangare S (2014) How do forestry codes affect access, use and management of protected indigenous tree species: evidence from West Africa Sahel, Occasional Paper 15.West Agroforestry Centre, Nairobi

CHAPTER 17

Off-farm participation decision and its impact on crop yield in Northern Ethiopia

Berihun Kassa Bihon, Kassa Kibrom Aregawi

[1]*College of Social Sciences and Humanities, Samara University, P. O. Box 132, Samara, Ethiopia.*

[2]*College of Business and Economics, Mekelle University, P. O. Box 450, Mekelle, Ethiopia.*

ABSTRACT

One of the most important facets of off-farm activities is providing employment opportunities and additional income for rural households and thereby accommodates the seasonal and fluctuating agricultural production. Given this, identifying the underlying determinants of off-farm participation and its impact on crop yield was found to be of the essence. Cross sectional data was collected through structured questionnaire administered on 384 randomly selected farmers. Descriptive statistics, probit and Ordinary Least Square (OLS) models as well as t and chi-square tests were used to analyze the data. The regression result revealed that off-farm participation was positively influenced by gender, education, working people, number of pack animals and credit access; while age and land size carried a negative sign. Off-farm participation was also found to have a negative and statistically significant impact on crop yield where non-participants were better producers; since majority of off-farm participants participate only on food-for-work which has nothing to add for crop yield rather than compromising farm activities. Hence, training on non-farm activities need to be given; the current adult education, being propagating, need to be strengthened; and there is a need to solve liquidity problem through credit access that could serve as startup capital.

Keywords: Off-farm, participation, smallholder farmers, probit, Ordinary Least Square (OLS).

1. INTRODUCTION

As engine of economic growth and poverty reduction, in developing countries, agriculture should be integrated with sectors that have direct or indirect linkages. According to Babatunde et al. (2010) financial capital appears to be the most limiting factor for farming, so that cash income from off-farm activities helps to expand farm production; increase household income and reduce risk of crop failure. Hence, off-farm is one among the activities whereby agriculture is believed to be integrated with. World Bank (2008) has reported that, in most developing countries, the importance of off-farm activities is increasing and estimated to account for 30 to 50% of rural incomes. According to Rios et al. (2008), the higher the off-farm income is, the larger capital endowments will be; and having higher capital endowments will in turn help to produce more and more and even to be productive. According to Asenso-Okyere and Samson (2012), Haggblade et al. (2010) and Diao and Nin Pratt (2007), modern agricultural inputs can result with ample production and productivity of marketable commodities that results with trade linkage; the requirement of agricultural inputs and marketing facilities by itself, then, induces off-farm activities.

As per the view of Haggblade et al. (2010), those who live in arid areas where production is sluggish tend to participate in off-farm activities and diversify their income sources. Similarly, in Burkina Faso, Zahonogo (2011) purport that a decline in farm income and farm production let farmers to participate in off-farm activities; and increment in farm production and income do the opposite. Furthermore, in areas where agriculture is grown, the rural off-farm economy was also been founded to be grown (Haggblade et al., 2009). While quantifying this direct linkage between agricultural growth and off-farm economy, Haggblade et al. (2009) reported that a one dollar value invested on to agriculture generates 0.3 and 0.5 additional dollar on to rural off-farm income in Africa and Latin America, respectively.

In Africa, off-farm economic activities, as a means of income diversification, are very much indispensable for improving the livelihood of rural poor

(Asenso-Okyere and Samson, 2012; Diao and Nin Pratt, 2007). Besides, it can serve as source of input supply for agricultural production and employment opportunity for those who do not have arable land and do not further want to rely on agriculture. Despite its vitality, in Africa, off-farm activities participation is low; and according to Haggblade et al. (2007), 37% of the rural households' income is really extracted from off-farm activities where surprisingly not more than 20% of the labor force is being participated. Likewise, in Nigeria, a study by Adewunmi et al. (2011) revealed that participation in off-farm activities more particularly wage employments of skilled and unskilled had resulted in a reduction of rural poverty by 11.02 and 10.68%, respectively; participants lessen poverty better than non-participants (Alaba and Kayode, 2011).

In Ethiopia, being part of developing regions in general and sub-Saharan Africa in particular, agriculture is the real backbone that enables 85% of Ethiopian people to walk upright and be able to feed the remaining 15% of the fortune coincidence urban residents. In line with this, Ministry of Agriculture and Rural Development (2010) acclaimed smallholder farmers in producing majority of the country's total production. Furthermore, World Bank (2010) inferred that 90% of the foreign earnings and 70% of the raw materials for industry are being sourced from agriculture. Hence, 85% of the Ethiopian people believe that agriculture is the way out to be food secured and even be luxuriant in the long run. Given this deep rooted belief, starting from recent past, arable land size and agricultural production are dwindling and thereby the society is being forced either switching to off-farm activities or stretch their hand for food aid basically for clothing the food longing mouths at home.

According to Merima and Peerlings (2012) and Abebe (2008), in Ethiopia, farmers tend to participate in off-farm activities to satisfy consumption needs by supplementing the sluggish agricultural income caused by erratic and seasonal rain fall; and due to crop failure and the resultant abundant labor force. Besides, farmers are vulnerable both for natural and manmade disasters and shocks that could be weather related, pests, death of livestock and others. As a means of relief, rural farm households prefer to participate in off-farm activities and diversify their income thereby (Berg and Kumbi, 2006; Dercon and Woldehanna, 2005). Hence, a certain farm household is tempted to participate in off-farm activities due largely to push factors, like drought. They further inferred that, Ethiopian rural poor prefer to participate in off-farm activities that require less entry costs like fire wood collection and charcoal production, while the most lucrative off-farm activities are left for the haves. Contrary to this, some rural farm households tend to participate in off-farm activities basically to widen their income sources given their ample agricultural production and the resultant accumulated wealth, as a pull factor.

In line with the above inclination, Rijkers et al. (2002) as cited in Woinshet (2010) purported that 25% of Ethiopian rural households own more than one labor intensive off-farm activities, although 23% of these rural households did their off-farm activities parallel with their agricultural practices. Hence, only 2% of Ethiopian rural households were those who exclusively rely on off-farm activities. Besides, according to Merima and Peerlings (2012) for the past eight years, back from 2012, not more than 25% of the rural households had engaged in off-farm activities which are minimal compared with 42% of the sub-Saharan Africa average; and its contribution for employment creation in Ethiopia is 1.14%. From this one can deduce that Ethiopian rural households' off-farm participation is insignificant and its effect onto crop production as well as its overall benefit has not yet been exploited. Hence, identifying determinants of off-farm participation decision and examining the impact of participation on crop yield is found to be imperative for a clear and sound policy indication whereby farm households would participate and diversify their income and cope up food security challenges.

2. MATERIALS AND METHODS

2.1 Site selection, sampling method and data collection

As far as the study area selection was concerned, compared with the rest districts of Southern zone, Raya-Azobo and Raya-Alamata are more of promising and lucrative with plenty resource reservations basically fertile land resource, ground water, flat plain land supported by runoff rain water from the surrounding hilly parts. Despite these endowments, since production is decreasing, these districts were reported to face recurrent drought and food insecurity; and the society is been engaged in off-farm activities basically tocurtail the food shortage. As a result of this researchers were intended to have a look on the impact of off-farm activities participation on crop yield.

Multi-stage sampling procedure was followed to reach on to final respondents. Firstly, Raya-Azebo and Raya-Alamata districts were selected from Southern Tigray; followed by the randomly selected four sub-districts, Bala-Ulaga, Kukufito, Lemeat and Tao. Thirdly, on the basis of Yamanes' sample size determination formula, as cited in Israel (1992), 384 sample respondents were calculated. Finally, 384 respondents were randomly selected from the list provided from each sub-district, as its proportion is shown in Table 1 and these people were considered to collect primary data mainly through structured questionnaire in 2014 cropping year. In-depth interview with crop production personnel and farmers who were on the verge to leave agriculture were interviewed. Besides, four focus group discussions with off-farm participants and non-participants, including female participants, were held to support the questionnaire based data. Interviewees and focus group discussants were selected purposively with the assumption that these people could provide us with tangible and pertinent information in the ground.

Table 1. Targeted sub-districts and villages with their total population and sample size taken (CSA, 2007; Own Computation, 2016).

S/N	Name of sub-district	Population size (N)	Sample size (n)
1	Bala-Ulaga	2164	89
2	Kukufito	3784	155
3	Lemeat	1007	70
4	Tao	1697	70
-	Total	9342	384

2.2 Method of data analysis and econometric model specification

To analyze the data, descriptive statistics like mean, percentage and standard deviations were used to assess socio-economic and demographic characteristics of sample respondents; while probit model was used to examine determinants of participation, Ordinary Least Square (OLS) model was employed to examine the impact of participation on crop yield. Besides, to test the relationship between dependent and independent variables vis-à-vis the two groups of individuals, t-test and Chi-square tests were used, respectively, for continuous and dummy variables. Data collected through interview and focus group discussion was basically analyzed qualitatively.

Regardless of the amount of money earned and the duration farmers relied on off-farm activity, in this article, a farmer was taken as off-farm participant if he/she did out of his/her farm land; while those who solely dependent on their farm land were considered as non-participants. Due to this dichotomous classification, the dependent variable (off-farm participation) has a binary nature taking the value of "1" for participants and "0" for non-participants that paved the way to employ binary outcome models particularly probit model. The reason behind employing probit model was due to its normal distribution assumption of error terms and farmers' unobserved or latent behavior by which households are assumed to make decisions pertaining to utility maximization; and hence in this model there is a latent or unobservable variable that takes all the values in

(-∞, +∞). As a result, probit model can be expressed by the following general formula:

$$\Pr(Y=1/X_i) = \Phi(\beta_1 X_1 + \varepsilon_{i1}) \quad (1)$$

The latent variable Y_i^* is not observable and is represented by its proxy Y_i taking a value one (1) for participants and zero (0) for non-participants.

$$Y_i = \begin{cases} 1, & Y_i^* > 0 \\ 0, & Y_i^* \leq 0 \end{cases}$$

$$Y_i^* = X_i \beta + \varepsilon_i \quad (2)$$

where $\varepsilon \mid x$ is a normally distributed error term.

Thus, for the household *i*, probability of participation is given by:

$$P(1) = \Phi(\beta X_i) \quad (3)$$

where P (1) is the probability of participation, Φ is the cumulative distribution function of the standard normal distribution, β is the parameters that are estimated by maximum likelihood, x' is a vector of exogenous variables which explains off-farm participation. Therefore,

$$OFP=\Phi(\beta_1 GENDER+\beta_2 AGE+\beta_3 EDUC+\beta_4 WORKING+\beta_5 DEPENDENT+\beta_6 LANDSZ+\beta_7 MRKTDIS+\beta_8 LOCATION+\beta_9 TLU+\beta_{10} PACKANIM+\beta_{11} CREDIT \quad (4)$$

Besides, to estimate magnitude of parameters or variables mainly to clearly put the percentage likelihood of participation, marginal effect of variables was calculated. Marginal effect of a variable is the effect of unit change of that variable on the likelihood of participation and it can be seen as *P(Y = 1|X = x)*, given that all other variables are constant. Hence, it is expressed as:

$$\frac{\partial P(Yi=1/Xi}{\partial Xi} = \frac{\partial E(Yi/Xi)}{\partial Xi} = \varphi(X_i'\beta)\beta \quad (5)$$

As far as examining the impact of participation on crop yield was concerned, Ordinary Least Square (OLS) regression model was employed due to the continuous nature of the dependent variable, crop yield measured in quintal. Furthermore, according to Gujarati (2006), with the assumption of classical linear model, OLS estimators are with unbiased linear estimators with minimum variance and hence they are Best Linear Unbiased Estimators. Hence, its specification is given as follows using the same independent variables used in probit model earlier.

$$Y = \beta_0 + \beta i\, Xi + U$$

where Y is the dependent variable (crop yield), Xi is a vector of explanatory variables, βi is a vector of estimated coefficient of the explanatory variables (parameters) and u_i indicates disturbance term which is assumed to satisfy all OLS assumptions (Gujarati, 2006).

$$Cropyld=\beta_0+\beta_1 GENDER+\beta_2 AGE+\beta_3 EDUC+\beta_4 WORKING+\beta_5 DEPENDENT+\beta_6 LANDSZ+\beta_7 MRKTDIS+\beta_8 LOCATION+\beta_9 TLU+\beta_{10} PACKANIM+\beta_{11} CREDIT \quad (6)$$

where Cropyld=Continuous dependent variable indicating crop yield measured in quintal.

Therefore, description of variables indicated earlier and their expected sign or hypothesis has shortly been summarized in Table 2.

Variable type	Character	Unit of measure	Expected sign
Sex of household head (GENDER)	Dummy	1 if Male, 0 if Female	+
Age of household head (AGE)	Continuous	Years	+/-
Education of household head (EDUC)	Dummy	1 if literate, 0 otherwise	+
Number of Working people (WORKING)	Continuous	Numbers	+/-
Number of Dependents (DEPENDENT)	Continuous	Number	+
Cultivated land size (LANDSZ)	Continuous	*Tsimad*[1]	+
Tropical Livestock Unit (TLU[2])	Continuous	Number	+/-
Number of pack animals (PACKANIM)	Continuous	Number	+
Distance to the nearest market (DISMRKT)	Continuous	Kilometers	-
Location effect (LOCATION)	Dummy	Across locations	+/-
Access to Credit (CREDIT)	Dummy	1 if Yes, 0 otherwise	+
Non-farm Trainings (NFTGS)	Dummy	1 if trained, 0 otherwise	+

[1]One *Tsimad* is equivalent with 0.25 hectare or one hectare is equivalent with 4 *Tsimad*; [2]TLU was calculated excluding pack animals basically to solve problem of multicollinearity. +, Positive; _, Negative.

3. RESULTS AND DISCUSSION

3.1 Socio-economic and demographic characteristics of sample households

Examining farm households' socio-economic and demo-graphic characteristics is believed to be an important indicator for probability of off-farm participation in such a way that farm households can be determined with. As a result of this, an attempt has been made to describe some socio-economic and demographic characteristics where differences between participants and non-participants were tested using t and Chi-square tests.

Table 3 clearly purports that majority of the sample respondents (73.44%) were off-farm participants. In line with this rate of participation, majority of participants (78%) were male headed households supported with a significant Chi-square test result. Compared with non-participants, participants were male headed households, younger in age, educationally better, large in number of working people, wealthier in pack animals possessed and farm income and then with better credit access. On the other hand, non-participants were owners of large arable land size and better crop producers. These all differences were supported with significant t and Chi-square tests for continuous and dummy variables, respectively.

Table 3. Descriptive statistics of variables used in regression analysis and important ones (Own Survey Result, 2016).

Variable	Average statistics (N=384)	Off-farm non-participants (N= 102)	Off-farm participants (N= 282)	t & chi-square tests (P-Value)
	Mean (Std.dev)	Mean (Std.dev)	Mean (Std.dev)	
Gender (1=Male)	0.786 (0.410)	0.706 (0.458)	0.816(0.388)	0.016**
Age	44.421 (11.481)	46.716 (12.846)	43.592 (10.851)	0.009***
Education (1=Literate)	0.349 (0.477)	0.245 (0.432)	0.386 (0.488)	0.030**
Number of Working people	3.086 (1.301)	2.931 (1.299)	3.142 (1.299)	0.038**
Number of Dependents	3.102 (1.462)	3.088 (1.695)	3.106 (1.372)	0.542
Cultivated Land Size	1.474 (0.693)	1.649 (0.884)	1.410 (0.599)	0.001***
Tropical Livestock Unit	4.853 (3.574)	4.779 (3.239)	4.879 (3.692)	0.596
Number of pack animals	0.602 (0.919)	0.324 (0.616)	0.702 (0.989)	0.074**
Distance to the nearest market	24.395 (13.023)	23.088 (9.057)	24.869 (14.172)	0.881
Access to credit (1=Yes)	0.365 (0.482)	0.127 (0.335)	0.450 (0.498)	0.000***
Non-farm Training (1= Trained)	0.099 (0 .298)	0.098 (0.297)	0.166 (0.408)	0.576
Farm income in Birr	39286.2 (29614.8)	32538.2 (29009.9)	46034.3 (29880.4)	0.041**
Crop yield in Quintal	8.522 (6.145)	9.946 (7.337)	8.369 (5.659)	0.024**

Std.dev: Standard deviations; ***, ** and * is significant at 1, 5 and 10% significance level.

3.2 Types of off-farm activities vis-à-vis place of work

Types of off-farm activities along with place of work were discussed in Table 4.

As it can be seen in Table 4, 65.6% of the total off-farm participants were engaged in food-for-work under the umbrella of Productive Safety Net Program (PSNP). These people were paid employees where one person was expected to do five days per month and will be awarded birr 19 in cash and 3 kg (0.03 quintal) in food/kind. As far as place of work was concerned, 76.95% of off-farm participants have been doing with in their respective sub-districts. To engage in food-for-work and transacting goat, sheep and cereals by buying and selling on market day time mainly on weekly and monthly basis, farmers were not forced to go out of their residence and thereby lose their social tie and family kinship.

Table 4 further reveal that, 67.02% of the total participants were instigated to participate due to the advent of and access to PSNP; followed by their residences' proximity to urban area that is believed to expose them to new and updated information as well as market identification mainly for petty trading activities like trading egg, coffee, honey, selling local beer (ale) and using cart.

Table 4 also indicated that, more than 57.84% of non- participants were seriously inhibited by absence and the resultant farness of off-farm opportunities coupled with shortage of startup capital. Had there been off-farm opportunities, they would not have been exploited opportunities due largely to shortage of startup capital. From this one can infer that, farm respondents were highly intended to do activities that requires startup capital like trading camel-the then costly animal being transacted and trading cereals across different regions. On the other hand, opportunities' farness highlights that, respondents were intended to do off-farm activities without compromising their farm activities, their family's love, affection and social ties.

Table 4. Off-farm activities and place of work (Own Survey Result, 2016).

Variable	Labels	Freq.	%
Off-farm activities being done	Trading goat, sheep and cereal crops	37	13.12
	Trading cereal crops	24	8.51
	Trading egg, coffee, honey and shopping	5	1.78
	Cart and Daily work	28	9.93
	Hair dressing and selling local beer (ale)	3	1.06
	Food for work	185	65.60
	Total	282	100
Location of employment	Within village	15	5.32
	Within sub-district	217	76.95
	Within district	22	7.8
	Neighboring region (Afar)	28	9.93
	Total	282	100
Off-farm motivating factors	Proximity to urban area	51	18.09
	Education level and Excess labor at home	12	4.25
	Small land size possessed	30	10.64
	Access to Productive Safety Net Program	189	67.02
	Total	282	100
Inhibiting factors	Needed on farm	5	4.9
	Retired	12	11.77
	Shortage of startup capital	26	25.49
	No opportunity and far away to find work	59	57.84
	Total	102	100

3.3 Estimation results of probit regression model

Before rushing to regression result display, pair-wise correlation matrix and the Brush Pagan test were used to test the problem of multicollinearity and hetroscedasticity, respectively (Table 5).

Table 5. Determinants of off-farm participation (Probit Estimates) and marginal effect (Own Estimation Result, 2016).

Explanatory variable	Coef.	Std. Err.	z	P>|z|	Marginal Effect (dy/dx)
-Cons	0.1731	0.4400	0.39	0.694	-
GENDER	0.5874	0.2034	2.89	0.004***	0.1790
AGE	-0.0169	0.0074	-2.31	0.021**	-0.0461
EDUC	0.5120	0.2085	2.46	0.014**	0.1275
WORKING	0.2508	0.0716	3.50	0.000***	0.2672
NUMBDEPEN	-0.1184	0.0850	-1.39	0.164	-0.0317
LANDSZ	-0.3050	0.1514	-2.01	0.044**	-0.0817
TLU	-0.0278	0.0320	-0.87	0.385	-0.0074
PACKANIM	0.2942	0.1469	2.00	0.045**	0.0788
DISTMRKT	-0.0028	0.0252	-0.11	0.910	-0.0010
Dummylocation1(Bala-Ulaga)	0.7205	0.3988	1.81	0.071*	0.1599
Dummylocation2 (Kukufito)	-0.7449	0.2419	-3.08	0.002***	-0.2105
Dummylocation3 (Lemeat)	-0.2008	0.3242	-0.62	0.536	-0.0568
CREDIT	0.5857	0.2130	2.75	0.006***	0.1455
NFTGS	0.0302	0.3784	0.08	0.936	0.0082

Log likelihood = -163.49267
Number of obs = 384
LR chi^2 (13) = 117.58
Prob > chi^2 = 0.0000
Pseudo R^2 = 0.2645

*,**,***Significant at 10, 5 and 1 %, respectively.

Implication of gender on participation decision is positive and statistically significant at 1% level. Male headed households, citrus paribus, have 17.9% higher probability of participation than female headed households who are more probable to be sandwiched with child rearing practices. In fact, in the study districts, letting females to be a household head is not yet well developed and recognized. Consequently, female headed households mostly are those who are widowed and divorced; and hence, challenged by districts' labor division culture. In line with this, the prior proposed hypothesis is not rejected at 1% significance level. The finding corroborates with the findings of Haggblade et al. (2010) in North Africa, Latin America and West Asia; Babatunde and Qaim (2009) in Kwara State of Nigeria and Abebe (2008) in Ethiopia where by female-headed households are less probable to participate and influenced by cultural influences. However, it strongly disagrees with the findings of Merima and Peerlings (2012) in Tigray, Amhara, Oromia and Southern Nations Nationalities and Peoples regions of Ethiopia; Alaba and Kayode (2011) in South West Zone of Nigeria; and Kaija (2007) in rural parts of Uganda where female-headed households were better off-farm participants unlike male headed households.

Effect of age on participation is negative and statistically significant at 5% level that lets not to reject the prior proposed hypothesis. By its very nature, off-farm participation does require physical strength and fitness whereby younger farmers are better than older ones. Keeping other things constant, as age increases by a year, probability of participation would decrease by 4.61%. Old aged farmers really have an accumulated farm experience like plot preparation and in clearly identifying what and when to sow crops. Besides, they are too stable and mostly could not think some other off-farm activities and migration. Due to this, they are more probable to invest their time on farm activities that participating in off-farm activities.

Contrary to old aged farmers, younger farmers are improbable to be stable and be motivated to do their farm activities. They really wonder here and there in search of some lucrative off-farm activities. Typically, inter and intra migrations were their features and hence they engage in farm activities consciously to buy time. The finding is consistent with the findings of Babatunde and Qaim (2009) in Kwara State of Nigeria and Abebe (2008) in Ethiopia; and Kaija (2007) in Uganda in such a way that probability of participation decreases as age increases. Nevertheless, the finding contradicts with the finding of Zahonogo (2011) in Burkina Faso and Berg and Kumbi (2006) in Ethiopia where older farmers tend to participate better than middle or younger farmers.

The probit estimation result also reveals that, the effect of education (literate and illiterate) is positive and statistically significant at 1% level. In fact, majority of the sample households have engaged in food-for-work, as their typical and best off-farm activity. Food-for-work, therefore, did not require education and for that pointing finger signature is enough basically to take the salary either in cash or in food. Despite this, the implication here is that, the more farmers become literate the higher will be their probability of searching off-farm work in non-agricultural sectors. Magnitude of the positive sign suggests that literate households, keeping other things constant, have 12.75% higher probability of participation unlike their counter parts. The finding confirms the findings of Zhu and Luo (2006) cited in Babatunde and Qaim (2009) in such a way that the more literate the household is, the more probable to search and participate in profitable off-farm activities. Hence, the prior hypothesized positive coefficient is accepted at 1% level of significance.

Off-farm participation decision is positively related with number of working people or active labor force at home; and statistically significant at 1% level. Presence of larger number of working people in a certain house can really do and manage both farm and off-farm activities without compromising each of the activities. A unit increase in number of working people, citrus paribus, would raise the probability of off-farm participation decision by 26.7%. From this inference can be made that households with large active labor force would more be probable to participate in such a way that labor division would prevail more; and besides, enough labor force at home would let family members to participate in off-farm activities. The finding is consistent with the findings of Zahonogo (2011) in Burkina Faso. As a result of this, the prior indeterminate hypothesis is liable to be positive and statistically significant at 1% level.

The regression result indicated that, the effect of land size on off-farm participation decision is negative and statistically significant at 5% level of significance. As land size increases by one *Tsimad*, keeping other things constant, the probability of participation would decrease by 8.17%. In the study districts, households who owned an arable land size ranged in between 1 and 2 *Tsimad* are young farmers while the larger land share was positioned on modal age group interval of 40 to 56 years and the preceding age group interval of 57 to 73 since they have had a great share from the last regional land redistribution held twenty five years back.

Young age groups, therefore, are more liable to meager land size transfer from their fathers' and mothers' lottery, a single dead bodies and if possible, expanding to the frontier. In the study region, Tigray, for instance, 4,005 hectare was given for 15,198 youngsters through inheritance; and 27,924 hectare was divided among 23.6 thousand youngsters in the frontiers (Bureau of Agriculture and Rural Development (BoARD), 2011). It is possible, therefore, to deduce that when we come across higher age groups, it is fortunate to find higher land holding size. This large land holding size is found to be imperative for producing a relatively higher crop production that would retain them to do farm activities and not to participate in off-farm activities. Hence, it is possible to infer that, participation in off-farm activities was in response to farm land constraints. The finding is in line with the findings of Asenso-Okyere and Samson (2012) in Africa at large; Alaba and Kayode (2011) in South West Zone of Nigeria; Babatunde and Qaim (2009) in Kwara State of Nigeria and Abebe (2008) in Ethiopia. The prior hypothesized positive coefficient is rejected at 5% level of significance.

Presence of pack animals has positive relationship with participation and statistically significant at 5% significance level. The finding reveals that a unit increase in draft animals would raise the probability of participation by 7.88%. Farm households who possess pack animals are more probable to participate in activities like trading cereals, doing with cart, transporting sand and stone for construction purpose, charcoal production, fire wood selling and the like. This finding is similar with the findings of Abebe (2008) and Berg and Kumbi (2006) in Ethiopia as if households who possess pack animals would participate better than those who did not possess. Nonetheless, the finding strongly contradicts with the finding of Rios et al. (2008) whereby pack animals would transport manure from home to farm land and thereby increase crop production that finally decreases the probability of off-farm participation decision. The prior hypothesized positive coefficient is accepted at 5% significance level.

As a liquidity factor, access to credit is one best option whereby smallholder farmers could be instigated in diversifying their economic base. Access to credit has a positive effect on participation and statistically significant at 5% level of significance. In line with this, credit non-rationed farm households, keeping other things constant, have 14.55% higher probability of participation unlike credit rationed farmers. Hence, access to credit and taking credit influences participation, indicating that the more farmers have access to source of finance, the more likely to decide and participate in off-farm activities. The finding agrees with the findings of Adebiyi and Okunlola (2013) in Oyo State of Nigeria and Abebe (2008) in Ethiopia; as if taking credit is imperative for solving liquidity problem and thereby increases the probability of participation. On the other hand, the finding contradicts with the findings of Babatunde and Qaim (2009) in Kwara State of Nigeria where liquidity problems, while intending to participate in off-farm activities, could not be solved by accessing and taking credit. The prior proposed positive coefficient is not rejected at 5% level of significance.

Location can serve as a proxy of rain fall availability, productive potential of districts, sub-districts and villages; and it is worth to note discrepancy of participation among sub-districts. Consequently, sub-district dummies have been created; *Tao* was taken as a reference group due to its middle crop production position. Relative to *Tao*, theprobability of participation is higher in *Balla-Ulaga* sub-district by 15.99%; a sub-district basically with higher crop yield and total farm income and it is statistically significant at 10% level of significance. Compared with *Tao*, the probability of participation is much lesser in *Kukufito* sub-district by 21.05% that relatively is food-deficit and a sub-district typically credit rationed due to rules of their holly Qur'an and it is statistically significant at 1% level. Farmers with relatively better crop yield and total farm income, therefore, tend to participate in lucrative off-farm activities unlike farmers with low crop yield and farm income. Due to this, the finding completely disagrees with the findings of Zahonogo (2011) in Sudanese, Sahelian, North and South-Guinean zones of Burkina Faso and Abebe (2008) who found as participation is higher and lower in food-deficit and surplus regions, respectively.

It is believed that providing non-farm training is imperative for instigating farmers to participate and thereby diversifying their economic and income base. Although, the probit estimation result is not statistically significant, non-farm training has a positive linkage with off-farm participation decision where it gives an insight to provide and strengthen trainings pertaining to off-farm activities. In the study districts, there is a cross generational task of preparing cultural manifestation using animal skin as well as a culture of weaving (to prepare cultural dresses). Hence, there is a need to invigorate this cross generational endowment via providing an inclusive training pertaining to the aforementioned focus areas.

3.4 Impact of off-farm participation on crop yield

Table 6 depicts the impact of off-farm participation on crop yield considering Ordinary Least Square (OLS) regression model as a measuring instrument.

Table 5. Determinants of off-farm participation (Probit Estimates) and marginal effect (Own Estimation Result, 2016).

Explanatory variable	Coef.	Std. Err.	z	P>|z|	Marginal Effect (dy/dx)
-Cons	0.1731	0.4400	0.39	0.694	-
GENDER	0.5874	0.2034	2.89	0.004^{***}	0.1790
AGE	-0.0169	0.0074	-2.31	0.021^{**}	-0.0461
EDUC	0.5120	0.2085	2.46	0.014^{**}	0.1275
WORKING	0.2508	0.0716	3.50	0.000^{***}	0.2672
NUMBDEPEN	-0.1184	0.0850	-1.39	0.164	-0.0317
LANDSZ	-0.3050	0.1514	-2.01	0.044^{**}	-0.0817
TLU	-0.0278	0.0320	-0.87	0.385	-0.0074
PACKANIM	0.2942	0.1469	2.00	0.045^{**}	0.0788
DISTMRKT	-0.0028	0.0252	-0.11	0.910	-0.0010
Dummylocation1(Bala-Ulaga)	0.7205	0.3988	1.81	0.071^{*}	0.1599
Dummylocation2 (Kukufito)	-0.7449	0.2419	-3.08	0.002^{***}	-0.2105
Dummylocation3 (Lemeat)	-0.2008	0.3242	-0.62	0.536	-0.0568
CREDIT	0.5857	0.2130	2.75	0.006^{***}	0.1455
NFTGS	0.0302	0.3784	0.08	0.936	0.0082

Log likelihood = -163.49267
Number of obs = 384
LR chi^2 (13) = 117.58
Prob > chi^2 = 0.0000
Pseudo R^2 = 0.2645

*,**,***Significant at 10, 5 and 1 %, respectively.

As it has already been discussed earlier, the t-test statistics result inferred that yield difference between off-farm participant and non-participant is significant where non-participants produce better than participants. Consistent with the descriptive statistics, regression result revealed that off-farm participation has a negative and statistically significant impact on crop yield, at 1% level of significance. Hence, both descriptive and econometric statistics results purport the negative impact of off-farm participation on crop yield. The regression result shows that, keeping other things constant, off-farm non-participants were much better to get 1.99

quintal than their non-participant counterparts. Hence, off-farm participants produce lesser than non-participants. This could basically be due to their participation in non-lucrative off-farm activities that could add nothing to their total production. In this case food-for-work was the most important off-farm activity on which majority of the participants engaged on. Having this activity does not mean that they could be productive rather they spent half of their time by compromising their farm activities; and at the end of a month they will be given only birr 19 in cash or 3 kg(0.03 quintal) in kind which is very negligible. On the contrary, off-farm non-participants are believed to spend their time on farm activities and thereby produce better.

4. CONCLUSION AND POLICY IMPLICATION

One of the most important facets of off-farm activities is providing employment opportunities and additional income for rural households more particularly during the slack time. Besides, off-farm participation has a multifaceted effect on agricultural production in such a way that it paves the way for ease access of inputs and on the other hand it negatively shares the time to be allocated for farm activities. By and large, it is one of the best means of risk minimization. This research paper has tried to examine the underlying determinants of off-farm participation and the impact of participation on crop yield in Tigray region, Ethiopia. The probit regression result revealed that off-farm participation was positively influenced by gender, education, working people, number of pack animals, credit access and dummy location one, while age, land size and dummy location two influenced off-farm participation negatively. In a nut shell, even though some people do participate in off-farm activities on the basis of their internal motivation or demand driven, majorly a push driven participation has been investigated. Ordinary Least Square (OLS) regression result revealed that off-farm participation has a negative and statistically significant impact on crop yield where off-farm non-participants were found to be better producers unlike their counter parts.

Without the shadow of doubt, instigating off-farm participation is the only way-out and the sole alternative that smallholder farmers need to follow basically to accommodate their seasonal and fluctuating agricultural production. In doing so, an inclusive non-farm activities training need to be given for the community like on weaving, leather and leather products training like shoe and mat making, leather strap; carpentry and some other activities. Hence, investment on human resource actually on to education is highly required (both short-term and long-term trainings). The current adult education, being propagating, need to be strengthened since literate households (those who simply can read and write) were found to be better participants unlike illiterate farmers. On the other hand, access to credit need to be propagated basically to solve liquidity problems and thereby serve as off-farm startup capital.

REFERENCES

1. Abebe DB (2008). Determinants of off-farm participation decision of farm households in Ethiopia. Agrekon 47:1.
2. Adebiyi S, Okunlola JO (2013). Factors Affecting Adoption of Cocoa Farm Rehabilitation Techniques in Oyo State of Nigeria. World J. Agric. Sci. 9(3):258-265.
3. Adewunmi IO, Awoyemi TT, Omonona BT, Falusi AO (2011). Non-farm income diversification and poverty among rural farm households in Southwest Nigeria. Eur. J. Soc. Sci. 21:1.
4. Alaba AO, Kayode SK (2011). Non-farm income diversification and welfare status of rural households in South West Zone of Nigeria. Conference paper presented for IFPRI.
5. Asenso-Okyere K, Jemaneh S (2012). Increasing Agricultural Productivity & Enhancing Food Security in Africa: New Challenges & Opportunities. Synopsis of an International Conference, International Food Policy Research Institute Washington, Dc.
6. Babatunde RO, Olagunju FI, Fakayode SB, Adejobi AO (2010). Determinants of Participation in Off-farm Employment among Small-holder Farming Households in Kwara State, Nigeria; PAT December. 6(2):1-14.
7. Babatunde RO, Qaim M (2009). The role of off-farm income diversification in rural Nigeria: Driving forces and household access. Paper Presented at the center for the study of African Economies (CSAE) conference, Economic Development in Africa, March 22-24, University of Oxford, United Kingdom.
8. Berg M, Kumbi G (2006). Poverty and the Rural Non-Farm Economy in Oromia, Ethiopia', International Association of Agricultural Economists Conference, Gold Coast, Australia, August 12-18. Int. Assoc. Agric. Econ. pp. 1-17.
9. BoARD (2011). Government of National Regional State of Tigray Five Years (2010/11 - 2014/15) Growth & Transformation Plan, Mekelle, Tigray, government printer.
10. CSA (2007). The Population and Housing Census of Ethiopia. Addis Ababa, Ethiopia government printer.
11. Dercon S, Woldehanna T (2005). Shocks and Consumption in 15 Ethiopian Villages, 1999-2004'. Afr. Econ. 14:559-585.
12. Diao X, Nin Pratt A (2007). Growth options and poverty reduction in Ethiopia –An economy-wide model analysis. International Food Policy Research Institute, Development Strategy and Governance Division, 2033 K St., NW, Washington, DC 20006, USA; Food Policy 32 (2007) 205–228 available online at www- siencedirect.com
13. Haggblade S, Hazell PBR, Reardon T (2007). Transforming the Rural Nonfarm Economy: Opportunities and Threats in the Developing World. (1st edn). Baltimore: International Food Policy Research Institute.
14. Haggblade S, Hazell PBR, Reardon T (2009). Transforming the Rural Non- Farm Economy: Opportunity and Threats in the Developing World', IFPRI Issue Brief No. 58. Washington DC. International Food Policy Research Institute.
15. Haggblade S, Hazell PBR, Reardon T (2010). The Rural Nonfarm Economy: Prospects for Growth and Poverty Reduction. World Dev. 38(10):1429-1441.
16. Israel GD (1992). Determining Sample Size. Program Evaluation and Organizational Development, IFAS PEOD-6, Series of Agricultural Education and Communication Department, Florida Cooperative Extension Service, Institute of Food and Agricultural Sciences, University of Florida; available at EIDS web site at

17. Kaija D (2007). Income Diversification and Inequality in Rural Uganda: The Role of Non-Farm Activities. A paper prepared for the Poverty reduction, Equity and Growth Network (PEGNeT) Conference, Berlin, September 6-7, 2007.
18. Merima A, Peerlings J (2012). Farm households and nonfarm activities in Ethiopia: does clustering influence entry and exit? Agric. Econ. 43:253-266.
19. MoARD (Ministry of Agriculture and Rural Development) (2010). Federal Democratic Republic of Ethiopia, Ethiopia's agriculture sector policy and investment framework: Ten year road map (2010-2020). Final Report
20. Rios AR, Masters WA, Shively GE (2008). Linkages between Market Participation and Productivity: Results from a Multi-Country Farm Household Sample. Prepared for presentation at the American Agricultural Economics Association Annual Meeting, Orlando, Florida, July 27-29, 2008.
21. Shumet A (2011). Analysis of technical efficiency of crop producing smallholder farmers in Tigray, Ethiopia. MPRA Paper No. 40461, posted 6. August 2012 13:41 UTC Available online at
22. Shumet A (2011). Analysis of technical efficiency of crop producing smallholder farmers in Tigray, Ethiopia. MPRA Paper No. 40461, posted 6. August 2012 13:41 UTC Available online at
23. World Bank (2008). The Growth Report: Strategies for Sustained Growth and Inclusive Development. Washington D.C., Commission on Growth and Development, World Bank or The International Bank for Reconstruction and Development / The World Bank
24. Woinshet A (2010). Participation into off-farm activities in rural Ethiopia: who earns more? MA Thesis, Graduate School of Development Studies, The Hague, The Netherlands.
25. Woldehanna T (2002). Rural farm/non-Farm Income Linkages in Northern Ethiopia', in B. Davis, T. Reardon, K. Stamoulis and P. Winters (eds) Rome: FAO pp 121-144.
26. Zahonogo P (2011). Determinants of non-farm activities participation decisions of farm households in Burkina Faso. J. Dev. Agric. Econ. 3(4):174-182.

CHAPTER 18

Determinants of small-scale mango farmers' market channel choices in Kenya: An application of the two-step Cragg's estimation procedure

Davis Nguthi Muthini*, Rose Adhiambo Nyikal and David Jakinda Otieno

Department of Agricultural Economics, University of Nairobi, P.O Box 29053-00625, Nairobi, Kenya.

ABSTRACT

The study estimates small-scale mango farmers' choice of market channels using the Cragg's two-step procedure where the farmer decides on the channel in the first step and the proportion sold to the selected channel in the second step. Cross section data was collected from a sample of 224 mango farmers selected through multistage sampling just after the mango season. The study was carried out in Makueni County in Eastern Kenya. The county is leading in production of mangoes in Kenya, having produced over 146,000 tonnes valued at over 18 million US dollars, in 2015. The data was analyzed using Cragg's two step regression model. The first step assessed factors that determine choice of a particular channel, while the second step assessed factors that influence the proportion of produce sold to the channel. Results show that socio-economic factors significant in the first stage are not necessarily significant in the second stage. In some cases, the direction of effect reverses. Factors such as distance to tarmac road, number of mango trees in the farm, membership in producer marketing groups, training in mango agronomy, and access to extension services affect choice of export market channel. Only membership to mango marketing groups significantly influences proportion sold. Household income, distance to tarmac, number of trees, market information, and gender significantly affect choice of the direct market channel. The direct market channel earns farmers the largest margins, followed by the export channel. However, majority of farmers sell to brokers followed by export channel. It was found that despite being aware that they could fetch higher prices through direct selling, they lacked financial capacity, transport resources, and information on market locations and requirements. Policies need to enhance financial capacity of farmers, as well as expand efforts to disseminate timely and accurate market information.

Keywords: Small-scale farmers, mango market channels, Kenya.

1. INTRODUCTION

Marketing plays a critical role both in stimulating production and accelerating the pace of economic development. In Kenya, marketing chains for agricultural commodities are generally not transparent and consist of many players, making them inefficient and unresponsive to producer needs (Government of Kenya, 2010). Despite recent urbanization and supermarket revolution which is creating market for horticulture farmers in Africa, Neven et al. (2009) found that majority of supermarket suppliers are not small scale poor farmers, but an emergent crop of educated farmers who own commercial medium sized farms. Olwande et al. (2015) found that market access in Kenya has improved over the last decade, but market participation has remained relatively stagnant in most sub-sectors, an evidence of subsistence agriculture.

There exist in literature a number of studies that analyze farmer choice of market channel. Bogiwe and Masuku, (2012) found that factors such as the age of the farmer, quantity produced, and education, significantly influenced choice of market channel among corn farmers in Swaziland. However, unlike the Swaziland market where the government is a player through the National Agricultural Marketing Board, the Kenyan mango market is fully liberalized, with the government only playing regulatory role. There exist mixed results on the impact of liberalization on access to markets by farmers (Kherallah et al., 2002), and thus not factual to generally apply the results.

In liberalized markets, individual farmers lose bargaining power, and thus are routinely exploited by buyers because they are price-takers in the absence of agreements (Koning and Anderson, 2007). It is on the basis of such that recent literature that support marking is biased towards farmer collective action to access high value markets for cash crops (Okello et al., 2007; Rao and Qaim, 2011; Fischer and Qaim, 2012). The possibility of accessing high value markets as a result of collective action is not universal; other factors significantly affect access to different markets.

The choice of channel to sell is not mutually exclusive; farmers sell to more than one channel within the same season for the same crop. Majority of the studies that assess market channel choice have one fundamental weakness in that they fail to acknowledge this common phenomenon of agricultural marketing in developing countries. Farmers would probably have a preferred main channel that they sell a larger proportion of the produce to. In such a case therefore, multinomial logit or tobit models commonly used would not be appropriate because they assume mutual exclusivity between the channels, and that the effect of the independent variables on choice of channel, and quantity of produce sold to the channel, is similar. This assumption is not true, as Katchova and Miranda (2004) found out, a variable that increases the probability of choosing a particular channel does not necessarily influence the quantity sold to that channel. In this study, we adopt a two-step procedure proposed by Katchova and Miranda (2004). The procedure captures farmer characteristics that influence choice of channel in a probit model in the first step, and characteristics that influence the quantity sold using a truncated regression model in the second step.

The study was carried out in Makueni County which is located in the semi-arid south eastern part of Kenya. The County experiences bi-modal rainfall, with the lower side receiving little rainfall ranging from 300 to 400 mm and the higher areas receiving. Similarly, the high altitude areas experience temperatures ranging from 20.2 to 24.6°C, while in the low-lying areas temperatures can exceed 30°. The Kenya Agriculture and Livestock Research Organization introduced improved mango varieties in in the area 15 years ago due to its climatic adaptability. Mangoes can thrive in low rainfall (500 to 1000 mm) and a wide range of temperatures (10 to 42°C) which makes it suitable even for the arid and semi-arid lands. Makueni County is one of the leading mango producing areas in Kenya, with an annual estimated value of Kshs. 1.2 billion (1USD$ = Kshs100 on average) (Agricultural Business Development, 2011).

The county however, has relatively high poverty levels at 64%, compared to a national average of approximately half of the population (Kenya National Bureau of Statistics, 2009). Improving market access for mango farmers is therefore critical to reducing poverty levels. Marketing of mangoes is not organized, it is estimated that margin to mango farmers is very low, at Kshs.

1.70 per fruit in some channels, while post-harvest losses could be up to 30%, which is a disincentive to production (Agricultural Business Development, 2011). The losses are exacerbated by the perishable nature of mangoes.

According to Tsourgiannisa et al. (2008), the marketing channel used has a bearing on the profit farmers may make. For most small scale farmers dealing with perishable products, a decision between selling to the most profitable channel, and having to sell to the easily available buyer to meet urgent financial needs or avoid post-harvest losses has to be made. It is not clear what drives the decisions on the choice of marketing channels and the economic implications for the farmers. The purpose of this study therefore was to evaluate the factors that influence the mango farmers' choice of market channel. This study contributes to the existing literature by providing new insights into how smallholder farmers decide on market channels and quantities sold in those channels, and how such decisions interact with factors such collective action, income, and perishability of agricultural produce.

2. MATERIALS AND METHODS

2.1 Sampling and data collection

Respondents were selected through multistage sampling techniques. In the first stage, 3 locations (Kilili, Mumbuni and Kilala) were selected purposively due to high volumes of mango produced. The villages, from which respondents were interviewed (Table 1), were selected based on two criteria; level of market organization and access to market, following Omiti et al. (2006).

The level of market organization was based on membership of the farmers in marketing groups, while distance to the nearest tarmac road was used as a proxy for access to market. Villages were classified as high on low regarding market organization and market access. The number of farmers interviewed from each village was based on the estimated total number of mango farmers in the respective village. Respondents were selected through systematic random sampling. Logistic regressions require larger samples than linear regression. According to Schwab (2002), the minimum number of cases per independent variable required in logistic regression is 10; the current study used 20 cases to 1. With 11 independent variables, a minimum 220 cases were required, the study proposed 240 cases to cater for non-response and incomplete questionnaires. Following Kothari (2004), systematic sampling was used to select the respondents. The nth farmer (where n = 3) was selected along the determined routes with a random start in each of the villages to give a total of 227 respondents as illustrated in Table 1. Data was collected using both qualitative and quantitative methods. Quantitative data was collected using both open and closed ended questionnaires administered by trained enumerators. Data was collected in the month of May, 2014, immediately after the peak mango harvesting season that spans December to March.

Table 1. Village selection matrix.

High organization and low access	No. of respondents	Location
Kathatu	13	Kilili
Kilumbu	20	Kilili
Kavuliloni	8	Kilili
Mulenyu	12	Kilili
Itaa	13	Mumbuni
Low organization and low access		
Kilanga	12	Mumbuni
Kisuu	12	Mumbuni
Kithiani	18	Mumbuni
Mboani	15	Kilili
Wee	20	Kilili
Low organization and high access		
Muselele	12	Kilala
Itangini	6	Kilala
Nduundune	12	Kilala
Kaseve	16	Mumbuni
High organization and high access		
Nzueni	16	Mumbuni
Kyumu	9	Kilala
Ngutw'a	13	Mumbuni

Source: Survey data (2014).

2.2 Theoretical framework

This study is based on the random utility model, which assumes that a decision maker, faced with a set of alternatives, will select the alternative that offers the highest utility (Greene, 2007). Suppose an individual *i* is faced with two choices *a* and *b* with utilities u_a and u_b respectively (Equation 1 and 2).

$$U_a = w'\beta_a + z_a'\gamma a + \varepsilon_a \quad (1)$$

$$U_b = w\beta_b + z_b'\gamma b + \varepsilon_b \quad (2)$$

Where, w represents the observable characteristics of the individual, such as age, income, and other demographics. The vector z denotes choice specific attributes of the two choices. The random terms, ε_a and ε_b, denote individual specific stochastic elements not be known to the researcher. If the individual's choice of alternative a is denoted by Y=1, then $u_a>u_b$, which follows:

$$\text{Prob}[Y=1|w,z_a,z_b]=\text{Prob}[u_a>u_b] \quad (3)$$

$$=\text{Prob}[x'\beta+\varepsilon>0|X] \quad (4)$$

Where $X'\beta$ are the observable elements of the difference of the two utility functions and ε represents the difference between the two random elements.

The choice of channel to sell is not mutually exclusive; farmers sell to more than one channel within the same season for the same crop. Majority of the studies that assess market channel choice have one fundamental weakness in that they fail to acknowledge this common phenomenon of agricultural marketing in developing countries. Farmers would probably have a preferred main channel that they sell a larger proportion of the produce to. In such a case therefore, multinomial logit or tobit models commonly used would not be appropriate because they assume mutual exclusivity between the channels, and that the effect of the independent variables on choice of channel, and quantity of produce sold to the channel, is similar. This assumption is not true, as Katchova and Miranda (2004) found out, a variable that increases the probability of choosing a particular channel does not necessarily influence the quantity sold to that channel. In this study, we adopt a two-step procedure proposed by Katchova and Miranda (2004). The procedure captures farmer characteristics that influence choice of channel in a probit model in the first step, and characteristics that influence the quantity sold using a truncated regression model in the second step. The study also applied the one step tobit model for comparison with the two step Cragg's procedure as discussed subsequently.

2.3 One-step Tobit model

Tobit model was used to analyze the effect of independent variables on the dependent variable as there were numerous zero occurrences and corner solutions where the respondent did not sell to a particular channel (Wooldridge, 2002). For a specific respondent therefore, given , the y^* is 0 if the farmer does not sell to that particular channel. Alternatively, if the farmer sells all produce to only one channel, then y^* is 1. The dependent variable is therefore censored from above (1) and below (0). Separate models were regressed for each channel.

Following Cogg (2000), the Tobit model is described as:

$$y_i^* = \begin{cases} a & if \quad y_i \leq a \\ y_i & if \; a < y_i < b \\ b & if \quad y_i \geq b \end{cases} \tag{5}$$

Given that *a* and *b* are the lower and upper limits respectively, which in this study is 0 and 1, then y_i*= 0 when $y_i < 0$; y_i*= $y_{i\ when}$ $0 < y_i < 1$; and Y_i* =1 when *yi* ≥1 The latent regression model can be expressed as $y = x\beta + \varepsilon$, whereby *y* is a continuous outcome variable and β is the change in the mean of the latent dependent variable, that is, $\beta - \partial E(y_i) / \partial x_{i.}$ The Tobit model assumes that the error term is normally distributed and homoscedastic; $\varepsilon \approx N(0, \delta^2)$ The marginal effects are reported at the means of the independent variable, for dummy variables the marginal effects represent net effect on the dependent variable when the independent variable changes from 0 to 1. Since the $E(y|x)$ is linear in *x*, then the marginal effect of *xj* on *y* is simply the corresponding β_j

2.4 Two-step procedure

The study used the Cragg's "two-tier" alternative to Tobit for corner-solution models (Cragg, 1971). The model applies the probit model in the first stage and the truncated normal regression in the second stage. The probit regression models a discrete decision, probability of a farmer choosing any particular channel, whether or not *y* is *0* or positive:

$$\text{Prob}(y > 0) = \Phi(x\beta)$$

Whereby Φ is the symbol for normal distribution.

After a farmer makes a decision to sell to a particular channel, then he/she has to decide what quantity to sell to that particular channel, the quantity sold was used to calculate the proportion for each channel. The truncated normal regression was used to assess the factors that influence the proportion sold to that particular channel for the farmers who sold a positive quantity:

$$E(y|y > 0) = x'\beta + \delta\lambda\left(\frac{x'\beta}{\delta}\right) \tag{6}$$

The term $\frac{x'\beta}{\delta}$ is an adjustment factor because the respondents with zero values have been dropped in the truncated model. The 2-step procedure is better than the one stage tobit because the effect of an independent variable on the probability of choosing the channel, and effect of the quantity sold to the particular channel are determined in separate processes. We also test the one step tobit against the Cragg's 2-step procedure as found in Katchova and Miranda (2004):

$$\lambda = 2(\ln L_{probit} + \ln L_{truncatedegression} - \ln L_{Tobit}) \tag{7}$$

Whereby is distributed as chi-square with K degrees of freedom. The Cragg's 2-step model is preferred if is greater than chi-square critical value.

3. RESULTS AND DISCUSSION

3.1 Sample characteristics

Some important features of mango farming are presented in Table 2. The average number of years spent in school by mango farmers was 10.6, thus a majority of respondents had gone through primary school (8 years). Education is expected to influence a household's understanding of markets, and consequently decisions on where to sell. Previous studies have found that majority of supermarket suppliers are educated farmers who own commercial medium sized farms (Neven et al. 2009; Sebatta et al., 2014).

Table 2. Summary of sample characteristics.

Variable (n=227)	Min	Max	Mean	Std. deviation
Number of years completed in school	0	18	10.06	3.72
Number of total family members	1	12	6.2	2.243
Distance to the nearest market center in km	0.1	22	4.933	5.36684
Distance to nearest tarmac road in km	0.1	27	11.324	7.58646
Total land size in acres owned	0.75	64	7.3427	8.34738
Land size in acres under mangoes per farmer	0.2	15	2.3379	2.1719
Age of the farmer in years	26	84	49.9	11.602
Years farmer has been growing mangoes	3	30	10.18	4.536
Total mango trees per farmer	12	1900	184.26	202.3
Percent of male respondents				86.8
Percent of farmers who had signed contract with buyers				12.8
Percent of farmers who are members of mango marketing group				45.8
Percent of farmers with access to market information				40.1
Percent of farmers visited by extension officer in last three years				47.1
Percent of farmers who have received training on mangoes				64.3

Source: Survey data (2014).

Majority of decision makers among mango farmers in the study area are male. This is more than proportion of households headed by females estimated by Kenya National Bureau of Statistics (2009). According to Agricultural Business Development (2011), the low number of women in mango farming is occasioned by the prevailing socio cultural factors where land and permanent crops are generally owned by men.

Only less than half of the farmers were members of any mango marketing groups. This shows a very disadvantaged position of the farmers considering the importance of collective action in marketing such as: Increasing farmers bargaining power, reducing transaction costs and improving information flows (Fafchamps and Hill 2005; Shiferaw et al., 2006; Kirsten et al., 2008).

Less than a quarter of the respondents had pre-agreed contracts with buyers, majority of whom sold to the export channel. Most of the contracts were agreed between groups and main buyer as opposed to individual farmers. Having a contract was howevermore of a consequence of the channel selected rather than a determining factor. Furthermore, the contracts did not seem to be binding, as the buyers would occasionally change prices and other terms of the contracts.

Nearly two thirds of farmers had received training on various aspects of mango farm management comprising grafting, pruning, pest and disease control, marketing, post-harvest management and record keeping. Only about a quarter of farmers had been able to access credit; among the different channels. The low figure was attributed to lack of information on the sources and availability of credit, and fear of risk.

More than half of the respondents mainly sold to middlemen, commonly known as brokers, while only slightly more than ten percent sold directly to the market. Only channels used as the main outlets are presented here. Some farmers would sell most of the mangoes to these main outlets and the rest, which is a negligible quantity, to rural retailers or brokers. Figure 1 presents distribution of farmers based on market channels targeted.

High prices were however not tantamount to high profitability because of the difference in costs incurred in accessing the channels. One way Analysis of Variance (ANOVA) was used to test for differences between the net prices offered by the channels, results are presented in Table 3. With $p = 0.000 < 0.05$ (Table 3), the hypothesis that there is no difference in prices offered by the different channels was rejected.

Results for gross margins across channels are presented in Table 4. Farmers who sold to the direct channel earned the highest gross margins, while those selling to brokers earned the lowest gross margins. Farmers who sold directly to the market incurred very high marketing costs in transport and storage, the price they earned was however sufficient to cover the cost and earn the farmer a significant profit. Farmers who sold to brokers incurred the lowest costs because brokers moved around in the farms in search of mangoes. These farmers incurred very low costs of searching for buyers, entering contracts, being in groups, transport, and monitoring as compared to those who sold to exporters and direct market. Despite the apparent advantage of selling to the direct market, only a minority of the farmers sold to this channel.

Table 3. One way ANOVA results for average net prices between and within groups.

Group	Sum of squares	Df	Mean Square	F	P-value.
Between groups	82.862	33	2.511	16.482	
Within groups	29.402	193	0.152		0.000***
Total	112.264	226			

***, **, * significance levels at 1, 5 and 10%, respectively.

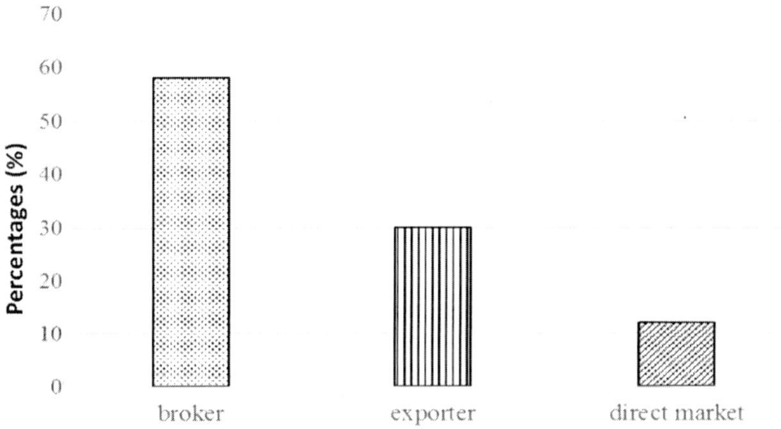

Figure 1. Distribution of farmers in market channels.Source: Survey data (2014).

3.2 Empirical results

Parameter estimates (marginal effects and coefficients) for the two-step Cragg's model is presented in Tables 4 to 6. The hypothesis for the log likelihood test for the Tobit versus the two-stage Cragg's model is rejected and thus the two stage model is preferred.

Table 4. Parameter estimates for the first-stage probit model

Variable	Export	Broker	Direct market
Formal education	-0.0004 (0.0340)	-0.0372 (0.0405)	0.0550 (0.0436)
Off-farm occupation (dummy)	0.2979 (0.2639)	-0.6202** (0.2959)	0.1726 (0.3235)
Household income	0.0608 (0.1145)	-0.0473 (0.1252)	0.3570** (0.1388)
Distance to tarmac road	0.0025*** (0.0005)	0.0004 (0.0007)	-0.0014* (0.0008)
Total mango trees	0.6423*** (0.1617)	0.1421 (0.1654)	0.3546* (0.1825)
Group membership (dummy)	0.4954** (0.2335)	-0.2537 (0.2698)	0.1666 (0.2889)
Market information (dummy)	-0.4017 (0.2495)	-0.5035** (0.2540)	0.7907*** (0.2784)
Agronomy training	0.1095** (0.0553)	0.0263 (0.0702)	-0.0671 (0.0733)
Extension	0.3967*** (0.1164)	0.1988 (0.1270)	-0.1167 (0.1355)
Own Vehicle (dummy)	-0.0447 (0.3268)	-0.3029 (0.3157)	0.4160 (0.3315)
Experience	0.0380 (0.0256)	-0.0000 (0.0297)	-0.0218 (0.0354)
Gender (dummy)	-0.1098 (0.3719)	0.8680*** (0.3328)	-0.7806** (0.3476)
Cons	-5.6808*** (0.9124)	1.9528** (0.8135)	-0.2629 (1.8524)
Log likelihood	-87.16	-67.24	-57.17
LR Chi^2	106.99	34.31	54.45
Prob > Chi^2	0.0000	0.0006	0.0000

The probit model results show that household income is significantly linked with a higher probability of choosing the direct market channel. This could be attributed to the ability of relatively wealthy farmers to pay for the transport and storage costs associated with direct marketing, as well as the risk of incurring loss. Fafchamps and Hill (2005) found mixed results among coffee farmers in Uganda; that wealthy farmers were less likely to sell directly to the market, except when they have large quantities. The truncated regression model results appear to corroborate Fafchamps and Hill (2005), although not significant, the negative effect means that increasingly smaller quantities are sold through the direct channel as incomes increase for farmers that select the channel. For farmers who sell to brokers, the quantity sold increases as household income, which is expected as poor farmers lack ability to target alternative channels that pay premium prices.

Contrary to expectations, distance from tarmac road is positively associated with a higher likelihood of farmers selling to export market and lower likelihood of selling directly to the market. This may be because farmers farther from the tarmac road were more likely to have large farms which exporters prefer because of economies of scale. Martey et al. (2012) found that farmers in Nigeria who were located far from the tarmac roads were more likely to participate in marketing cooperatives, which exporters prefer to buy from as opposed to individual farmers. A Pearson correlation test proved that there was a significant positive relationship between distance to tarmac road and membership to mango marketing group. Farmers close to the tarmac road were more likely to sell to direct market as expected. The direction of effect does not change for the truncated regression model. As distance from the tarmac increases, the quantity sold to brokers decreases, an indication that brokers venture only to a certain extent in the rural areas, possibly to reduce the associated transport and bulking costs.

Table 5. Parameter estimates for the second stage truncated regression model.

Variable	Export		Broker		Direct market	
	Truncated regression	Tobit	Truncated regression	Tobit	Truncated regression	Tobit
Formal education	0.0103 (0.0116)	0.0095 (0.0179)	0.0045 (0.0056)	-0.0079 (0.0196)	-0.0682* (0.0359)	0.0731 (0.0719)
Off-Farm occupation (dummy)	0.0371 (0.0876)	0.1542 (0.1343)	0.0158 (0.0440)	-0.2271 (1467)	0.0878 (0.1578)	0.4756 (0.5117)
Household Income	-0.0030 (0352)	0.0120 (0.0568)	-0.0620*** (0.0185)	-0.1375** (0.0629)	-0.03[illegible]6 (0.0835)	0.4929** (0.2396)
Distance to tarmac road	0.0002 (0.0002)	0.0012*** (0.0003)	-0.0003*** (0.0000)	-0.0006* (0.0003)	-0.00[illegible]0 (0.0006)	-0.0027* (0.0014)
Total Mango trees	0.0239 (0.0524)	0.3109*** (0.0814)	-0.0269 (0.0231)	-0.1061 (0.0809)	-0.0464 (0.1123)	-0.5234* (0.3038)
Group Membership (dummy)	0.1348 (0.0854)	0.1776 (0.1220)	-0.0062 (0.0380)	-0.1160 (0.1335)	0.1048 (0.1806)	0.2335 (0.4636)
Market Information (dummy)	0.1114 (0.0833)	-0.1376 (0.1241)	-0.0542 (0.0378)	-0.3217** (0.1314)	0.2073 (0.1998)	1.2669** (0.5204)
Agronomy training	0.0259 (0.0203)	0.0708** (0.0293)	-0.0205** (0.0094)	-0.0328 (0.0330)	-0.0458 (0.0492)	-0.1053 (0.1190)
Extension	-0.0074 (0.0401)	0.1327** (0.0580)	-0.0388** (0.0176)	-0.3011 (0.0622)	-0.2225*** (0.0852)	-0.2911 (0.2252)
Own Vehicle (dummy)	-0.0076 (0.1028)	-0.0832 (0.1614)	-0.0002 (0.0533)	-0.1876 (0.1745)	0.3545** (0.1812)	0.7076 (0.5426)
Experience	0.0066 (0.0077)	0.0262** (0.0128)	-0.0093** (0.0042)	-0.0131 (0.0144)	-0.0428* (0.0221)	-0.0530 (0.0585)
Gender (dummy)	-0.0657 (0.1164)	-0.0924 (0.1823)	-0.0020 (0.0620)	0.4680** (0.1938)	0.1453 (0.1805)	-1.1954** (0.6049)
Cons	-0.1722 (0.3113)	-2.90611*** 0.4880)	1.3257*** (0.1226)	2.2883*** (0.4301)	2.2296*** (0.5342)	-0.0560 (1.3787)
Log likelihood	-4.114	-122.20677	3.004	-181.34805	-1.9666	-74.64
Wald Chi2/LR Chi^2	11.75	101.42	92.32	68.50	15.75	56.4
Prob > Chi2	0.4663	0.0000	0.0000	0.0000	0.2030	0.0000
Log likelihood test for tobit vs truncated regression test	**38.816**		**100.059**		**23.525**	

Table 6. Marginal effects for the second-stage truncated regression model.

Variable	Export	Broker	Direct market
Formal education	0.0090 (0.0102)	0.0045 (0.0055)	-0.0605** (0.0311)
Off-farm occupation (dummy)	0.0325 (0.0767)	0.0157 (0.0438)	0.0779 (0.1404)
Household income	-0.0027 (0.0308)	-0.061*** (0.0184)	-0.0280 (0.0741)
Distance to tarmac road	0.0002 (0.0001)	-0.0003*** (0.0000)	-0.0009 (0.0006)
Total mango trees	0.0210 (0.0459)	-0.0268 (0.0230)	-0.0411 (0.0999)
Group membership (dummy)	-0.1181 (0.0745)	-0.0061 (0.0378)	0.0930 (0.1606)
Market Information (dummy)	0.0975 (0.0727)	-0.0539 (0.0376)	0.1839 (0.1762)
Agronomy training	0.0226 (0.0177)	-0.0204** (0.0094)	-0.0407 (0.0431)
Extension	-0.0065 (0.0350)	-0.0386** (0.0175)	-0.1974*** (0.0725)
Own vehicle (dummy)	-0.0067 (0.0900)	-0.0002 (0.0530)	0.3145** (0.1567)
Experience	0.0058 (0.0068)	-0.0092** (0.0042)	-0.0380** (0.0194)
Gender (dummy)	-0.0575 (0.1019)	-0.0020 (0.0616)	0.1289 (0.1594)

Standard errors are shown in parenthesis. ***, **, * significance levels at 1, 5 and 10%, respectively. Source: Survey data (2014).

The number of mango trees is significantly associated with higher probability of selling to the export channel and less to direct market. Apart from being a perishable product, mangoes are very seasonal, buyers prefer unripe mangoes to avoid losses hence those with large quantities avoid transporting to the market because of their low capacity to sell at the market which is not as organized as farm-gate buyers. Martey et al. (2012) further explains that farmers who produce more prefer selling at their immediate market (farm gate) to avoid post-harvest loss, especially if the product is highly perishable, as is the case with mangoes. In addition, Shilpi and Umali-Deininger (2008) noted that farmers incur very high transaction cost selling to the market as compared to traders especially if the physical market infrastructure is poor. Poor market infrastructure results in long waiting periods, which will affect farmers more negatively than traders who may have better access to facilities even in congested markets, thus farmers with large quantities may be discouraged to sell directly in the market. As found by this study, they will sell at the farm gate but they are more likely to sell to exporters at the farm gate and not brokers. Although not significant, the direction of the effect does not change for the truncated regression model.

Membership to a mango marketing group is associated with increased likelihood of a farmer selling to the export channel as opposed to the broker channel. The importance of collective action has been emphasized in the literature (Kirsten et al., 2008; Fischer and Qaim, 2012). Farmers in groups have the advantage of bulking hence gaining economies of scale. It is also easier and cheaper for exporters to enforce quality and grade requirements of the export market through reaching farmers in groups rather than individual farmers.

Access to market information is significantly associated with farmers selling to direct market channel. Conversely, farmers who lacked market information were more likely to sell to brokers. A significant proportion of farmers in the study area were found not to actively search for market and market information, and are thus price-takers. However, based on the results of the truncated regression model, access to market information does not affect the quantity sold to the direct market.

Having access to training is significantly associated with higher probability of selling to the export channel. As explained earlier, the export channel only buys grade one mangoes of high quality. Only farmers who had gone through training were able to attain these quality requirements. Similarly increased contact with extension service providers increased farmer's likelihood of selling to the export channel. The results from the two variables can be attributed to the farmer's ability to produce quality mangoes not damaged by pests and diseases. Similarly, lack of access to training and extension services significantly affected positively the quantity of produce sold to brokers.

Although ownership of a vehicle is not significantly associated with a higher probability of selecting any of the market channels, it is positive and significant for the direct channel in the truncated regression model. Owning a vehicle increased the quantity sold directly sold to the market by 31 percent, which is substantial. Lack of transport to the market and the associated cost and risk even where available was a major deterrent in accessing the direct channel. As observed by Panda and Sreekumar (2012), farmers' own vehicles allow them to access marketing centers located far off at a lower cost and within a shorter period as compared to their colleagues who had no transport means.

The number of years that a farmer had been growing mangoes does not significantly affect choice of market channel. However, for farmers selling to brokers, less experience is associated with selling a larger proportion to brokers. This implies that these farmers who were presumably new in the sector had not created market networks with other buyers. The proportion of produce sold to brokers increased by 0.93% for every year of experience less.

Farmers with an off farm source of income were less likely to sell to brokers. This is because these farmers were probably not extremely cash-constrained and therefore could delay sales and seek for better prices from other channels. As discussed earlier, poor farmers sell to brokers at low prices to meet urgent financial needs. Female farmers were more likely to sell directly to the market and less likely to brokers. Women engaged more in marketing activities of mangoes in markets, especially transporting to and selling small quantities to nearby markets. However, the direction of the effect reverses for the truncated regression model results, implying that quantities sold to this channel increased when the farmer was male.

The Wald chi-square, which is used to test the hypothesis that at least one of the predictors' regression coefficient is not equal to zero, is less than the degrees of freedom for the export channel. The Prob>chi^2 is not significant thus as a whole, the model is statistically not significant. In other terms all the coefficients in the model are equal to zero and so none of the independent variables influences proportion of produce sold to export channel.

CONCLUSIONS AND POLICY IMPLICATIONS

This study analyzed choice of market channel as occurring in two steps; the farmer first decides on the channel to sell to, and in the second stage decides on the quantity to sell to the particular channel. The log likelihood test indicates that the two-stage Cragg's is preferred over the one-stage Tobit model. The two-stage model shows that interaction of variables is different between the first stage binary choice of channel and second stage with continuous dependent variable. Socio-economic factors that affect the first-stage do not necessarily affect the second stage. In some instances the direction of effect reverses in the second stage. The results therefore contrast the results when choice of market channel is modelled as one step, and provides more insights into the decision process.

For instance, whereas factors such as distance to tarmac, number of trees, membership in producer marketing groups, training, and access to extension services affect choice of export market channel, only membership to mango marketing groups significantly influences proportion sold. Household income, distance to tarmac, number of trees, market information, and gender significantly affect choice of the direct market channel. Variables that were found to be significant for choice of brokers channel are off farm income, market information, and gender. The proportion of produce sold to brokers increases with lower household income and experience, poor access to training and extension services, and increasing distance to tarmac road. Ownership of a vehicle positively influences the proportion of produce sold directly to the market.

The study found that unlike recent bias for farmers to participate in groups, collective action has not enabled participation in the value chain through selling directly at the market. Rather, factors such as income, access to information, and ownership of a means of transport. The ability of farmer groups to fill the gaps in financing, information access, and bulking is intricate. Thus, for effective market access, the drive to have farmers participate in collective action should be combined with interventions that ensure the resulting producer marketing groups improve access to financial assistance, information, and common transport. In addition, access to the export market depends on capacity to attain the quality requirements of the market. It was found that access to training and extension services improve ability of farmers to meet quality requirements.

It is therefore clear that policies should focus on improving quality of produce, by increasing the geographical coverage of especially extension and training services to farmers. Market information for perishable horticultural products is not available to farmers, and therefore price discovery is biased against farmers. In addition to providing market information, facilitating farmers to acquire affordable means of transport would assist in reducing reliance on brokers and middlemen. Policy initiatives aimed at reducing costs of transport for farmers, as well as storage and export would enable farmers to participate in the value chain and earn higher margins.

This study is limited in that it focused more on the producer and discussions are based only on producer characteristics and needs. There is need for future research to focus on the whole value chain. Further research on the opportunities and constraints faced by buyers will help in coming up with broad based all-inclusive interventions.

REFERENCES

1. Agricultural Business Development (2011). The Mango Sector in Eastern Kenya: Eastern Region Mango Census and Baseline Survey. Final report, Agricultural Business Development, Nairobi.
2. Bogiwe G, Masuku M (2012). Factors Affecting the Choice of Marketing Channel by Vegetable Farmers in Swaziland. Sustain. Agric Res. 2(1).
3. Cragg JG (1971). Some statistical models for limited dependent variables with application to the demand for durable goods. Econometrica 39:829-844
4. Fafchamps M, Hill R (2005). Selling at the Farm gate or Traveling to Market. Am. J. Agric. Econ. 87(3):717-734.
5. Fischer E, Qaim M (2012). Linking Smallholders to Markets: Determinants and Impacts of Farmer Collective Action in Kenya. World Dev. 40(6):1255-1268.
6. Government of Kenya (2010). Agricultural Sector Development Strategy 2010–2020. Government Printer, Nairobi, Kenya.
7. Greene WH (2003). Econometric analysis. Fifth edition. New Jersey, Prentice Hall.
8. Katchova AL, Miranda MJ (2004). Two-Step Econometric Estimation of Farm Characteristics Affecting Marketing Contract Decisions. Am. J. Agric. Econ. 86(1):88-102.
9. Kenya National Bureau of Statistics (2009). Makueni District Multiple Indicator Cluster Survey, 2008. Nairobi, Kenya.
10. Kherallah M, Delgado C, Gabre-Madhin E, Minot N, Johnson N (2002). The road half traveled: agricultural market reform in Sub-Saharan Africa. OCL 9(6):416-425.
11. Kirsten J, Doward A, Poulton C, Vink N (2008). Institutional Economics, Perspectives on African Agricultural Development. International Food Policy Research Institute, Washington
12. Koning N, Andersen P (2007). Agricultural Trade Liberalization and the Least Developed Countries. First Edition. Wageningen, Netherlands, Springer.
13. Kothari CR (2004). Research Methodology; Methods and Techniques. Second Edition. New Delhi, New Age International Publishers.

14. Martey E, Annin K, Nimo AW, Attoh C (2012). Does Access to Market Information Determine the Choice of Marketing Channel among Smallholder Yam Farmers in the Brong Ahafo Rgion of Ghana ? A Multinomial Logit Regression Analysis. J. Econ. Sustain. Dev. 3(12):18-28.
15. Neven D, Odera MM, Reardon T, Wang H (2009). Kenyan supermarkets, emerging middle-class horticultural farmers, and employment impacts on the rural poor. World Dev. 37(11):1802-1811.
16. Okello JJ, Narrod C, Roy D (2007). Food safety requirements in African green bean exports and their impact on small farmers. IFPRI discussion paper 00737. Washington, DC.
17. Olwande J, Smale M, Mathenge MK, Place F, Mithöfer D (2015). Agricultural marketing by smallholders in Kenya: A comparison of maize, kale and dairy. Food Policy 52:22-32.
18. Omiti J, McCullough E, Otieno D, Madelon M, Nyanamba T, Murage A (2006). Participatory Prioritization of Issues in Smallholder Agricultural Commercialization in Kenya. KIPPRA Discussion Paper No. 64. Nairobi, Kenya.
19. Panda RK, Sreekumar (2012). Marketing Channel Choice and Marketing Efficiency Assessment in Agribusiness. J. Int. Food Agribus. Mark. 24(3):213-230.
20. Rao EJ, Qaim M (2011). Supermarkets, farm household income, and poverty: Insights from Kenya. World Dev. 39(5):784-796.
21. Schwab JA (2002). Multinomial logistic regression: Basic relationships and complete problems.
22. Sebatta C, Mugisha J, Katungi E, Kashaaru A, Kyomugisha H (2014). Smallholder Farmers' Decision and Level of Participation in the Potato Market in Uganda. 11th African Crop Science Conference Proceedings, 10 -13 October 2013, Entebbe, Uganda.
23. Shiferaw B, Obare G, Muricho G (2006). Rural institutions and producer organizations in imperfect markets: Experiences from Producer Marketing Groups in semi-arid eastern Kenya. CAPRi Working Paper 60. Also jointly released as ICRISAT Socioeconomics and Policy Working Paper Series No. 23. Patancheru, Andhra Pradesh: International Crops Research Institute for the Semi-Arid Tropics 40 p.
24. Shiferaw B, Obare G, Muricho G (2006). Rural institutions and producer organizations in imperfect markets: Experiences from Producer Marketing Groups in semi-arid eastern Kenya. CAPRi Working Paper 60. Also jointly released as ICRISAT Socioeconomics and Policy Working Paper Series No. 23. Patancheru, Andhra Pradesh: International Crops Research Institute for the Semi-Arid Tropics 40 p.
25. Shilpi F, Umali-Deininger D (2008). Market facilities and agricultural marketing: evidence from Tamil Nadu, India. Agric. Econ. 39(3):281-294.
26. Tsourgiannis L, Eddison J, Warren M (2008). Factors affecting the marketing channel choice of sheep and goat farmers in the region of east Macedonia in Greece regarding the distribution of their milk production. Small Rumin. Res. 79(1):87-97.
27. Wooldridge JM (2002). Econometric Analysis of Cross Section and Panel Data. Cambridge, MA: MIT Press.

INDEX

A

Agroforestry, 323, 324, 330, 365, 366, 367

C

Chronic, 282, 302, 303

Common Agricultural Policy, 247, 248, 250, 252, 271

Consumer behaviour, 36

crop production, 103, 108, 118, 144, 167, 305, 306, 319, 320, 351, 352, 359, 372, 380, 381

D

Determinants, 35, 47, 50, 59, 60, 66, 69, 74, 75, 98, 112, 321, 378, 382, 384, 385, 387, 402

Dietary diversity, 50, 53, 74

Direct marketing, 2

E

Egyptian orange exports, 78, 79, 80, 81, 95, 96, 97

Entrepreneurial orientation, 2, 6, 31, 32, 33

environmental valuation, 144

F

Farm diversification, 248

Fertilizer, 210, 220

Food expenditure, 50

food security, 35, 36, 42, 47, 48, 49, 50, 51, 52, 53, 54, 55, 325, 353, 359, 362, 364, 365, 366, 372

Food security, 37, 50, 53, 75, 102, 353, 366

G

Granger causality, 173, 175, 176, 178, 179, 186, 187

growth enhancement scheme, 210

H

household, 35, 36, 38, 41, 43, 44, 46, 47, 50, 51, 52, 53, 54, 354, 361, 362, 364, 370, 371, 374, 378, 379, 384, 394, 397, 401, 403

human capital, 288, 305, 306, 307, 310, 311, 313, 314, 318, 319, 320, 321

K

Korean beef prices, 190, 191, 193

L

labor, 144, 282, 288, 291, 298, 301, 302, 305, 306, 307, 309, 310, 311, 313, 314, 315, 318, 319, 371, 372, 378, 379

Livelihoods, 324, 364

Local conventions, 324, 332, 344

M

mango market channels, 388

Market integration, 173, 174, 175, 176, 188

Market orientation, 2, 5, 29, 30, 32, 33

Marketing innovation, 2

mid-scale farms, 221, 222, 223, 227, 229, 236, 237, 240, 241, 242, 243

Mozambique, 140, 281, 282, 284, 285, 290, 291, 294, 296, 298, 301, 302, 303

N

Natural regeneration Sahel, 324

Nepalese agriculture, 144, 168

O

Off-farm, 369, 370, 377, 379, 384

Ordinary Least Square, 212, 369, 370, 373, 374, 382, 383

P

precautionary storage, 102

price risk, 101, 102, 103, 104, 109, 119, 175

probit, 63, 121, 122, 125, 126, 127, 129, 130, 131, 133, 135, 212, 213, 369, 370, 397

probit Ghana, 122

processed cocoyam, 35, 36, 37, 39, 41, 43, 44, 45, 47

Product innovation, 2

R

ricardian approach, 144

Rotterdam model, 78, 87, 88, 95, 97

rural migration, 306, 307, 318

Russia Import demand, 78

S

small scale farms, 222

smallholder farmers, 118, 121, 122, 123, 136, 138, 141, 211, 325, 326, 370, 371, 381, 383, 385, 390

Smallholder farmers, 121, 122

spatial panel data analysis, 144

specialty foods, 222

Structural change, 190

T

transient, 281, 282, 283, 284, 285, 286, 287, 288, 289, 290, 296, 297, 298, 301, 302

V

values-based supply chains, 221, 222, 223, 226, 244